River Ecological Restoration and Groundwater Artificial Recharge II

River Ecological Restoration and Groundwater Artificial Recharge II

Editors

Yuanzheng Zhai
Jin Wu

MDPI

Basel • Beijing • Wuhan • Barcelona • Belgrade • Novi Sad • Cluj • Manchester

Yuanzheng Zhai
College of Water Sciences
Beijing Normal University
Beijing
China

Jin Wu
Faculty of Architecture, Civil
and Transportation Engineering
Beijing University of Technology
Beijing
China

Editorial Office
MDPI AG
Grosspeteranlage 5
4052 Basel, Switzerland

This is a reprint of the Special Issue, published open access by the journal *Water* (ISSN 2073-4441), freely accessible at: www.mdpi.com/journal/water/special_issues/artificial_recharge2.

For citation purposes, cite each article independently as indicated on the article page online and using the guide below:

Lastname, A.A.; Lastname, B.B. Article Title. *Journal Name* **Year**, *Volume Number*, Page Range.

ISBN 978-3-7258-1408-4 (Hbk)
ISBN 978-3-7258-1407-7 (PDF)
https://doi.org/10.3390/books978-3-7258-1407-7

Contents

About the Editors

Yuanzheng Zhai

Yuanzheng Zhai, PhD, has been a professor and faculty member of Beijing Normal University since 2011. He performs research in hydrogeology and teaches undergraduate and postgraduate classes on applied hydrogeology and modern hydrogeochemistry.

Jin Wu

Jin Wu, PhD, has been an associate professor and faculty member of Beijing University of Technology since 2017. He performs research on ecological risk assessment and teaches undergraduate and postgraduate classes on water environment models.

Preface

The goal of this Special Issue is to highlight the research frontier of river ecological restoration and groundwater artificial recharge, considering worldwide river shrinkage and groundwater depletion. After peer review, organized by the journal, and the selection of our Special Issue, 15 papers were finally selected and published in this Special Issue. The papers can generally be classified according to the following topic areas: (1) identifying and characterizing the complex hydrological processes in the river ecological restoration or groundwater artificial recharge; (2) characterizing the impacts of the river ecological restoration or groundwater artificial recharge on the physiochemical and biological evolution of and the ecological risk to aquatic environments; and (3) demonstrating the redistribution of water resources and the safety of water quality in the river ecological restoration or groundwater artificial recharge. We believe that these high-quality papers have important reference value for the sustainable management of water resources and the ecological protection of water.

Yuanzheng Zhai and Jin Wu
Editors

water

Editorial

River Ecological Restoration and Groundwater Artificial Recharge II

Yuanzheng Zhai [1] and Jin Wu [2,*]

[1] College of Water Sciences, Beijing Normal University, Beijing 100875, China; zyz@bnu.edu.cn
[2] Faculty of Architecture, Civil and Transportation Engineering, Beijing University of Technology, Beijing 100124, China
* Correspondence: wujin@bjut.edu.cn

The depletion of rivers and groundwater caused by climate change and human activity is threatening water security and ecosystems. In order to mitigate this trend, some initiatives have been implemented, including the ecological restoration of rivers and the artificial recharge of groundwater [1,2]. For instance, the South-to-North Water Diversion Project [3,4], the ecology water replenishment of the Yongding River in Beijing, and the comprehensive treatment of groundwater overexploitation [5] have been instrumental in alleviating the depletion of water resources in North China. However, these actions can change the natural connections between river water and groundwater, affecting their hydrological characteristics and the exchange of materials between them [2]. Currently, the responses of rivers and groundwater to these behaviors and the mechanisms behind them are not fully understood. As a result, the Special Issue "River Ecological Restoration and Groundwater Artificial Recharge" was created in March 2022 to review and present advanced methodologies, recent progress and challenges, and future opportunities in this field. The first Special Issue comprised 10 papers that discussed the impacts of river ecological replenishment and groundwater recharge on watershed ecology and groundwater quality, as well as the sustainable utilization of water resources at the regional and basin scale. Unfortunately, due to the limited capacity of the publication, many excellent studies could not be included. In response to the requests of the authors and readers, a second Special Issue was created to further enrich our understanding of this issue.

This Special Issue comprises fifteen papers encompassing three interlinked research fields. Three papers focused on revealing the hydrobiogeochemical cycles existing in riverbank filtration. Seven papers focused on the sources, distribution, and transformation of various types of pollutants in river–groundwater systems. Five papers focused on the impacts of climate change and human activities on groundwater dynamics.

Riverbank filtration (RBF) is an important part of the surface water–groundwater cycle, and it intercepts and retains many pollutants present in rivers. Understanding the material cycling process is of paramount importance for the comprehension and implementation of RBF. During groundwater recharge using RBF, pollutants such as ammonium and COD enter the aquifer and change the hydrogeochemical processes and microbial community structure, which in turn causes the release of elements such as Fe, Mn, and As (described in contributions 1–3).

The effects of human activity on the quantity and quality of water in river–groundwater systems can change the migration and transformation behaviors of pollutants in river water and groundwater [6]. Research on the physicochemical behavior of pollutants in river–groundwater systems is crucial for understanding their risks and their subsequent control [7,8]. The behavior of different types of pollutants is determined by different factors such as their sources and properties. The anthropogenic influence on inorganic contamination is mainly seen in the release of poor-quality primary components from aquifers

Citation: Zhai, Y.; Wu, J. River Ecological Restoration and Groundwater Artificial Recharge II. *Water* **2024**, *16*, 1328. https://doi.org/10.3390/w16101328

Received: 25 April 2024
Accepted: 29 April 2024
Published: 7 May 2024

(especially in mining areas) caused by changes in hydrodynamic conditions, such as fluorine contamination from red mud pits (contribution 4) and uranium from sandstone-type mines (contribution 5). The behavior of insoluble organic contaminants (e.g., petroleum hydrocarbons) in the zones of groundwater level fluctuation is sensitive to changes in the water level caused by artificial recharge, especially under freezing and thawing cycles (contributions 6–7). In addition, ecological replenishment strongly alters the hydrodynamic conditions and chemical composition of surface water, which in turn causes the secondary release of pollutants, such as heavy metals (contribution 8), pharmaceuticals, and personal care products (PPCPs) (contribution 9), from sediments. With different types of contaminants and distinct aquifer conditions, the choice of water treatment also needs to be assessed comprehensively from a multi-dimensional perspective to prevent potential pollution risks (contribution 10).

Groundwater dynamics, which are under the influence of climate change and human activity, is a coupled natural–human system problem, and numerical simulation is an effective tool for studying it. Authors have combined fuzzy mathematics, random forests, and climate models with groundwater models to study groundwater dynamics under different scenarios, such as ecological recharge (contribution 11), riverbank filtration (contribution 12), and artificial recharge (contribution 13). The impact of climate change on the sustainable utilization of water resources is also discussed and studied in detail (contributions 14–15).

These published papers provide useful scientific evidence that could lead to a better understanding of the relationship between river water and groundwater impacted by human activity and climate change. We believe that these high-quality papers have important value as references for the sustainable management of water resources and the protection of water ecological security.

We thank all the authors for contributing to this Special Issue and making it a success.

Author Contributions: Y.Z. and J.W.: conceptualization and writing—original draft preparation, review, and editing. All authors have read and agreed to the published version of the manuscript.

Funding: This research received no external funding.

Institutional Review Board Statement: Not applicable.

Informed Consent Statement: Not applicable.

Conflicts of Interest: The authors declare no conflicts of interest.

List of Contributions:

1. Xia, X.; Teng, Y.; Zhai, Y. Effects of Ammonium and COD on Fe and Mn Release from RBF Sediment Based on Column Experiment. *Water* **2023**, *15*, 120. https://doi.org/10.3390/w15010120
2. Lu, S.; Yang, Y.; Yin, H.; Su, X.; Yu, K.; Sun, C. Microbial Community Structure of Arsenic-Bearing Groundwater Environment in the Riverbank Filtration Zone. *Water* **2022**, *14*, 1548. https://doi.org/10.3390/w14101548
3. Zhang, B.; Chen, L.; Li, Y.; Liu, Y.; Li, C.; Kong, X.; Zhang, Y. Impacts of River Bank Filtration on Groundwater Hydrogeochemistry in the Upper of Hutuo River Alluvial Plain, North China. *Water* **2023**, *15*, 1343. https://doi.org/10.3390/w15071343
4. Qi, Y.; Zhou, P.; Wang, J.; Ma, Y.; Wu, J.; Su, C. Groundwater Pollution Model and Diffusion Law in Ordovician Limestone Aquifer Owe to Abandoned Red Mud Tailing Pit. *Water* **2022**, *14*, 1472. https://doi.org/10.3390/w14091472
5. Zheng, F.; Teng, Y.; Zhai, Y.; Hu, J.; Dou, J.; Zuo, R. Geo-Environmental Models of In-Situ Leaching Sandstone-Type Uranium Deposits in North China: A Review and Perspective. *Water* **2023**, *15*, 1244. https://doi.org/10.3390/w15061244
6. Huang, J.; Zhong, R.; Lyu, H. Investigating the Change Pattern in Adsorption Properties of Soil Media for Non-Polar Organic Contaminants under the Impact of Freezing and Thawing. *Water* **2023**, *15*, 2515. https://doi.org/10.3390/w15142515

7. Zhou, J.; Pan, M.; Chang, C.; Wang, A.; Wang, Y.; Lyu, H. Migration Law of LNAPLs in the Groundwater Level Fluctuation Zone Affected by Freezing and Thawing. *Water* **2022**, *14*, 1289. https://doi.org/10.3390/w14081289
8. Liu, T.; Zhang, D.; Yue, W.; Wang, B.; Huo, L.; Liu, K.; Zhang, B.-T. Heavy Metals in Sediments of Hulun Lake in Inner Mongolia: Spatial-Temporal Distributions, Contamination Assessment and Source Apportionment. *Water* **2023**, *15*, 1329. https://doi.org/10.3390/w15071329
9. Pei, S.; Li, B.; Wang, B.; Liu, J.; Song, X. Distribution and Ecological Risk Assessment of Pharmaceuticals and Personal Care Products in Sediments of North Canal, China. *Water* **2022**, *14*, 1999. https://doi.org/10.3390/w14131999
10. Li, J.; Wen, Y.; Jiang, J.; Tan, W.; Zhang, T. A Multi-Dimensional Comprehensive Assessment (MDCA) Method for the Prioritization of Water Pollution Treatment Technologies in China. *Water* **2023**, *15*, 751. https://doi.org/10.3390/w15040751
11. Nan, T.; Cao, W. Effect of Ecological Water Supplement on Groundwater Restoration in the Yongding River Based on Multi-Model Linkage. *Water* **2023**, *15*, 374. https://doi.org/10.3390/w15020374
12. He, Z.; Kang, B.; Tao, Y.; Qin, L. Mining Scheme for Small Rivers near Water Sources—A Case Study of Liuan River in Linquan County, China. *Water* **2022**, *14*, 1921. https://doi.org/10.3390/w14121921
13. Asmael, N.; Dupuy, A.; McLachlan, P.; Franceschi, M. Hydro-Geochemical Characteristics of the Shallow Alluvial Aquifer and Its Potential Artificial Recharge to Sustain the Low Flow of the Garonne River. *Water* **2023**, *15*, 2972. https://doi.org/10.3390/w15162972
14. Qi, X.; Li, W.; Zheng, Y.; Cui, H.; Kang, W.; Liu, Z.; Shao, X. Coupling Simulation and Prediction of Sustainable Utilization of Water Resources in an Arid Inland River Basin under Climate Change. *Water* **2023**, *15*, 3232. https://doi.org/10.3390/w15183232
15. Hou, X.; Yang, H.; Cao, J.; Feng, W.; Zhang, Y. A Review of Advances in Groundwater Evapotranspiration Research. *Water* **2023**, *15*, 969. https://doi.org/10.3390/w15050969

References

1. Wohl, E.; Lane, S.N.; Wilcox, A.C. The science and practice of river restoration. *Water Resour. Res.* **2015**, *51*, 5974–5997. [CrossRef]
2. Dillon, P.; Stuyfzand, P.; Grischek, T.; Lluria, M.; Pyne, R.D.G.; Jain, R.C.; Bear, J.; Schwarz, J.; Wang, W.; Fernandez, E.; et al. Sixty years of global progress in managed aquifer recharge. *Hydrogeol. J.* **2019**, *27*, 1–30. [CrossRef]
3. Nong, X.; Shao, D.; Zhong, H.; Liang, J. Evaluation of water quality in the South-to-North Water Diversion Project of China using the water quality index (WQI) method. *Water Res.* **2020**, *178*, 115781. [CrossRef] [PubMed]
4. Long, D.; Yang, W.; Scanlon, B.R.; Zhao, J.; Liu, D.; Burek, P.; Pan, Y.; You, L.; Wada, Y. South-to-North Water Diversion stabilizing Beijing's groundwater levels. *Nat. Commun.* **2020**, *11*, 3665. [CrossRef] [PubMed]
5. Yang, H.; Cao, W.; Zhi, C.; Li, Z.; Bao, X.; Ren, Y.; Liu, F.; Fan, C.; Wang, S.; Wang, Y. Evolution of groundwater level in the North China Plain in the past 40 years and suggestions on its overexploitation treatment. *Geol. China* **2021**, *48*, 1142–1155. [CrossRef]
6. Wei, Y.; Chen, Y.; Cao, X.; Xiang, M.; Huang, Y.; Li, H. A Critical Review of Groundwater Table Fluctuation: Formation, Effects on Multifields, and Contaminant Behaviors in a Soil and Aquifer System. *Environ. Sci. Technol.* **2024**, *58*, 2185–2203. [CrossRef] [PubMed]
7. Min, L.; Liu, M.; Wu, L.; Shen, Y. Groundwater Storage Recovery Raises the Risk of Nitrate Pollution. *Environ. Sci. Technol.* **2022**, *56*, 8–9. [CrossRef] [PubMed]
8. Zang, Y.; Hou, X.; Li, Z.; Li, P.; Sun, Y.; Yu, B.; Li, M. Quantify the effects of groundwater level recovery on groundwater nitrate dynamics through a quasi-3D integrated model for the vadose zone-groundwater coupled system. *Water Res.* **2022**, *226*, 119213. [CrossRef] [PubMed]

water

MDPI

Article

Effects of Ammonium and COD on Fe and Mn Release from RBF Sediment Based on Column Experiment

Xuelian Xia, Yanguo Teng * and Yuanzheng Zhai *

Engineering Research Center of Groundwater Pollution Control and Remediation of Ministry of Education of China, College of Water Sciences, Beijing Normal University, Beijing 100875, China
* Correspondence: ygteng@bnu.edu.cn (Y.T.); zyz@bnu.edu.cn (Y.Z.); Tel.: +86-010-58802738 (Y.T.); +86-010-58802736 (Y.Z.)

Abstract: Riverbank filtration (RBF) is an important part of the surface water–groundwater cycle, and it intercepts and retains many pollutants in rivers. However, RBF affects the biogeochemical process which enables aquifer sediments to release iron (Fe) and manganese (Mn). In this study, column experiments were performed to investigate the effects of ammonium ions and organic matter on Fe and Mn release from anaerobic RBF sediments. In addition, high-throughput sequencing technology was used to characterize the microbial community. The results showed that the ammonium ions (NH_4^+) and organic matter (COD) in groundwater promote the release of Fe and Mn from aquifer sediments. The trends of Fe and Mn release were similar during the leaching process. The maximum concentrations of Fe and Mn were 0.32 and 40 μg/L, respectively. The structural diversity and abundance of the microbial communities in the groundwater were closely related to the Fe/Mn content. *Actinobacteriota*, *Proteobacteria*, *Acidobacteriota*, *Bacteroidota*, and *Chloroflexi* were the dominant phyla, while *Rhodococcus*, *Ochrobactrum*, and *Pseudarthrobacter* were the dominant genera. These functional microbes are actively involved in the biogeochemical cycling of Fe, Mn, and N. In summary, contaminants and the microbial-community structure have dual effects on the release of Fe and Mn from RBF aquifers.

Keywords: RBF; column experiment; groundwater; Fe; Mn; ammonia; organic matter; microbial community

check for updates

Citation: Xia, X.; Teng, Y.; Zhai, Y. Effects of Ammonium and COD on Fe and Mn Release from RBF Sediment Based on Column Experiment. *Water* **2023**, *15*, 120. https://doi.org/10.3390/w15010120

Academic Editor: Glen R. Walker

Received: 27 November 2022
Revised: 22 December 2022
Accepted: 26 December 2022
Published: 29 December 2022

1. Introduction

As an important interface and natural filter of surface water and groundwater systems, riverbank filtration (RBF) is an active biogeochemical transition zone [1]. The presence of certain pollutants in natural RBF systems enhances the heavy-metal-release process in sediments [2]. Iron (Fe) and manganese (Mn) are widely distributed in aquifer sediments, and they can enter groundwater through leaching and become common water-chemistry components [3]. High Fe and Mn concentrations in groundwater affect the purification efficiency of RBF. The concentrations of Fe and Mn in groundwater are highly spatially variable, which is related to the geochemical processes of Fe and Mn in aquifer sediments [4,5]. The superposition of exogenous contaminants and chemical components in the aquifer may complicate hydrogeochemical processes [6]. Therefore, it is necessary to understand the flow and release of Fe and Mn from aquifers.

Fe and Mn in groundwater are mainly derived from aquifer sediments and anthropogenic activities [7]. This not only pollutes groundwater to varying degrees, but it also alters the environmental conditions of the groundwater, thus affecting the natural environmental processes of Fe and Mn in groundwater [8,9]. In recent years, it has been shown that the Fe and Mn concentrations are usually higher in groundwater with higher ammonia and organic-matter contents, for example, in the aquifers in the United States [10], China [11], and some other regions [12]. It is worth noting that the premise of the above study was the observation under homogeneous redox zones, and since redox zones are usually inhomogeneous in natural environments, a newer method of aquifer redox zoning was proposed to compare the redox zoning and morphology of Fe and Mn in aquifers by fusing the Honduras aquifer data; as a result, anomalous concentrations of Fe and Mn were also detected [13]. In addition, at the site scale, such as in landfills, the Fe and Mn concentrations in the groundwater significantly increase after a period of operation, while the background values in the groundwater are not high [11]. Various organic and nitrogen-containing substances are among the main pollutants in these sites [14]. Agriculture, urbanization and industrialization determine the nitrogen pollution of soils, surface water and groundwater, and hydrological and biological processes between rivers and aquifers can control the concentration and fate of nitrogen compounds. In groundwater multilayer systems, the interaction between groundwater and surface water is influenced by changes in river flow and morphology, which can have an impact on the global nitrogen cycle system [15]. In addition, the detection of high concentrations of Fe and Mn in groundwater due to the presence of organic matter is usually associated with the occurrence of redox processes. In the hydrogeological context, multi-layered alluvial aquifers are directly connected to surface waters heavily affected by eutrophication, and redox processes involving organic loads in groundwater control the distribution and reaction rates of Fe and Mn [16]. Thus, formation of Fe- and Mn-containing groundwater is the result of a combination of a series of water–rock interactions and the hydrogeochemistry.

The release of Fe and Mn from aquifers is influenced by the physicochemical properties and microbial response [17]. Microbes can change the morphology and potential parameters of Fe and Mn in aquifers, such as through redox reactions, which affect the circulation and distributions of Fe and Mn in groundwater [18]. Native iron-reducing bacteria (IOB) and manganese-reducing bacteria (MnOB) play important roles in the migration processes of Fe and Mn in aquifers [19]. Fe-reducing bacteria (IRB) and Mn-reducing bacteria (MRB) can reduce iron and manganese oxides through metabolic activities, making them susceptible to migration in groundwater [20]. In addition, sulfate-reducing bacteria can indirectly reduce Fe/Mn oxides, and sulfate-reducing bacteria can stimulate the reaction of SO_4^{2-} in sediments with Fe/Mn minerals to form synthetic minerals such as Fe-S/Mn-S, which may also inhibit some Fe/Mn migration [21]. However, the effects of the microbial community structure and function on Fe and Mn release from RBF sediment and their potential mechanisms are not clear. Therefore, the effect of the coupled processes of exogenous pollutant species and microbes on the behavior of Fe and Mn in the actual RBF system requires further investigation.

In this study, an anaerobic column experiment simulating the RBF system was developed to investigate the effects of ammonium ions and organic matter (COD) on the mobility of Fe and Mn in aquifer sediments. The objectives of this study were as follows: (1) to determine the enhancement effects of ammonium and COD in groundwater on the release of Fe and Mn from aquifer sediments through time-series analysis, (2) to analyze the structure and dynamics of the response processes of the microbial community based on Illumina MiSeq high-throughput sequencing, and (3) to reveal the mechanisms behind these effects. This study will aid in better understanding the range of biogeochemical effects of contaminants on Fe in groundwater systems, and we hope that the results of this study will lead to more effective control of groundwater pollution.

2. Materials and Methods

2.1. Sediment Sampling and Geochemical Analysis

The aquifer sediments were sampled in Northeast China at a specific point (45°50′ N, 126°40′ E) located east of the Songnen Plain, which belongs to the Songhua River Basin. One hole was drilled in the study area in July 2021 to collect aquifer sediment samples from 0 to 2 m, 5 to 6 m, 10 to 11 m, 15 to 16 m, and 20 to 21 m below the surface, and the lithological variation (sand, clay, or silty clay) was recorded. The physicochemical properties of the sediments are given in Table 1, and the test methods and detection limits for the physical and chemical properties of the sediments are given in Table 2. The in situ sediment samples were stored in sterile tubes, kept frozen in a cold environment, and transported to the laboratory within 3 days for storage in a −70 °C refrigerator for microbial-community analysis.

Table 1. Physical and chemical properties of the aquifer sediments used for the experiment.

Sampling Site (m)	Depth * (m)	Sediment Texture	pH	Fe (g/Kg)	Mn (g/Kg)	TOC (%)	NH_4^+ (mg/kg)
0–2	1–2	Clay	5.25	7.36	0.42	21	1.5
5–6	5–6	Clay	7.87	15.40	0.38	52	1.3
10–11	10–11	Clay and sand	7.86	9.6	0.29	12	0.9
15–16	15–16	Sand	6.3	12.8	0.20	8	1.3
20–21	20–21	Clay and sand	6.61	5.40	0.18	7	1.2

Note(s): * Refers to the buried depth of the sampling site underground.

Table 2. Methods for determining the physicochemical properties of the sediment and the detection limits.

Item (Unit)	Detection Method	Detectable Limit
pH (/)	Glass electrode method	–
NH_4^+ (mg/L)	Gas phase molecular absorption spectrometry	0.15 mg/kg
TOC (mg/L)	Combustion method	–
Fe (mg/L)	ICP-AES (PerkinElmer Optima 8000)	0.06 mg/kg
Mn (mg/L)	ICP-AES (PerkinElmer Optima 8000)	0.02 mg/kg

Note(s): "–" means that there is no detection limit.

2.2. Column Experiment

A one-dimensional column system was constructed to simulate the roles of ammonium and COD in the release of Fe and Mn from the sediments to groundwater. The sediments were air-dried and sieved to a particle size of less than 2.0 mm before the column experiments. Three polyvinyl chloride columns with an inner diameter of 8 cm and a height of 80 cm were designed (Figure 1), and called column A (blank control), column B (ammonium solution), and column C (COD solution). A total of five outlets were designed, with each column filled to a height of approximately 60 cm with a sediment weight of approximately 1.2 kg. The top and bottom of the column were lined with 1 cm of quartz sand, and a layer of glass wool was placed at the bottom of the quartz sand at the lower end of the column. The sediment was tapped several times with a pounding stick after each filling. After filling, the outer wall of the sediment column was wrapped with a shade cloth, which was used to simulate the original light-proof environment of the underground aquifer.

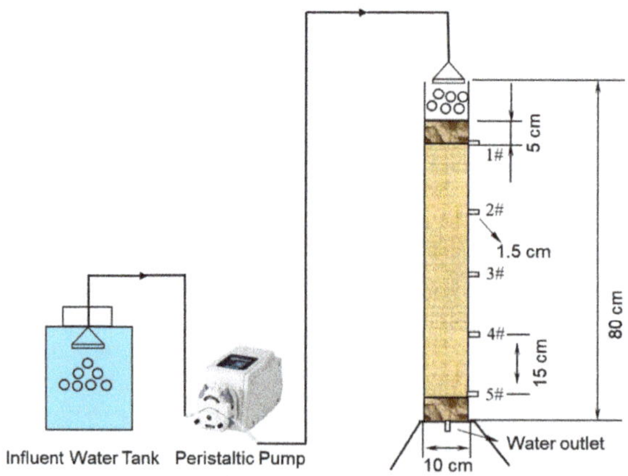

Figure 1. Schematic diagram of the column system.

The column experiments were conducted at room temperature (20 ± 3 °C) and protected from light to analyze the migration patterns of sediment Fe and Mn during contaminant fixed-head drenching tests. Before the experiment was started, ultrapure water (column A), 50 mg/L ammonium solution (column B), and 50 mg/L COD solution (column C) were injected into the columns at a flow rate of 2.5×10^{-5} cm/s, and the columns were protected from light for treatment. Ammonium and COD for the leaching test were artificially prepared with NH_4Cl and potassium hydrogen phthalate, respectively. The sampling times were 0 h, 3 h, 5 h, 9 h, 18 h, 20 h, 24 h, 30 h, . . . , 742 h (31 days). The samples from the sampling points were analyzed, and the experiments were stopped when the pollutant concentrations at the outlets stabilized. The detected parameters were the pH, oxidation-reduction potential (Eh), dissolved oxygen (DO) concentration, ammonium concentration, Fe concentration, and Mn concentration. Owing to the special characteristics of DO, the DO concentration was measured immediately after sample collection. The other parameters were measured after storing the sample in a low-temperature environment. The solution used to determine the Fe and Mn concentrations was stored with a protective agent.

2.3. Aqueous-Phase Analysis

Approximately 5 mL of the existing water solution was collected from sampling ports 1–5 on the column with a sediment sampler. The samples were filtered through a 0.45-μm membrane, and the solutions for measurement of the Fe and Mn concentrations were preserved with acid. The pH, DO concentration, and Eh were measured by a water quality parameter meter (TE-600Plus, USA). The Fe and Mn concentrations were determined by inductively coupled plasma atomic emission spectrometry (Optima 8000, PerkinElmer, Waltham, MA, USA) [22]. Determination of the Fe(II) concentration was performed by the 1,10-*o*-phenanthroline method with a detection limit of 5 μM [23]. The difference between the total Fe concentration and Fe^{2+} concentration was used as the concentration of Fe^{3+} in the aqueous phase of the effluent. The ammonium concentration was measured by ultraviolet spectrophotometry.

All of the reagents were analytical grade. All of the glassware was carefully treated before use to ensure that the data were of high quality. The glassware was immersed in 10% HCl for 4 h, rinsed three times with ultrapure water, and then dried at 180 °C for 4 h.

2.4. Microbial Community Analysis

Sediment samples at five depths (20, 40, 60, 80, and 100 cm) of columns A–C were collected, which are called A1–A5, B1–B5, and C1–C5, respectively. Total DNA extraction from the microbial community was performed according to the instructions of the E.Z.N.A. soil DNA kit (Omega Bio-tek, Norcross, GA, USA), the quality of DNA extraction was determined by 1% agarose gel electrophoresis, and the DNA concentration and purity were determined by a NanoDrop2000 spectrophotometer. The 16S rRNA gene V3–V4 variable region was amplified by polymerase chain reaction (PCR) using 338F (5′-ACTCCTACGGGAGGCAGCAG-3′) and 806R (5′-GGACTACHVGGGTWTCTAAT-3′). PCR amplification of the variable region was performed as follows: 95 °C pre-denaturation for 3 min, 27 cycles of 95 °C denaturation for 30 s, 55 °C annealing for 30 s, and 72 °C extension for 30 s, 72 °C stable extension for 10 min, and finally storage at 4 °C (ABI GeneAmp 9700 PCR system, ABI, USA). The PCR reaction system was 4 μL of 5× TransStart FastPfu buffer, 2 μL of 2.5 mM dNTPs, 0.8 μL of upstream primer (5 μM), 0.8 μL of downstream primer (5 μM), 0.4 μL of TransStart FastPfu DNA polymerase, and 10 ng of template DNA (made up to 20 μL). There were three replicates per sample.

The PCR products from the same sample were mixed and recovered using 2% agarose gel, purified using an AxyPrep DNA gel extraction kit (Axygen Biosciences, Union City, CA, USA), detected by 2% agarose gel electrophoresis, and quantified with a Quantus fluorometer (Promega, Madison, WI, USA). Library construction was performed using a NEXTFLEX Rapid DNA-Seq kit with the following steps: (1) splice linkage, (2) removal of the splice self-linked fragments by magnetic-bead screening, (3) enrichment of the library template by PCR amplification, and (4) recovery of the PCR products by magnetic beads to obtain the final library. Sequencing was performed using Illumina's Miseq PE300/NovaSeq PE250 platform (Shanghai Meiji Biomedical Technology Co., Ltd., Shanghai, China).

3. Results and Discussion

3.1. Changes of pH, DO and Eh

In the control experiment (column A) (Figure 2, Table S1), the average leachate pH was 7.1, and the pH first decreased and then slowly increased with increasing infiltration time. The pH decreased to 5.92 at sampling port 3, and then the pH basically stabilized. With continuous introduction of ultrapure water, the pH of the column increased after 576 h. In the ammonium test (column B) (Figure 2), the average pH of the leachate was 6.2. With increasing infiltration time, the pH first decreased and then slowly increased, and the pH fluctuated before 432 h. The pH increased to 7.12 at 48 h of the experiment, and it then decreased to 5.8 at 144 h. During this period, ammonium underwent a hydrolysis reaction to produce H^+, so the pH briefly decreased. After 432 h of the experiment, the reaction in the column tended to be stable, so there was no significant change in the pH in the later stage. In the COD test (column C) (Figure 2), the average pH was 6.2, and the pH fluctuated and then reached a stable state after 430 h. The pH increased at 300 h. The pH was the lowest at the beginning of the test, because the COD solution was acidic (pH = 5–6). Overall, compared with the ultrapure water group, the pH values in the ammonium and COD tests were lower, the pH values of the sampling points were similar, and they all reached a stable state after 430 h.

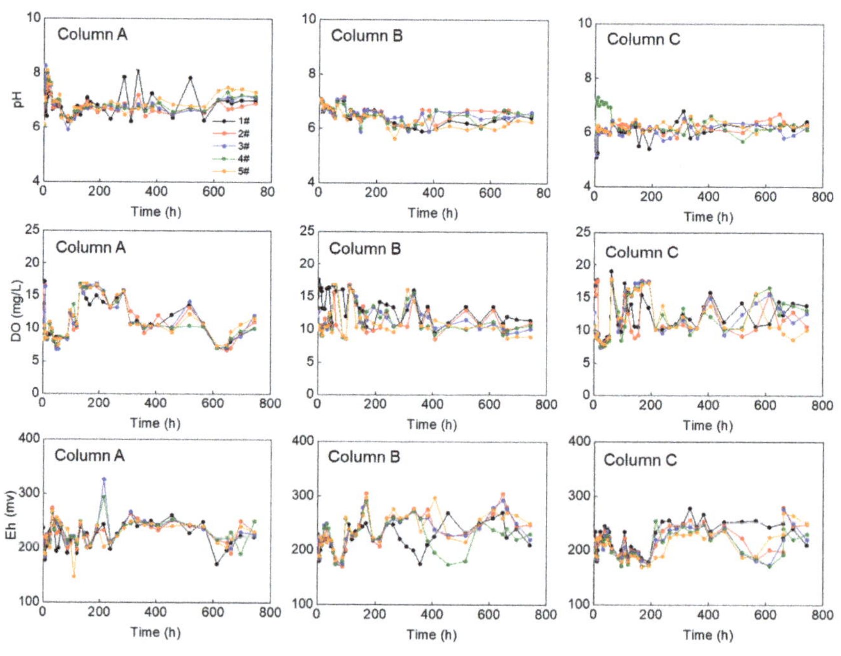

Figure 2. Changes of the pH, DO concentration, and Eh with time.

In column A (Figure 2), the average DO concentration in the leachate was 10.2 mg/L. The DO concentration first decreased, then increased, and then decreased, and the DO concentration fluctuated in the different water-extraction holes. Before 96 h, the DO concentration showed a decreasing trend, during which the column consumed O_2. After 300 h, the DO concentration in the column began to stabilize. After 528 h, the DO concentration greatly fluctuated, which may be due to a change in the DO concentration caused by contact with air at the bottom of the column. In column B (Figure 2), the average DO concentration was 12.8 mg/L. The DO concentration fluctuated before 144 h, indicating a strong reaction in the column. In column C (Figure 2), the average DO concentration was 13.2 mg/L, and the DO concentration showed a downward trend before 48 h. During this period, aerobic microbes were relatively active in the column. After 48 h, the DO concentration in the column began to decrease and was then relatively constant at 300 h, indicating that the reaction was not intense during this period. The DO concentrations in the ammonium and COD tests were higher than that in the control test.

In column A (Figure 2), the average Eh of the leachate was 210 mV. Eh showed the trend of first decreasing, then increasing, and then decreasing. Eh is highly susceptible to the influence of redox substances. Eh greatly fluctuated before 324 h, and complex reactions occurred in the column during this period. After 324 h, Eh tended to a relatively stable state, indicating that the column had reached a relatively balanced state. In column B (Figure 2), the average Eh was 222 mV. Before 276 h, Eh greatly fluctuated, with obvious biochemical effects and strong microbial metabolic processes. After 276 h, the column tended to a relatively stable state. In column C (Figure 2), the average Eh was 212 mV. The lowest Eh was 178 mV at 96 h, the highest Eh occurred between 200 and 400 h, and Eh tended to be stable at 670 h.

3.2. *Changes of Fe and Mn*

With the increase in infiltration time, the concentration of Fe and Mn in leachate increased to the highest value, and then decreased gradually (Figure 3, Table S2). The concentration trends of Fe and Mn in different leaching depths were similar, but the concentration values were different. In column A, Fe was released faster before 48 h (Figure 3), and the maximum Fe concentration was 0.051 mg/L at 60 h for sampling point 1. For sampling points 2–5, the maximum Fe concentrations of 0.06, 0.07, 0.081, and 0.078 mg/L were reached at 140 h, respectively, and they then gradually decreased. The concentration of Fe in the leachate increased with increasing infiltration distance, and the mean Fe concentrations for sampling points 1–5 were 0.033, 0.039, 0.049, 0.05, and 0.053 mg/L, respectively. At 140 h, the Mn concentrations reached the maximum concentrations of 4, 4.5, 4.5, 4.1, and 5.9 μg/L, respectively (Figure 3). The Mn concentrations then gradually decreased. The average Mn concentrations for sampling points 1–5 were 2, 2, 3, 4, and 4 μg/L, respectively, and the Mn concentrations increased with increasing infiltration distance.

In column B, Fe was released at a fast rate before 50 h (Figure 3), and the concentration of Fe released at sampling point 1 reached a maximum of 0.12 mg/L at 144 h. At sampling points 2–5, the maximum Fe concentrations of 0.2, 0.28, 0.22, and 0.28 mg/L occurred at 240 h, respectively, and they then gradually decreased and reached a stable state at 576 h. The average Fe concentrations for sampling points 1–5 were 0.073, 0.11, 0.156, 0.131, and 0.148 mg/L, respectively, and the Fe concentrations increased with increasing infiltration distance. The Mn release rate was slow before 144 h (Figure 3), and the Mn concentration did not significantly change. The maximum Mn concentration was 0.03 mg/L at 525 h for sampling point 1. For sampling points 2 and 5, Mn showed the maximum concentrations of 0.068 and 0.08 mg/L at 550 h, respectively. The average Mn concentrations for sampling points 1–5 were 11, 16, 14, 13, and 10 μg/L, respectively, and the Mn concentrations decreased with increasing infiltration distance.

In column C, Fe was released at a fast rate before 48 h (Figure 3), and the maximum Fe concentrations for sampling points 1, 2, and 3 were 0.2, 0.28, and 0.34 mg/L at 150 h, respectively. The maximum Fe concentrations for sampling points 4 and 5 were 0.31 and 0.28 mg/L at 200 h, respectively, and they then gradually decreased and reached a stable state at 380 h. The average concentrations of Fe for sampling points 1–5 were 0.086, 0.131, 0.122, 0.114, and 0.084 mg/L, respectively. The concentration of Mn fluctuated before 144 h (Figure 3). For sampling point 1, the maximum Mn concentration of 25 μg/L occurred at 190 h. For sampling point 4, the maximum Mn concentration of 35 μg/L occurred at 240 h.

The maximum Mn concentration for sampling point 5 was 30 μg/L at 576 h, and the Mn concentration then gradually decreased and reached a stable state at 380 h. The mean Mn concentrations for sampling points 1–5 were 11, 14, 14, 19, and 13 μg/L, respectively.

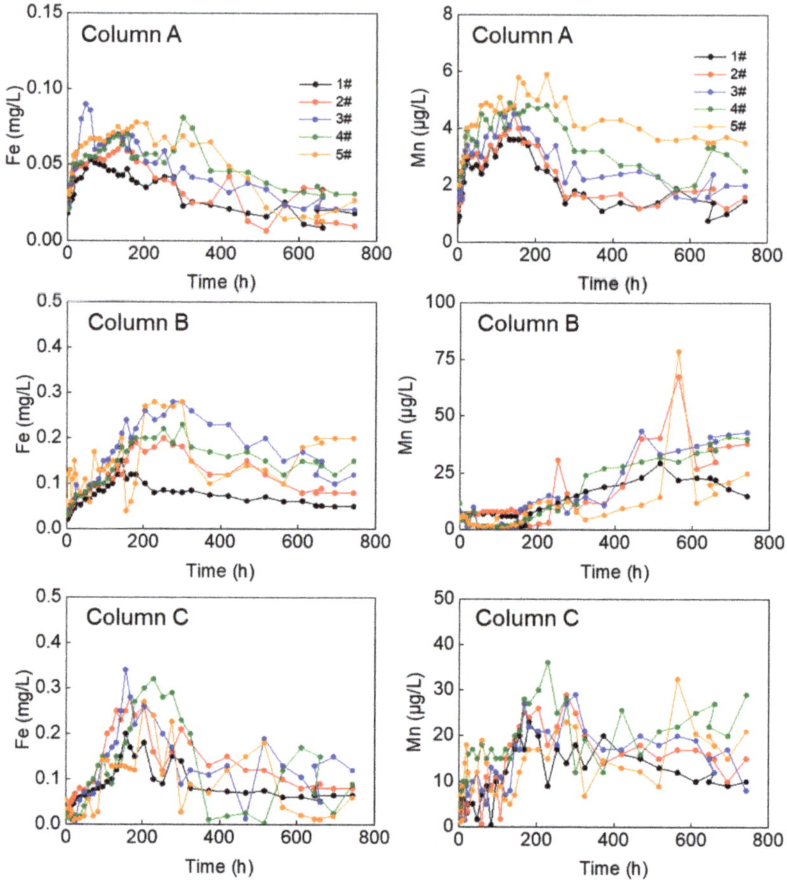

Figure 3. Changes of Fe and Mn concentrations.

The leaching trends of Fe and Mn in the three columns were similar. The concentrations of Fe and Mn rapidly increased and reached peaks in the early stage of leaching, and they then gradually decreased until they reached a stable state in the middle to late stages of leaching. Ammonium and COD can enhance the release of Fe/Mn in aqueous media. Compared with the ultrapure water group, the average release times of Fe/Mn were 2.8/10 and 3/7 times higher for columns A and B, respectively. By comparing the five sampling points, the leaching concentration order of Fe and Mn was 5 > 4 > 3 > 2 > 1.

Areas with high Fe and Mn contents in groundwater are accompanied by ammonium pollution, which has attracted great attention. For example, the chemical data of 43,334 wells in the United States were tested [24], and the results showed that high Mn concentration (>300 μg/L) was present in the groundwater. The potential value of the redox reaction of Fe, Mn, and ammonium is related to the order of the reaction speed. Therefore, under limited conditions, Fe can first undergo an oxidation reaction [25]. In the reaction process, the ammonium reaction is incomplete, resulting in the accumulation of nitrite ions. Ammonium solutions are weakly acidic (pH = 5.5–6.5), and the sediment will dissolve during leaching. In addition, the migration processes of Fe and Mn in the column are controlled by the adsorption–desorption equilibrium, and this equilibrium moves towards desorption when the concentration increases.

The ammonium concentration of the whole sediment column first decreased and then increased (Figure 4). The decrease in the ammonium concentration can be attributed to heterotrophic denitrification and dissimilatory nitrate reduction [26]. Ammonium-coupled Fe^{2+} oxidation is the main pathway of nitrate reduction under oxidizing conditions [27,28]. The Fe^{3+} content in the column was higher than the Fe^{2+} content. In addition to rapid oxidation of Fe^{2+} to Fe^{3+} owing to the instability of Fe^{2+}, Fe^{3+} can also form by the reaction between Fe^{2+} and ammonium in the column. Mn and Fe have similar oxidation characteristics. We speculate that nitrification driven by Mn^{2+} oxidation is the main reason for the decrease of the ammonium concentration, which is confirmed by the fact that dissolved reduced Mn can be used as a reducing agent for nitrogen oxide species [29]. The increase in the ammonium concentration may be due to the fact that the reaction in the sediment column tends to be stable in the later stage of the experiment, and the content of pollutants reacting with Fe and Mn oxides in the sediment becomes less. Compared with the change in the ammonium concentration, the COD concentration did not significantly change. Previous studies have shown that COD has certain effects on the dissolution of minerals [30], and COD can act as a reducing agent or complex formal ligand. Because COD is a good metal-chelating agent, it can provide electrons to promote reduction and dissolution of minerals [31]. COD creates a more suitable environment for the presence of reducing substances (such as Fe and Mn) by decreasing Eh in groundwater [32].

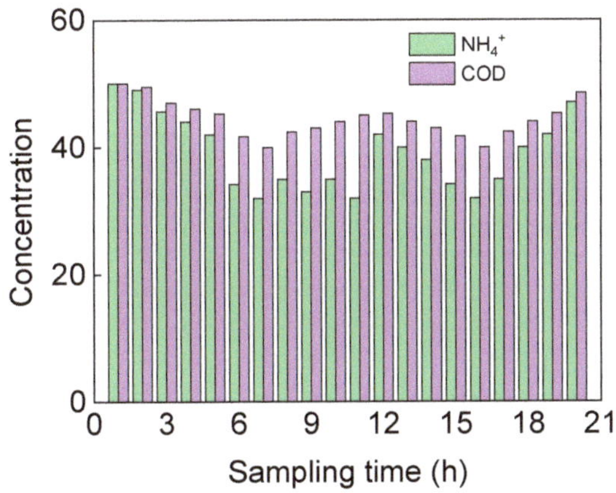

Figure 4. Changes of the concentrations of ammonium and COD in the column.

3.3. *Response of Microbial Community*

Microbial community sequence analysis revealed a series of groups associated with Fe-, Mn-, ammonium-, and organic-related processes. At the phylum level (Figure 5, Table S3), in column A, *Actinobacteriota*, *Proteobacteria*, *Acidobacteriota*, and *Chloroflexi* were the dominant phyla, with relative abundances of 23.1–40.5%, 11.4–45.5%, 1.4–15.2%, and 2.7–14.1%, respectively. Among the top ten bacterial groups, *Actinobacteriota*, *Proteobacteria*, *Acidobacteriota*, *Firmicutes*, and *Bacteroidota* were the common IOB and MOB. Their cumulative abundance was 64.9%. In column C, *Proteobacteria*, *Actinobacteriota*, *Bacteroidota*, and *Acidobacteriota* were the dominant phyla with relative abundances of 16.7–54.4%, 15.2–42.1%, 3.8–7.0% and 2.8–6.3%. The cumulative relative abundance of IOB and MOB increased by 74.7%, and organic-degrading bacteria (1.3–3.6%) appeared. Compared with column A, the cumulative relative abundance of IOB and MOB in columns B and C increased by 19.5% and 15.1%, respectively, indicating that ammonium and COD promote the release of iron. At the genus level, *Rhodococcus*, *Ochrobactrum*, and *Pseudarthrobacter* were common IOB and MOB in column A, with a cumulative abundance of 20.8%. The relative abundances of IOB and MOB in columns B and C were 23.3% and 30.3%, respectively, which were 12.0% and 45.6% higher than that in column A, respectively. This verified that ammonium and COD promote the release of Fe and Mn.

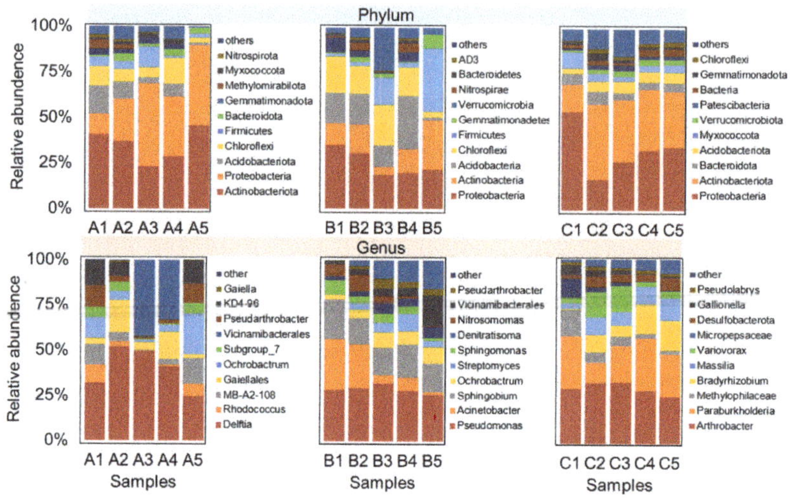

Figure 5. Histogram of the species abundance in the microbial community.

3.4. Fe and Mn Release Mechanism

The changes of the Fe and Mn concentrations in the sediments in ammonium and COD solution, and the structure of the microbial community indicate that microbe-mediated reductive dissolution of Fe/Mn oxides may still be the main mechanism of Fe/Mn release in the system (Figure 6). With the addition of a large number of pollutants, the DO was consumed in large quantities, and the microbial community in the system was mainly composed of anaerobic/supplementary anaerobic bacteria, in which IRB/MRB played an important role. The metabolic activity of *Acidobacteriota* affects the pH of the environment and the dissolution of Fe and Mn minerals through a change in the environmental conditions, and Fe and Mn minerals are then reduced under the action of IRB and MRB. In the presence of ammonium, *Nitrospirae*, *Denitratisoma*, and *Nitrosomonas* enhance the process of Fe release by nitrification and denitrification of NH_4^+ [33,34]. Through the reductive dissolution of Fe/Mn minerals, Fe and Mn attached to the surface of iron and manganese oxides are released and reduced to Fe^{2+} and Mn^{2+} under the action of IRB and MRB in the aqueous environment [35]. In the presence of COD, IRB and MRB enhance the Fe- and Mn-release processes by using organic matter, and organic-matter-degradation bacteria can enhance the Fe-release process [36]. In addition, *Desulfobacterota* (an anaerobic microorganism) was detected in the COD test, and S can be used as a strong reducing agent to participate in the reaction of iron oxide in the system, accelerating the reduction and release of iron/manganese oxides [37,38], while the Fe/Mn and S elements coexist. The S^{2-} or HS^- ions produced by reduction can combine with Fe^{2+} in water to form secondary minerals, which can inhibit the increase in the Fe/Mn concentration in groundwater.

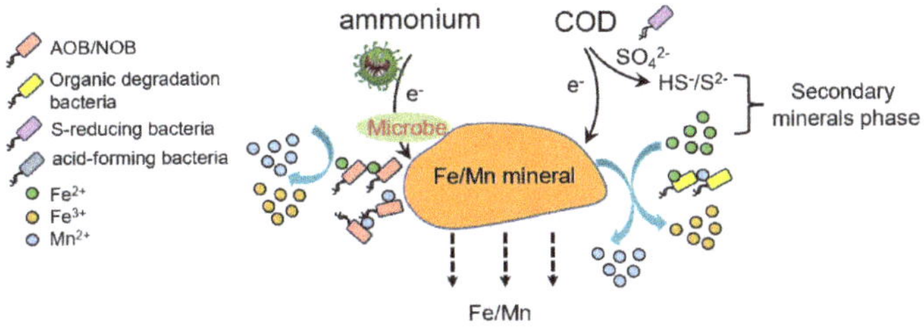

Figure 6. Effects of microbes, ammonium, and COD on the release of Fe and Mn from sediments.

4. Conclusions

The effect of anthropogenic pollutants on the release of Fe and Mn from sediments has been investigated by column experiments. The results showed that ammonium and COD significantly enhance the release of Fe and Mn from sediments to groundwater. The leached concentrations of Fe and Mn were similar, and the maximum leached concentrations were 0.32 and 40 μg/L, respectively. The diversity and abundance of the microbial community structure was closely related to the contents of Fe and Mn. *Actinobacteriota*, *Proteobacteria*, *Acidobacteriota*, *Bacteroidota*, and *Chloroflexi* were the dominant phyla, and *Rhodococcus*, *Ochrobactrum*, and *Pseudarthrobacter* were the dominant genera in the sediments. These microbes not only promote the release of Fe/Mn by reducing iron and manganese minerals, but they also actively participate in the biogeochemical cycles of Fe and Mn.

This study provides a good explanation for why the concentrations of Fe and Mn in groundwater spatially vary in the natural environment. The results show that the discharge of ammonium and COD can increase the concentration of Fe and Mn in groundwater. Therefore, more attention should be paid to the importance of ammonium ions and organics in the geochemical processes of Fe and Mn in sediments, which will facilitate the further development of groundwater pollution control measures.

Supplementary Materials: The following supporting information can be downloaded at: https://www.mdpi.com/article/10.3390/w15010120/s1, Table S1: The average value of pH, DO concentration and Eh over time; Table S2: The average value of Fe and Mn concentration over time; Table S3: The proportion of microbial community abundance in sediment samples of different soil columns.

Author Contributions: Methodology, Y.T.; formal analysis, X.X.; investigation, Y.Z.; data curation, Y.T.; writing—original draft, X.X.; writing—review and editing, X.X.; supervision, Y.T.; funding acquisition, Y.Z. All authors have read and agreed to the published version of the manuscript.

Funding: This work was supported by the National Natural Science Foundation of China (No. 41877355, 42077170 and 41831283), Beijing Advanced Innovation Program for Land Surface Science, and the 111 Project of China (No. B16020).

Data Availability Statement: Not applicable.

Conflicts of Interest: The authors declare no conflict of interest.

References

1. Abdelrady, A.; Bachwenkizi, J.; Sharma, S.; Sefelnasr, A.; Kennedy, M. The fate of heavy metals during bank filtration: Effect of dissolved organic matter. *J. Water Process Eng.* **2020**, *38*, 101563. [CrossRef]
2. Liu, R.; Ma, T.; Qiu, W.; Du, Y.; Liu, Y. Effects of Fe oxides on organic carbon variation in the evolution of clayey aquitard and environmental significance. *Sci. Total Environ.* **2020**, *701*, 134776. [CrossRef] [PubMed]
3. Wu, B.; Amelung, W.; Xing, Y.; Bol, R.; Berns, A.E. Iron cycling and isotope fractionation in terrestrial ecosystems. *Earth Sci. Rev.* **2019**, *190*, 323–352. [CrossRef]
4. Ying, S.C.; Schaefer, M.V.; Cock-Esteb, A.; Li, J.; Fendorf, S. Depth Stratification Leads to Distinct Zones of Manganese and Arsenic Contaminated Groundwater. *Environ. Sci. Technol.* **2017**, *51*, 8926–8932. [CrossRef]
5. Moon, J.W.; Paradis, C.J.; Joyner, D.C.; Majumder, E.L.; Dixon, E.R.; Podar, M.; Ge, X.; Walian, P.J.; Smith, H.J. Characterization of subsurface media from locations up- and down-gradient of a uranium-contaminated aquifer. *Chemosphere* **2020**, *255*, 126951. [CrossRef]

6. Wallis, I.; Prommer, H.; Berg, M.; Siade, A.J.; Sun, J.; Kipfer, R. The river-groundwater interface as a hotspot for arsenic release. *Nat. Geosci.* **2020**, *13*, 288–295. [CrossRef]
7. Xia, X.L.; Teng, Y.G.; Zhai, Y.Z. Influencing factors and mechanism by which DOM in groundwater releases Fe from sediment. *Chemosphere* **2022**, *300*, 134524. [CrossRef]
8. Yang, Y.; Hu, M.; Zhou, D.; Fan, W.; Wang, X.; Huo, M. Bioremoval of Cu^{2+} from CMP wastewater by a novel copper-resistant bacterium Cupriavidus gilardii CR3: Characteristics and mechanisms. *RSC Adv.* **2017**, *7*, 18793–18802. [CrossRef]
9. Bryant, S.R.; Sawyer, A.H.; Briggs, M.A.; Saup, C.M.; Nelson, A.R.; Wilkins, M.J.; Christensen, J.N.; Williams, K.H. Seasonal manganese transport in the hyporheic zone of a snowmelt-dominated river (East River, Colorado, USA). *Hydrogeol. J.* **2020**, *28*, 1323–1341. [CrossRef]
10. Zhai, Y.Z.; Han, Y.F.; Xia, X.L.; Li, X.D.; Lu, H.; Teng, Y.G.; Wang, J.S. Anthropogenic Organic Pollutants in Groundwater Increase Releases of Fe and Mn from Aquifer Sediments: Impacts of Pollution Degree, Mineral Content, and pH. *Water* **2021**, *13*, 1920. [CrossRef]
11. Papp, D.C.; Cociuba, I.; Baciu, C.; Cozma, A. Origin and Geochemistry of Mine Water and its Impact on the Groundwater and Surface Running Water in Post-mining Environments: Zlatna Gold Mining Area (Romania). *Aquat. Geochem.* **2017**, *23*, 247–270. [CrossRef]
12. Nguyen, B.T.; Nguyen, T.M.T.; Bach, Q.-V. Assessment of groundwater quality based on principal component analysis and pollution source-based examination: A case study in Ho Chi Minh City, Vietnam. Environ. *Monit. Assess.* **2020**, *192*, 395. [CrossRef] [PubMed]
13. Di Curzio, D.; Rusi, S.; Signanini, P. Advanced redox zonation of the San Pedro Sula alluvial aquifer (Honduras) using data fusion and multivariate geostatistics. *Sci. Total Environ.* **2019**, *695*, 133796. [CrossRef] [PubMed]
14. Xia, X.L.; Teng, Y.G.; Zhai, Y.Z. Influence of DOM and microbes on Fe biogeochemistry at a riverbank filtration site. *Environ. Res.* **2022**, *216*, 114430. [CrossRef]
15. Caschetto, M.; Barbieri, M.; Galassi, D.M.; Mastrorillo, L.; Rusi, S.; Stoch, F.; Petitta, M. Human alteration of groundwater–surface water interactions (Sagittario River, Central Italy): Implication for flow regime, contaminant fate and invertebrate response. Environs. *Earth Sci.* **2014**, *71*, 1791–1807. [CrossRef]
16. Di Curzio, D.; Palmucci, W.; Rusi, S.; Signanini, P. Multidisciplinary approach to assess the seasonal effect on redoc processes occurring in a tropical alluvial aquifer. In Proceedings of the 3rd National Meeting on Hydrogeology, Cagliari, Italy, 14–16 June 2017; pp. 11790–11805. [CrossRef]
17. Xiu, W.; Lloyd, J.; Guo, H.; Dai, W.; Nixon, S.; Bassil, N.M.; Ren, C.; Zhang, C.; Ke, T.; Polya, D. Linking microbial community composition to hydrogeochemistry in the western Hetao Basin: Potential importance of ammonium as an electron donor during arsenic mobilization. *Environ. Int.* **2020**, *136*, 105489. [CrossRef]
18. Yang, L.; Li, X.K.; Chu, Z.R.; Ren, Y.H.; Zhang, J. Distribution and genetic diversity of the microbe in the biofilter for the simultaneous removal of arsenic, iron and manganese from simulated groundwater. *Bioresour. Technol.* **2014**, *156*, 384–388. [CrossRef]
19. Yamamura, S.; Kurasawa, H.; Kashiwabara, Y.; Hori, T.; Aoyagi, T.; Nakajima, N.; Amachi, S. Soil microbial communities involved in reductive dissolution of arsenic from arsenate–laden minerals with different carbon sources. *Environ. Sci. Technol.* **2019**, *53*, 12398–12406. [CrossRef]
20. Ibekwe, A.; Ma, J.; Murinda, S. Bacterial community composition and structure in an urban river impacted by different contaminant sources. *Sci. Total Environ.* **2016**, *566–567*, 1176–1185. [CrossRef]
21. Tong, M.; Yuan, S.; Wang, Z.; Luo, M.; Wang, Y. Electrochemically induced oxidative removal of As (III) from groundwater in a dual-anode sand column. *J. Hazard. Mater.* **2016**, *305*, 41–50. [CrossRef]
22. Xie, S.; Yuan, S.; Liao, P.; Tong, M.; Gan, Y.; Wang, Y. Iron-anode enhanced sand filter for arsenic removal from tube well water. *Environ. Sci. Technol.* **2017**, *51*, 889–896. [CrossRef] [PubMed]
23. Bai, J.; Su, X.; Wang, J.; Lyu, H.; Gao, R.; Lu, S. Multi–isotope constraints on biogeochemical processes during bank filtration: A case study of the Liao River. Northeast China. *Appl. Geochem.* **2020**, *122*, 104762. [CrossRef]
24. Peter, B.M.; Kenneth, B.; James, E.R.; Tyler, D.J. Elevated manganese concentrations in united states groundwater, role of land surface-soil-aquifer connections. *Environ. Sci. Technol.* **2019**, *53*, 29–38. [CrossRef]
25. Gao, Z.P.; Jia, Y.F.; Guo, H.M.; Zhang, D.; Zhao, B. Quantifying geochemical processes of arsenic mobility in groundwater from an inland basin using a reactive transport model. *Water Resour. Res.* **2020**, *56*, e2019WR025492. [CrossRef]
26. Xin, J.; Liu, Y.; Chen, F.; Duan, Y.; Wei, G.; Zheng, X.; Li, M. The missing nitrogen pieces: A critical review on the distribution, transformation, and budget of nitrogen in the vadose zone-groundwater system. *Water Res.* **2019**, *165*, 114977. [CrossRef]
27. Robertson, W.D.; Moore, T.A.; Spoelstra, J.; Li, L.; Elgood, R.J.; Clark, I.D.; Schiff, S.L.; Aravena, R.; Neufeld, J.D. Natural attenuation of septic system nitrogen by anammox. *Groundwater* **2012**, *50*, 541–553. [CrossRef]
28. Rivett, M.O.; Buss, S.R.; Morgan, P.; Smith, J.W.N.; Bemment, C.D. Nitrate attenuation in groundwater: A review of biogeochemical controlling processes. *Water Res.* **2008**, *42*, 4215–4232. [CrossRef]
29. Mogollón, J.M.; Mewes, K.; Kasten, S. Quantifying manganese and nitrogen cycle coupling in manganese-rich, organic carbon-starved marine sediments: Examples from the clarion-clipperton fracture zone. *Geophys. Res. Lett.* **2016**, *43*, 7114–7123. [CrossRef]

30. Wang, J.W.; Teng, Y.G.; Zhang, C.X.; Liao, X.P.; Zhai, Y.Z.; Zuo, R. Activation of manganese dioxide with bisulfite for enhanced abiotic degradation of typical organophosphorus pesticides: Kinetics and transformation pathway. *Chemosphere* **2019**, *226*, 858–864. [CrossRef]
31. Liu, W.; Wang, Y.; Li, J.; Qian, K.; Xie, X. Indices of the dual roles of OM as electron donor and complexing compound involved in As and Fe mobilization in aquifer systems of the Datong Basin. *Environ. Pollut.* **2020**, *262*, 114305. [CrossRef]
32. Amarathunga, U.; Diyabalanage, S.; Bandara, U.G.C.; Chandrajith, R. Environmental factors controlling arsenic mobilization from sandy shallow coastal aquifer sediments in the Mannar Island, Sri Lanka. *Appl. Geochem.* **2019**, *100*, 152–159. [CrossRef]
33. Liu, S.; Chui, T. Numerical modelling to evaluate the nitrogen removal rate in hyporheic zone and its implications for stream management. *Hydrol. Process.* **2019**, *33*, 3084–3097. [CrossRef]
34. Wu, J.; Gu, L.; Hua, Z.; Liang, Z.; Chu, K.; He, X. Per-, poly-fluoroalkyl substances (PFASs) pollution in benthic riverine ecosystem: Integrating microbial community coalescence and biogeochemistry with sediment distribution. *Chemosphere* **2021**, *281*, 130977. [CrossRef] [PubMed]
35. Zhang, Y.; Ji, G.; Wang, R. Functional gene groups controlling nitrogen transformation rates in a groundwater–restoring denitrification biofilter under hydraulic retention time constraints. *Ecol. Eng.* **2016**, *87*, 45–52. [CrossRef]
36. Lyu, H.; Su, X.; Wang, Y.; Dai, Z.; Liu, M. Effectiveness and mechanism of natural attenuation at a petroleum-hydrocarbon contaminated site. *Chemosphere* **2018**, *206*, 293–301. [CrossRef]
37. Xia, X.; Zhai, Y.; Teng, Y. Microbial response to biogeochemical profile in a perpendicular riverbank filtration site. *Ecotoxicol. Environ. Saf.* **2022**, *244*, 114070. [CrossRef]
38. Lu, S.; Yang, Y.M.; Yin, H.L.; Su, X.S.; Yu, K.N.; Sun, C. Microbial community structure of arsenic-bearing groundwater environment in the riverbank filtration zone. *Water* **2022**, *14*, 1548. [CrossRef]

water

Article

Microbial Community Structure of Arsenic-Bearing Groundwater Environment in the Riverbank Filtration Zone

Shuai Lu [1,2,3], Yimeng Yang [1], Hanling Yin [1], Xiaosi Su [4,*], Kaining Yu [1,*] and Chao Sun [3,*]

1 Hebei Province Collaborative Innovation Center for Sustainable Utilization of Water Resources and Optimization of Industrial Structure, Hebei Center for Ecological and Environmental Geology Research, Hebei Province Key Laboratory of Sustained Utilization and Development of Water Resources, School of Water Resources and Environment, Hebei GEO University, Shijiazhuang 050031, China; lushuai@hgu.edu.cn (S.L.); yymeng21@126.com (Y.Y.); hlyins@126.com (H.Y.)

2 Key Laboratory of Groundwater Contamination and Remediation of Hebei Province and China Geological Survey, The Institute of Hydrogeology and Environmental Geology, Chinese Academy of Geological Sciences, Shijiazhuang 050061, China

3 Hebei Key Laboratory of Geological Resources and Environment Monitoring and Protection, Shijiazhuang 050021, China

4 Institute of Water Resources and Environment, Jilin University, Changchun 130021, China

* Correspondence: suxs16@163.com (X.S.); yukn2000@163.com (K.Y.); hkyschao@163.com (C.S.)

Abstract: Arsenic (As) contamination of groundwater is a global public health problem. Microorganisms have a great effect on the migration and transformation of arsenic. Studying the effect of microbial community structure and function on arsenic release in the groundwater environment of the riverbank filtration zone has important theoretical and practical significance. In this paper, in-situ monitoring technology and molecular biology technology were used to study the microbial community in the process of river water infiltration in the Shenyang Huangjia water source, China. The results showed that the structure, diversity and abundance of the microbial community in groundwater were closely related to the arsenic content. Proteobacteria was the dominant phylum in groundwater of the study area, and *Acinetobacter*, *Pseudomonas*, *Sulfuritalea*, *Sphingomonas* and *Hydrogenophaga* etc. were the main dominant bacterial genera. In addition to reducing and oxidizing arsenic, these functional microorganisms also actively participated in the biogeochemical cycle of elements such as iron, manganese, nitrogen and sulfur. There was a significant correlation between dominant bacteria and environmental factors. Fe/Mn had a significant positive correlation with As, which brought potential danger to the water supply in high iron and manganese areas.

Keywords: microbial community structure; arsenic-bearing groundwater environment; riverbank filtration zone

check for updates

Citation: Lu, S.; Yang, Y.; Yin, H.; Su, X.; Yu, K.; Sun, C. Microbial Community Structure of Arsenic-Bearing Groundwater Environment in the Riverbank Filtration Zone. *Water* **2022**, *14*, 1548. https://doi.org/10.3390/w14101548

Academic Editor: Dimitrios E. Alexakis

Received: 19 April 2022
Accepted: 10 May 2022
Published: 12 May 2022

1. Introduction

Driven by riverside exploitation, strong interactions occur between river water and groundwater [1–3]. The riverbank filtration zone, as the key zone of interaction, affects the migration and transformation of pollutants in the process of river water infiltration [4–7]. Arsenic is a typical toxic carcinogen. Long-term drinking of arsenic-bearing groundwater will do great harm to human health, which can lead to neurasthenia, cardiovascular disease, liver cancer, kidney cancer and so on [8–10]. In view of the high risk of arsenic, the World Health Organization (WHO), the European Union and many countries (including China and the United States) stipulate that the concentration of arsenic in drinking water should not exceed 10 μg/L [11,12]. Once arsenic contamination occurs, the impact on human beings is very serious [13–16].

The formation of arsenic-bearing groundwater is the result of a series of water-rock interactions, hydrogeochemistry and geological microorganisms. The release mechanisms are mainly as follows: desorption, reductive dissolution of arsenic-bearing minerals, oxidation of arsenic-bearing minerals and the direct reduction of arsenic [17–22]. There are mainly two forms of arsenic in nature, inorganic arsenic and organic arsenic, and its main valence states are As^{3+} and As^{5+}. These two valence states of arsenic are the main existing forms in groundwater and soil. Arsenic mainly exists in the form of arsenite or arsenate in groundwater, and As^{3+} has stronger mobility and biological toxicity than As^{5+} [23–25]. Most of these studies are based on traditional hydrogeochemical or mineralogical methods. With the continuous deepening of research, scholars found that microorganisms affect the geochemical process of arsenic in groundwater [26–28]. Microorganisms with an arsenic metabolism function can directly promote the redox and methylation of arsenic, which affects the migration and transformation of arsenic [29–31]. Indigenous iron-reducing bacteria can promote the reduction of iron hydroxides and release the adsorbed arsenic [32–34]. The reduced sulfur produced by microbial reduction can form arsenic-bearing sulfide precipitation with dissolved arsenic in groundwater. If the arsenic–sulfur complex is formed in this process, it will promote the migration of arsenic in groundwater, because it is not easy to be adsorbed [35]. Microbial activities in the groundwater system, especially the microbial-mediated cycling and metabolism of elements such as iron and manganese, affect the migration and transformation of arsenic in the arsenic-bearing groundwater environment [36–39].

Previous studies have shown that arsenic enrichment during riverbank filtration is closely related to microorganisms, and most of them focus on the effect of a certain bacterial species on arsenic. However, the effects and potential mechanisms of microbial community structure and function on arsenic release in the groundwater environment of the riverbank filtration zone are still unclear. Therefore, an in-depth study on the microbial community composition and functional genes of the arsenic-containing groundwater environment is of great scientific significance for understanding the arsenic release and the formation of arsenic-bearing groundwater under the influence of microorganisms.

Shenyang Huangjia water source is located in the north of Shenyang City, China. The content of iron, manganese and arsenic in groundwater exceeds the standard seriously [40]. At present, the source and cause of arsenic in the groundwater of this water source are not completely clear. Therefore, this paper uses in-situ monitoring technology and molecular biology technology to analyze the microbial community structure characteristics of the arsenic-bearing groundwater environment in the riverbank filtration zone, and it reveals the impact of the microbial community on the biogeochemical process of arsenic in groundwater. This study can deeply understand the formation law of groundwater chemical composition and reveal the formation of arsenic-bearing groundwater under the influence of microorganisms, which is of great significance to the monitoring and protection of groundwater.

2. Materials and Methods

2.1. Study Area Description

The study area is located in the Shenyang Huangjia water source, and the Liao River flows from east to west in the north of the study area (Figure 1). Groundwater is stored in an unconsolidated phreatic Quaternary aquifer, with a buried depth of 1–4 m. The aquifer is about 50 m thick and is mainly composed of fine sand, medium coarse sand and gravel. Twelve mining wells are arranged along the river, with a total mining volume of 30,000 m^3/d. The groundwater receives lateral recharge from the Liao River all the year round.

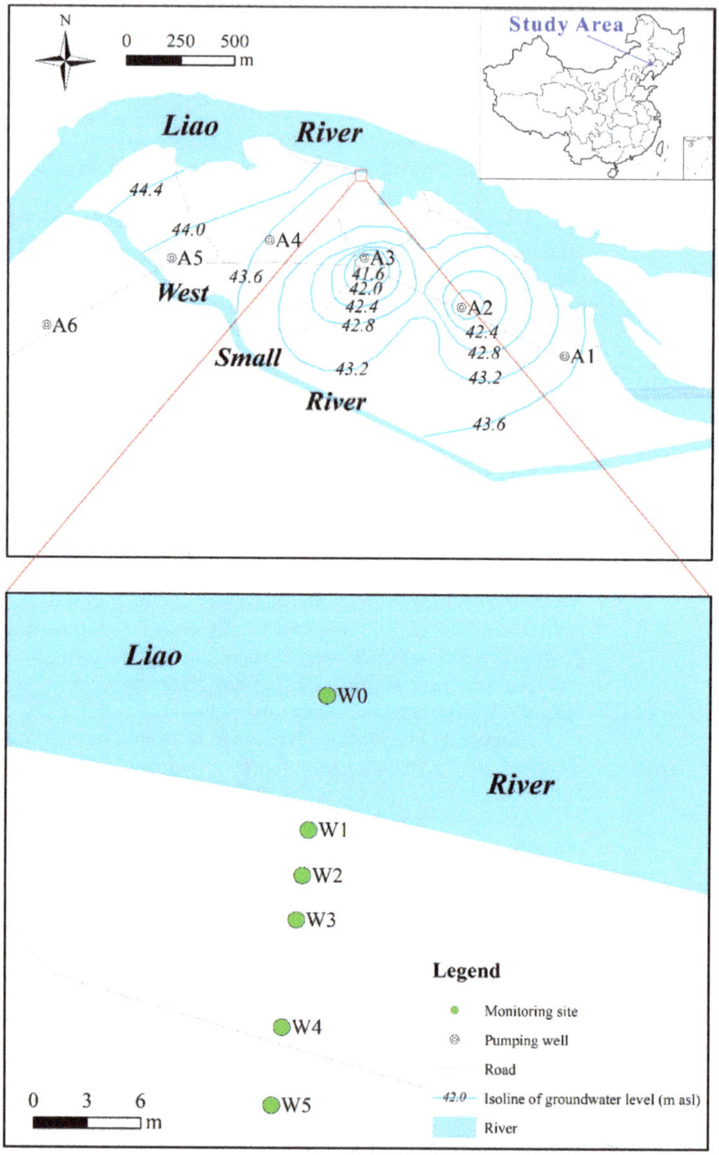

Figure 1. Location of the study area and distribution of monitoring sites.

2.2. Sampling and Testing

Along the groundwater flow direction, five monitoring wells (numbered W1, W2, W3, W4 and W5) were set up at 1.5 m, 4 m, 6.5 m, 12.5 m and 17 m away from the south bank of the Liao River, while these five wells were each screened to 10.0 m below the ground surface. The river water sampling point W0 was arranged about 6 m away from the south bank. We carried out sample collection in August and collected five groundwater samples and one river water sample. Before collecting groundwater samples, at least 3–5-times the volume of the monitoring well pipe was extracted, and a portable water quality analyzer (HQ40d, Hach Company, Loveland, CO, USA) was used to measure the groundwater temperature (T), pH, oxidation-reduction potential (ORP), dissolved oxygen (DO) and other indicators on site. Samples should have been taken after the readings were stable. The river water samples were collected about 20 cm below the water surface. The collected water samples were sealed with sealing film without headspace after adding a protective agent according to the preservation requirements [41].

Fe^{2+} and Mn^{2+} were measured in situ by a spectrophotometer (DR2800, Hach Company, Loveland, CO, USA). Conventional anions and cations were determined by ion chromatography (881 Compact IC, Metrohm AG, Herisau, Switzerland), and HCO_3^- was determined by acid–base titration. After the water samples were filtered by a GF/F membrane (1825-047 Whatman Glass Microfiber Filters, Cytiva Company, Buckinghamshire, UK), the DOC was determined by a TOC instrument (TOC-L CPH CN200, Shimadzu Corporation, Kyoto, Japan). As was determined by inductively coupled plasma mass spectrometer (7500C ICP-MS, Agilent Technologies, Santa Clara, CA, USA). The above tests were completed in the Key Laboratory of Groundwater Resources and Environment, Ministry of Education, Jilin University. The microbial community composition and arsenic functional genes were analyzed by 16S rRNA high-throughput sequencing, which was completed by Beijing Allwegene Technology Co., Ltd., Beijing, China.

3. Results and Discussion

3.1. Hydrochemical Characteristics

The pH of the river water in the study area was 8.17 (Table 1). The pH of groundwater ranged from 7.16 to 7.52, which was weakly alkaline on the whole, and it gradually decreased along the groundwater flow direction (Figure 2). The ORP and DO content of the river water were 35.02 mV and 9.34 mg/L, respectively, indicating that it was in a state of oxidation. The ORP of groundwater ranged from −135.23 mV to −11.54 mV, which decreased to −135.23 mV at 17 m away from the riverbank, showing strong reducibility. Along the groundwater flow direction, the ORP and DO content gradually decreased, reflecting that the infiltrated river water experienced a transition from an oxidation environment to a reduction environment. The DOC content in river water was relatively high. Along the groundwater flow direction, the DOC content in groundwater gradually decreased, reflecting the strong redox reaction between DOC and electron acceptors in groundwater.

Table 1. Physico-chemical parameters in river water (W0) and groundwaters (W1–W5).

Sample ID	W0	W1	W2	W3	W4	W5
Distance to the riverbank (m)	−6	1.5	4	6.5	12.5	17
T (°C)	22.24	14.71	13.17	12.21	10.94	10.49
pH	8.17	7.52	7.36	7.33	7.27	7.16
ORP (mV)	35.02	−11.54	−69.88	−94.45	−118.79	−135.23
DO (mg/L)	9.34	7.56	2.07	2.11	1.93	1.90
DOC (mg/L)	18.76	9.79	9.34	8.16	7.27	6.69
K^+ (mg/L)	6.32	5.17	3.39	3.41	3.75	8.31
Na^+ (mg/L)	31.27	49.66	43.87	39.85	32.37	30.00
Ca^{2+} (mg/L)	50.46	91.96	87.42	101.46	116.97	129.05
Mg^{2+} (mg/L)	20.90	28.17	27.18	32.64	40.43	109.35
Cl^- (mg/L)	41.80	56.56	57.26	60.11	69.78	27.57
HCO_3^- (mg/L)	169.00	378.41	348.80	431.06	480.42	957.55
NO_3^- (mg/L)	7.69	2.36	1.77	0.59	0.33	0.30
SO_4^{2-} (mg/L)	75.50	68.63	68.55	61.01	62.58	54.79
As (μg/L)	2.21	37.04	40.69	46.95	48.80	51.34
Fe^{2+} (mg/L)	0.11	13.35	17.02	19.93	21.89	25.93
Mn^{2+} (mg/L)	0.14	2.00	5.42	6.14	7.56	7.42

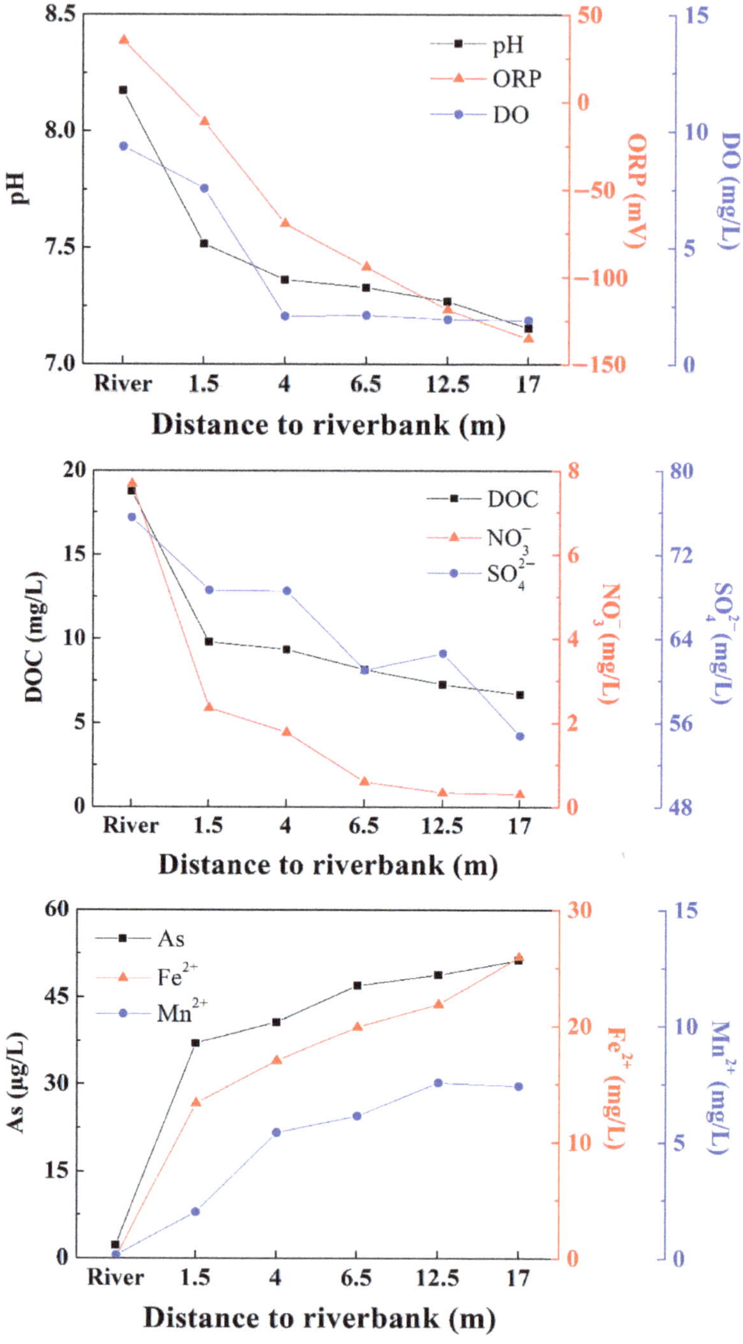

Figure 2. Variations in hydrochemical parameters along the groundwater flow direction.

The hydrochemical type of river water was a Ca-Mg-Na-HCO_3-SO_4-Cl type. The hydrochemical type of groundwater was mainly a Ca-Mg-HCO_3 type, and the SO_4^{2-} content decreased, which was affected by reduction (Figure 3). Along the groundwater flow direction, the SO_4^{2-} content decreased from 75.50 mg/L at the river water to 54.79 mg/L at W5, showing the consumption of SO_4^{2-} in groundwater, indicating the gradual enhancement of SO_4^{2-} reduction, pointing to the possible activity of sulfate-reducing bacteria in groundwater. As a sensitive redox component, the NO_3^- content gradually decreased to 0.30 mg/L along the groundwater flow direction, reflecting the strong denitrification effect of NO_3^- as an electron acceptor.

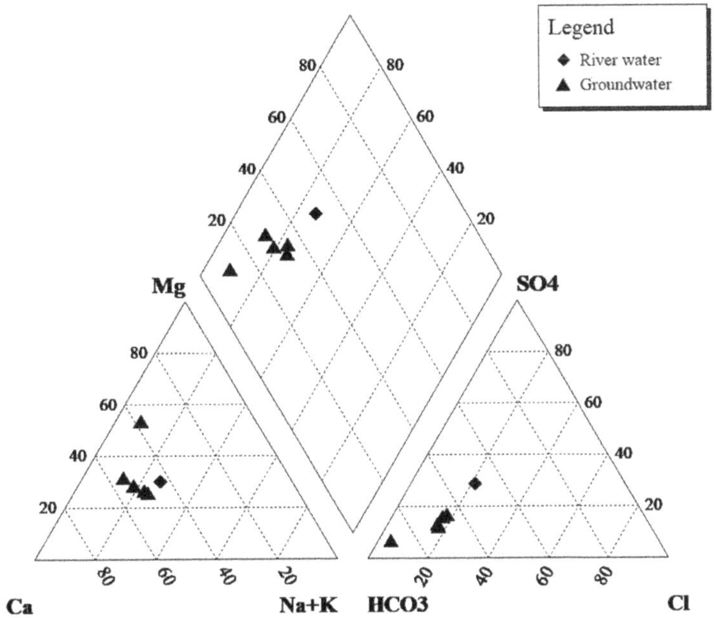

Figure 3. Piper diagram of river water and groundwater samples.

The content of Fe^{2+} and Mn^{2+} in the river water was low, both being less than 0.2 mg/L. Along the groundwater flow direction, the content of Fe^{2+} and Mn^{2+} in groundwater gradually increased. At W5, the content increased to 25.93 mg/L and 7.42 mg/L, respectively, indicating that the reductive dissolution of iron and manganese minerals occurred in the gradually reducing environment, which increased the content of Fe^{2+} and Mn^{2+} in groundwater. Similarly, the As content in the river water was low, which was far below the 10.0 μg/L set by the WHO for water quality standards, indicating that the river water may not be polluted by arsenic. Along the groundwater flow direction, the As content in groundwater gradually increased. At W5, the As content increased to 51.34 μg/L, which was about 25-times the As content in river water. At the initial stage of river infiltration, some arsenic-bearing minerals were oxidized and dissolved to release As. With the transformation of groundwater to the reducing environment, Fe/Mn minerals underwent reductive dissolution, and As adsorbed on the mineral surface, and the lattice was released.

As was significantly positively correlated with Fe^{2+} and Mn^{2+} and negatively correlated with NO_3^- and SO_4^{2-} (Table 2). NO_3^- and SO_4^{2-} also had a negative correlation with Fe^{2+} and Mn^{2+}, indicating that with the infiltration of river water, Fe^{3+}, Mn^{4+}, NO_3^- and SO_4^{2-} were reduced in the process of increasing the arsenic concentration. There was a significant correlation between the components, indicating that there was a close relationship between iron, manganese, nitrogen and sulfur and the release of arsenic.

Table 2. Correlations between As, Fe^{2+}, Mn^{2+}, nitrate and sulfate.

Component	As	Fe^{2+}	Mn^{2+}	NO_3^-	SO_4^{2-}
As	1.000				
Fe^{2+}	1.000 **	1.000			
Mn^{2+}	0.943 **	0.943 **	1.000		
NO_3^-	−1.000 **	−1.000 **	−0.943 **	1.000	
SO_4^{2-}	−0.943 **	−0.943 **	−0.829 *	0.943 **	1.000

Notes: ** correlation is significant at a confidence level of 0.01; * correlation is significant at a confidence level of 0.05.

3.2. Microbial Species Abundance and Diversity

Due to different physical and chemical conditions, microbial communities have different structures and compositions. The interactions among various communities and the collection of various enzymatic reactions play a significant role in the biogeochemical process of arsenic. The samples were classified by Operational Taxonomic Units (OUT) species, and a total of 2330 OTUs were obtained, belonging to 40 phyla, 106 classes and 302 genera. The microbial abundance of river water and groundwater was different, showing strong spatial heterogeneity. The species abundance of river water was relatively high, 1245 OTUs, and that of groundwater was relatively low, 157–288 OTUs (Figure 4). Along the groundwater flow direction, the OTUs decreased gradually on the whole.

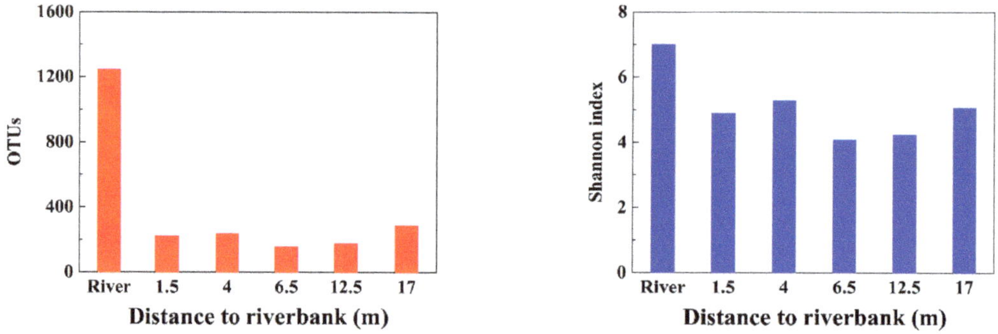

Figure 4. Microbial abundance and diversity.

The Shannon index of the river water was high, indicating that the river water had a high species diversity and rich species. Along the groundwater flow direction, the microbial diversity in groundwater generally decreased gradually, while the corresponding arsenic content in groundwater was high, reflecting that higher arsenic content had a strong inhibitory effect on microorganisms in groundwater, resulting in fewer species [42–44].

It can be seen from Figure 5 that the PC1 principal axis could explain 44.36% of the sample composition differences, and the PC2 principal axis could explain 13.79%. In the Principal Co-ordinates Analysis (PCoA) diagram, the closer the distance between two samples, the higher the similarity of species [45]. The microbial species composition of the five groundwater samples was similar, which was distributed in the upper left part of the PCoA diagram, while the river water sample was distributed in the lower right part, showing the difference in microbial species composition between river water and groundwater samples.

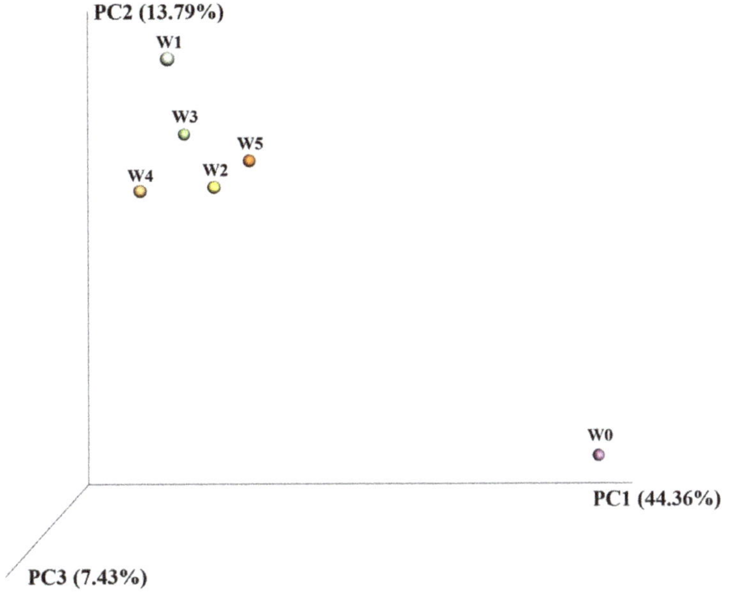

Figure 5. Principal Co-ordinates Analysis 3D map.

W1, which was spatially closer to the river water, had an obvious deviation from the river water (W0) in the PCoA diagram. With the increasing distance from the riverbank, the deviation between groundwater and river water decreased, indicating that W1 and river water species were less similar than other groundwater, and there were certain differences in the communities. There was a high physical and chemical gradient between W1 and the river water; the microbial activity in the groundwater was high, and the biogeochemical effect was relatively strong. With the further infiltration of river water, the species abundance decreased, and the gradient of environmental conditions decreased relatively.

3.3. Microbial Species Composition

Proteobacteria (38.4–85.4%) was the dominant bacterial phylum, followed by Bacteroidetes (4.1–18.8%) and Cyanobacteria (1.2–23.1%) (Figure 6). Proteobacteria was the most common and dominant bacterial phylum in arsenic-contaminated environments, with numerous arsenic-resistant bacteria [46], and its abundance generally increased first and then decreased along the groundwater flow. Among the Proteobacteria, *Alphaproteobacteria* was the dominant bacterial class, accounting for 15.2–33.7%, followed by *Betaproteobacteria* (8.7–34.9%), *Gammaproteobacteria* (1.9–30.3%) and *Deltaproteobacteria* (0.3–22.5%).

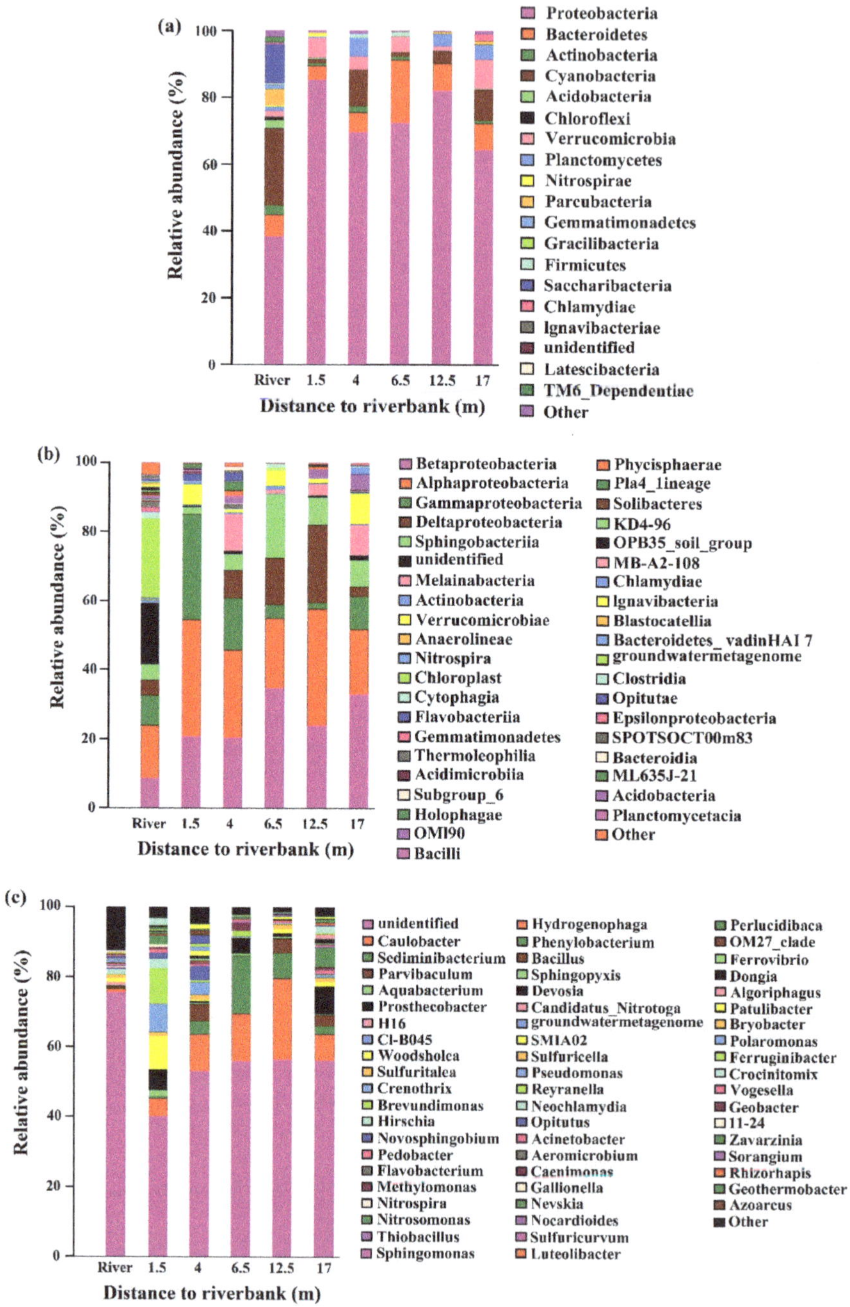

Figure 6. Microbial community composition: (**a**) phylum-level; (**b**) class-level; (**c**) genus-level (Other: the summation of the bacteria, of which the relative abundance was considerably below 2%).

Alphaproteobacteria included *Sphingomonas* and *Brevundimonas*, which have the ability of arsenic reduction and iron reduction [47].

Betaproteobacteria included many aerobic or facultative bacteria, usually with variable degradability. The study area mainly included *Vogesella*, *Aquabacterium*, *Polaromonas*, *Gallionella*, *Hydrogenophaga*, *Sulfuritalea*, etc. *Aquabacterium* is an iron-reducing bacterium. At the same time, as a facultative aerobic bacterium, *Aquabacterium* can use nitrate as an electron acceptor for nitrate reduction, accounting for a relatively high proportion in W1, which is related to the rapidly decreasing NO_3^- content along the groundwater flow, indicating that its existence affects the change of nitrate content in groundwater. *Gallionella* and *Hydrogenophaga* are ubiquitous in groundwater and can catalyze the reduction of As(V) to As(III), reduce part of arsenite and also catalyze the reduction and dissolution of iron manganese oxide bound iron, releasing arsenic in the lattice into groundwater [48].

The species abundance of *Gammaproteobacteria* was significantly different. The study area mainly included *Acinetobacter*, *Pseudomonas*, *Perlucidibaca*, *Aeromicrobium* and *Flavobacterium*. *Acinetobacter* is a typical bacterium with an arsenic metabolism ability, which is widely distributed in the arsenic-contaminated environment. Some strains have significant arsenic resistance compared with other bacteria, and some strains can reduce or oxidize arsenic. In addition, some strains also show an iron reduction and sulfur metabolism capacity [49–51]. *Pseudomonas* is mostly facultative anaerobic bacteria, which can use nitrate as an electron acceptor for metabolism and participate in denitrification. At present, many *Pseudomonas* strains that have been found are arsenic-resistant bacteria, which is closely related to the release and migration of arsenic [52–54]. At the same time, Drewniak et al. found that some *Pseudomonas* can produce a compound bound to iron to promote the migration of arsenic from natural minerals [55,56].

Most *Deltaproteobacteria* can utilize various ions as electron acceptors and widely participate in sulfate reduction and iron reduction in the environment, such as sulfate-reducing bacteria and sulfur-reducing bacteria, etc. [48]. The high presence of sulfate-reducing bacteria in a certain location of the aqueous medium indicates that the source of arsenic in the groundwater may not be the reduction of iron oxides, but the pyrite provides the substrate for sulfate-reducing bacteria to metabolize, resulting in the release of arsenic.

It can be seen that the species and genetic diversity of Proteobacteria are extremely rich, covering a wide range of physiological metabolism types, and they are widely present in groundwater in the study area. Microbial bacteria show certain metabolic functions of arsenic, iron, manganese, sulfur and nitrogen; actively participate in the redox process and have a great effect on the migration and transformation of various elements.

3.4. The Impact of Environmental Factors on Microbial Communities

Environmental factors affect the composition of the microbial community in groundwater. A Redundancy Analysis (RDA) was used to analyze the microbial community and environmental factors. The length of the arrows of environmental factors indicates the degree of its impact on the microbial community. As, Fe^{2+}, Mn^{2+}, NO_3^- and SO_4^{2-} affect the microbial community (Figure 7). The angles between the arrows of environmental factors represent positive and negative correlations. As was positively correlated with Fe^{2+} and Mn^{2+} environmental factors and negatively correlated with NO_3^- and SO_4^{2-}, which was consistent with the results of the Spearman correlation analysis. As, Fe^{2+} and Mn^{2+} pointed to W3, W4 and W5, indicating that they affected the microbial community structure of the groundwater sample in these three reducing-environment areas, corresponding to the high concentrations of arsenic, iron and manganese in these groundwater samples. *Sphingomonas* and *Hydrogenophaga*, which have the ability to metabolize arsenic and iron, accounted for a high proportion. *Sulfuritalea* is not only an arsenic dissimilatory reducing bacteria, but it is also a sulfur oxidizing bacteria and denitrifying bacteria; *Pseudomonas* is not only an arsenic-resistant bacteria, but it also can participate in denitrification; *Acinetobacter* not only has the ability of arsenic metabolism, but it also shows the ability for iron reduction and sulfur metabolism. These three bacteria had a good correlation with NO_3^- and SO_4^{2-}.

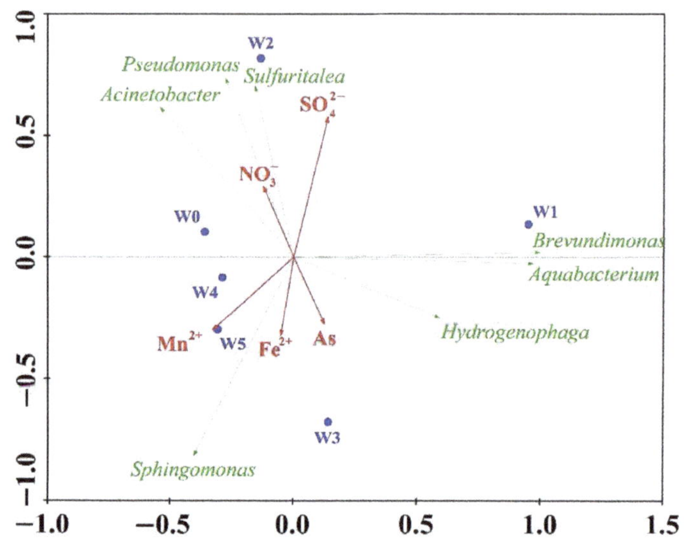

Figure 7. Redundancy analysis in the genus-level.

3.5. Arsenic Functional Genes

There were bacteria containing the As(V)-reducing gene *arr*A and the As(III)-oxidizing genes *aio*A and *arx*A in the river water. However, in the groundwater, only bacteria containing the As(V)-reducing gene *arr*A, but no As(III)-oxidizing genes, were found (Table 3).

Table 3. OTUs of river water and groundwater.

Gene Groups	W0	W1	W2	W3	W4	W5
*arr*A	1081	1162	1016	1082	1024	905
*aio*A	18	-	-	-	-	-
*arx*A	351	-	-	-	-	-

The bacteria containing *arr*A were mainly Proteobacteria and Actinobacteria, including a great quantity of reducing bacteria, such as *Geobacter*, *Sulfuritalea*, *Desulfuromonas*, etc. (Figure 8). Dissimilatory Arsenate-Reducing Prokaryotes (DARPs) existing in anaerobic environments can use As(V) as an electron acceptor for redox reactions to obtain energy for cell growth. This reduction mechanism is called the respiratory reduction mechanism [57–59]. The functional gene *arr*A is a reliable marker for the reduction of As(V) by DARPs, and microorganisms containing *arr*A have a great effect on the biogeochemical cycle of arsenic [60–62]. DARPs can reduce As(V) in amorphous iron oxide, alumina or arsenic-containing minerals; release arsenic in sediments and increase the arsenic content, especially the As(III) content in groundwater [63–65]. Although microorganisms containing the *arr*A gene are mostly detected in anaerobic environments, some scholars have cloned the gene in the rhizosphere soil of rice, indicating that the expression of the *arr*A gene may not be strictly anaerobic, so the *arr*A gene was also detected in the river water in the study area. [66,67].

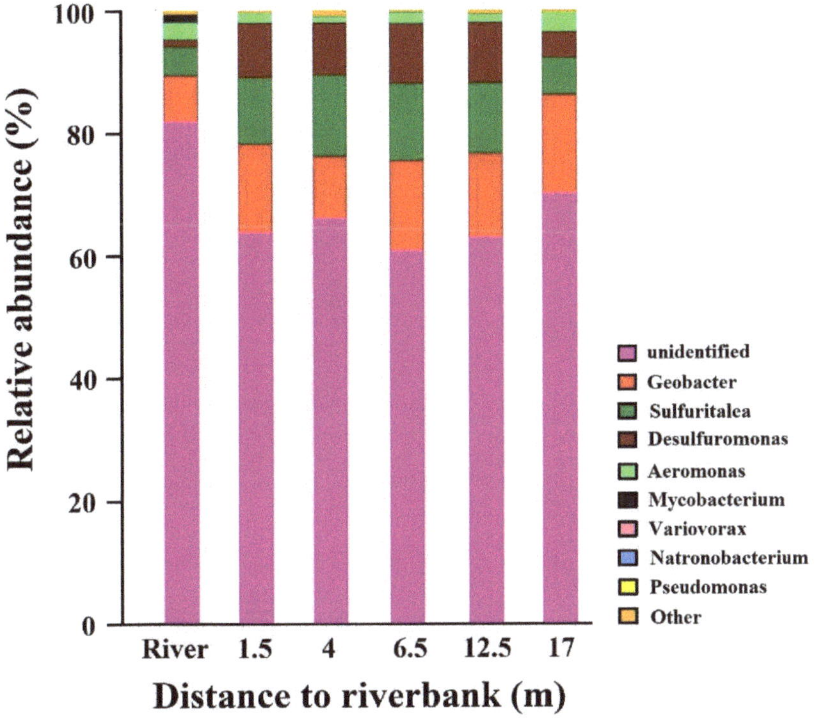

Figure 8. Species composition containing the As(V)-reducing gene *arr*A in genus-level (Other: the summation of the bacteria, of which the relative abundance was considerably below 1%).

The bacteria containing *aio*A were mainly Proteobacteria, including *Xanthobacter* and *Defluviimonas* (Figure 9). Bacteria containing *arx*A were mainly Proteobacteria and Actinobacteria, including *Parvibaculum*, *Rhodobacter*, *Maricaulis* and so on. Under aerobic conditions, As(III) is oxidized to As(V) under the action of microorganisms, and the oxidation of As(III) is achieved by oxidase. Currently, the As(III) oxidase is mainly divided into AioA and ArxA. The process by which microorganisms oxidize As(III) into As(V) to reduce toxicity is arsenic oxidation by microorganisms. There are mainly autotrophic arsenic-oxidizing bacteria and heterotrophic arsenic-oxidizing bacteria that can carry out this process [68–70]. These species were found only in the river water. The oxidation of As(III) occurs in the periplasm and is accompanied by electron transfer during the oxidation process [71–73].

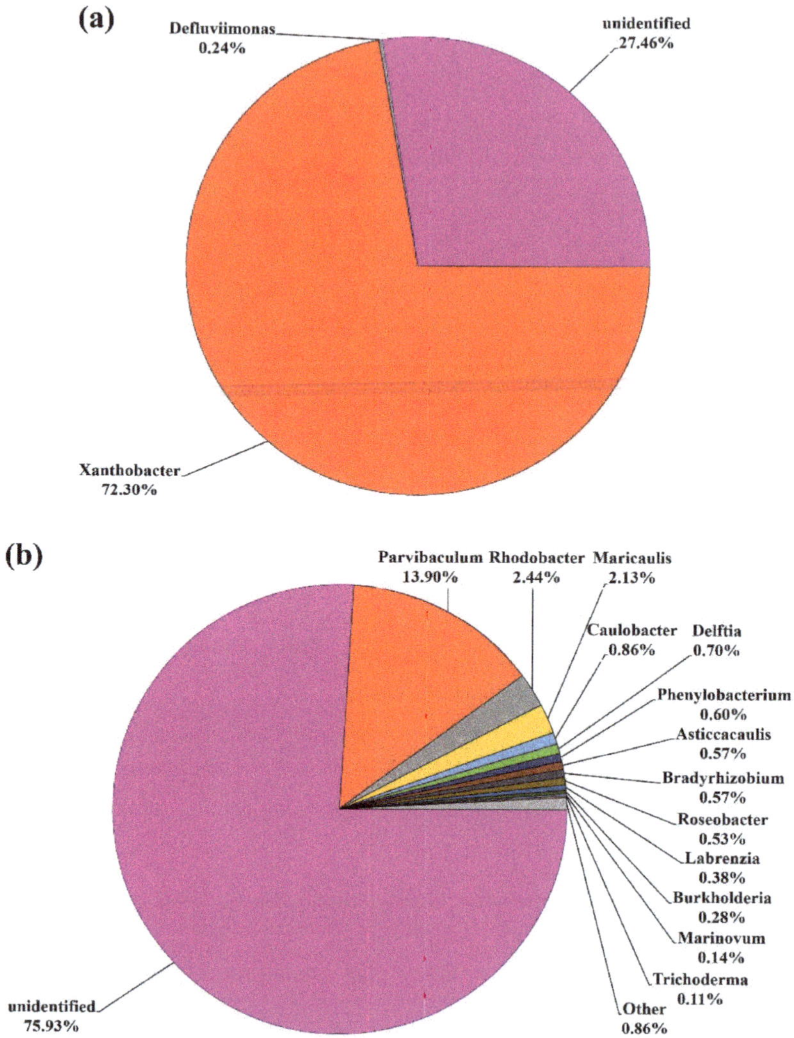

Figure 9. Species composition containing the As(III) oxidizing gene (**a**) *aio*A and (**b**) *arx*A in the genus-level (Other: the summation of the bacteria, of which the relative abundance was considerably below 0.1%).

4. Conclusions

The river water in the study area was not polluted by arsenic. Along the groundwater flow direction, the oxidation environment gradually transitioned to the reduction environment. The content of arsenic, iron and manganese in groundwater gradually increased, while NO_3^- and SO_4^{2-} showed a decreasing trend, indicating that Fe^{3+}, Mn^{4+}, NO_3^- and SO_4^{2-} were reduced in the process of increasing arsenic concentration.

The species abundance of river water in the study area was relatively high, and that of groundwater was relatively low. There was a significant negative correlation between microbial diversity and arsenic content, and microbial diversity was higher in groundwater with a relatively low arsenic content. Proteobacteria was the dominant phylum in the groundwater, and *Acinetobacter*, *Pseudomonas*, *Sulfuritalea*, *Sphingomonas* and *Hydrogenophaga* etc. were the main dominant bacterial genera. In addition to reducing and oxidizing arsenic, these functional microorganisms also actively participated in the biogeochemical cycle of elements such as iron, manganese, nitrogen and sulfur. Through the functional gene analysis, it could be seen that there were mainly DARPs in the groundwater. There was a significant correlation between dominant bacteria and environmental factors. Fe/Mn had a significant positive correlation with As, which brought potential danger to the water supply in high iron and manganese areas. It can provide reference for the site selection and construction of the water source in the future and promote the development of groundwater arsenic pollution control technology.

Author Contributions: Conceptualization, X.S.; methodology, S.L.; investigation, S.L.; resources, X.S.; data curation, Y.Y. and H.Y.; writing—Original draft preparation, S.L. and K.Y.; writing—Review and editing, S.L. and C.S.; supervision, C.S.; funding acquisition, S.L. and X.S. All authors have read and agreed to the published version of the manuscript.

Funding: This research was funded by the National Natural Science Foundation of China, grant number 42107100 and 41372238; the Hebei Province Water Conservancy Research and Promotion Plan Project, grant number 2021-37; the Hebei Province Professional Degree Postgraduate Teaching Case Construction Project, grant number KCJSZ2021098; the Open Fund for the Hebei Province Collaborative Innovation Center for Sustainable Utilization of Water Resources and Optimization of Industrial Structure, grant number XTZX202114; the Open Fund Project of the Key Laboratory of Groundwater Contamination and Remediation of Hebei Province and China Geological Survey (Innovation Center), grant number SK202103KF02; the Open Topic Project of the Hebei Key Laboratory of Geological Resources and Environment Monitoring and Protection, grant number JCYKT202003; the Special Project for the Cultivation of Scientific and Technological Innovation Ability of College and Middle School Students in Hebei Province, grant number 22E50150D, and the Science and Technology Innovation Team Project of Hebei GEO University, grant number KJCXTD-2021-14.

Institutional Review Board Statement: Not applicable.

Informed Consent Statement: Not applicable.

Data Availability Statement: Not applicable.

Conflicts of Interest: The authors declare no conflict of interest.

References

1. Zhang, Q.; Liu, H.; Zhu, P.; Xu, W.; Su, X. Evaluation of the influence of river bank infiltration on groundwater in an inland alluvial fan using spectral analysis and environmental tracers. *Hydrogeol. J.* **2021**, *29*, 1117–1128. [CrossRef]
2. Groeschke, M.; Frommen, T.; Taute, T.; Schneider, M. The impact of sewage-contaminated river water on groundwater ammonium and arsenic concentrations at a riverbank filtration site in central Delhi, India. *Hydrogeol. J.* **2017**, *25*, 2185–2197. [CrossRef]
3. Bai, J.; Su, X.; Wang, J.; Lyu, H.; Gao, R.; Lu, S. Multi-isotope constraints on biogeochemical processes during bank filtration: A case study of the Liao River, Northeast China. *Appl. Geochem.* **2020**, *122*, 104762. [CrossRef]
4. Zhu, Y.; Zhai, Y.; Teng, Y.; Wang, G.; Du, Q.; Wang, J.; Yang, G. Water supply safety of riverbank filtration wells under the impact of surface water-groundwater interaction: Evidence from long-term field pumping tests. *Sci. Total Environ.* **2020**, *711*, 135141. [CrossRef]
5. Zhu, Y.; Zhai, Y.; Du, Q.; Teng, Y.; Wang, J.; Yang, G. The impact of well drawdowns on the mixing process of river water and groundwater and water quality in a riverside well field, Northeast China. *Hydrol. Process.* **2019**, *33*, 945–961. [CrossRef]
6. Jiang, Y.; Zhang, J.; Zhu, Y.; Du, Q.; Teng, Y.; Zhai, Y. Design and Optimization of a Fully-Penetrating Riverbank Filtration Well Scheme at a Fully-Penetrating River Based on Analytical Methods. *Water* **2019**, *11*, 418. [CrossRef]
7. Yin, W.; Teng, Y.; Zhai, Y.; Hu, L.; Zhao, X.; Zhang, M. Suitability for developing riverside groundwater sources along Songhua River, Northeast China. *Hum. Ecol. Risk Assess.* **2018**, *24*, 2088–2100. [CrossRef]
8. He, X.; Li, P.; Ji, Y.; Wang, Y.; Su, Z.; Elumalai, V. Groundwater Arsenic and Fluoride and Associated Arsenicosis and Fluorosis in China: Occurrence, Distribution and Management. *Expos. Health* **2020**, *12*, 355–368. [CrossRef]
9. Abdul, K.S.M.; Jayasinghe, S.S.; Chandana, E.P.; Jayasumana, C.; De Silva, P.M.C. Arsenic and human health effects: A review. *Environ. Toxicol. Pharmacol.* **2015**, *40*, 828–846. [CrossRef]
10. Smith, A.H.; Lopipero, P.A.; Bates, M.N.; Steinmaus, C.M. Arsenic epidemiology and drinking water standards. *Science* **2002**, *296*, 2145–2146. [CrossRef]
11. Lu, S.; Su, X.; Feng, X.; Sun, C. Study on the formation and influencing factors of arsenic in nearshore zone during river water infiltration. *Earth Sci. Front.* **2021**. (In Chinese with English Abstract) [CrossRef]

12. Su, X.; Lu, S.; Yuan, W.; Woo, N.C.; Dai, Z.; Dong, W.; Du, S.; Zhang, X. Redox zonation for different groundwater flow paths during bank filtration: A case study at Liao River, Shenyang, northeastern China. *Hydrogeol. J.* **2018**, *26*, 1573–1589. [CrossRef]
13. Jadhav, S.V.; Bringas, E.; Yadav, G.D.; Rathod, V.K.; Ortiz, I.; Marathe, K.V. Arsenic and fluoride contaminated groundwaters: A review of current technologies for contaminants removal. *J. Environ. Manag.* **2015**, *162*, 306–325. [CrossRef] [PubMed]
14. Podgorski, J.; Berg, M. Global threat of arsenic in groundwater. *Science* **2020**, *368*, 845–850. [CrossRef]
15. Wallis, I.; Prommer, H.; Berg, M.; Siade, A.J.; Sun, J.; Kipfer, R. The river–groundwater interface as a hotspot for arsenic release. *Nat. Geosci.* **2020**, *13*, 288–295. [CrossRef]
16. He, X.; Li, P.; Wu, J.; Wei, M.; Ren, X.; Wang, D. Poor groundwater quality and high potential health risks in the Datong Basin, northern China: Research from published data. *Environ. Geochem. Health* **2021**, *43*, 791–812. [CrossRef]
17. Guo, H.; Liu, Z.; Ding, S.; Hao, C.; Xiu, W.; Hou, W. Arsenate reduction and mobilization in the presence of indigenous aerobic bacteria obtained from high arsenic aquifers of the Hetao basin, Inner Mongolia. *Environ. Pollut.* **2015**, *203*, 50–59. [CrossRef]
18. Lu, S.; Feng, X.; Su, X. Geochemical characteristics of arsenic in groundwater during riverbank filtration: A case study of Liao River, Northeast China. *Water Supply* **2020**, *20*, 3288–3300. [CrossRef]
19. Xu, N.; Gong, J.; Tao, X.; Liu, L. Hydrogeochemical Processes and Potential Exposure Risk of Arsenic-Rich Groundwater from Huaihe River Plain, China. *Water* **2022**, *14*, 693. [CrossRef]
20. Zhi, C.; Cao, W.; Wang, Z.; Li, Z. High-Arsenic Groundwater in Paleochannels of the Lower Yellow River, China: Distribution and Genesis Mechanisms. *Water* **2021**, *13*, 338. [CrossRef]
21. Cao, W.; Guo, H.; Zhang, Y.; Ma, R.; Li, Y.; Dong, Q.; Li, Y.; Zhao, R. Controls of paleochannels on groundwater arsenic distribution in shallow aquifers of alluvial plain in the Hetao Basin, China. *Sci. Total Environ.* **2018**, *613–614*, 958–968. [CrossRef] [PubMed]
22. Zhang, Y.; Cao, W.; Wang, W.; Dong, Q. Distribution of groundwater arsenic and hydraulic gradient along the shallow groundwater flow-path in Hetao Plain, Northern China. *J. Geochem. Explor.* **2013**, *135*, 31–39. [CrossRef]
23. Tufano, K.J.; Fendorf, S. Confounding Impacts of Iron Reduction on Arsenic Retention. *Environ. Sci. Technol.* **2008**, *42*, 4777–4783. [CrossRef] [PubMed]
24. Wang, Y.; Liu, X.; Si, Y.; Wang, R. Release and transformation of arsenic from As-bearing iron minerals by Fe-reducing bacteria. *Chem. Eng. J.* **2016**, *295*, 29–38. [CrossRef]
25. Carlin, D.J.; Naujokas, M.F.; Bradham, K.D.; Cowden, J.; Heacock, M.; Henry, H.F.; Lee, J.S.; Thomas, D.J.; Thompson, C.; Tokar, E.J.; et al. Arsenic and Environmental Health: State of the Science and Future Research Opportunities. *Environ. Health Perspect.* **2016**, *124*, 890–899. [CrossRef] [PubMed]
26. Xiu, W.; Lloyd, J.; Guo, H.; Dai, W.; Nixon, S.; Bassil, N.M.; Ren, C.; Zhang, C.; Ke, T.; Polya, D. Linking microbial community composition to hydrogeochemistry in the western Hetao Basin: Potential importance of ammonium as an electron donor during arsenic mobilization. *Environ. Int.* **2020**, *136*, 105489. [CrossRef]
27. Wang, L.; Yin, Z.; Jing, C. Metagenomic insights into microbial arsenic metabolism in shallow groundwater of Datong basin, China. *Chemosphere* **2020**, *245*, 125603. [CrossRef] [PubMed]
28. Zheng, T.; Deng, Y.; Wang, Y.; Jiang, H.; O'Loughlin, E.J.; Flynn, T.M.; Gan, Y.; Ma, T. Seasonal microbial variation accounts for arsenic dynamics in shallow alluvial aquifer systems. *J. Hazard. Mater.* **2019**, *367*, 109–119. [CrossRef]
29. Wang, P.; Sun, G.; Jia, Y.; Meharg, A.A.; Zhu, Y. A review on completing arsenic biogeochemical cycle: Microbial volatilization of arsines in environment. *J. Environ. Sci.* **2014**, *26*, 371–381. [CrossRef]
30. Ye, J.; Rensing, C.; Rosen, B.P.; Zhu, Y. Arsenic biomethylation by photosynthetic organisms. *Trends Plant Sci.* **2012**, *17*, 155–162. [CrossRef]
31. Hamamura, N.; Itai, T.; Liu, Y.; Reysenbach, A.; Damdinsuren, N.; Inskeep, W.P. Identification of anaerobic arsenite-oxidizing and arsenate-reducing bacteria associated with an alkaline saline lake in Khovsgol, Mongolia. *Env. Microbiol. Rep.* **2014**, *6*, 476–482. [CrossRef] [PubMed]
32. Burton, E.D.; Johnston, S.G.; Kocar, B.D. Arsenic mobility during flooding of contaminated soil: The effect of microbial sulfate reduction. *Environ. Sci. Technol.* **2014**, *48*, 13660–13667. [CrossRef] [PubMed]
33. O'Day, P.A.; Vlassopoulos, D.; Root, R.; Rivera, N. The influence of sulfur and iron on dissolved arsenic concentrations in the shallow subsurface under changing redox conditions. *Proc. Natl. Acad. Sci. USA* **2004**, *101*, 13703–13708. [CrossRef] [PubMed]
34. Wang, Y.; Li, P.; Jiang, D.; Li, B.; Dai, X.; Jiang, Z.; Wang, Y. Vertical distribution of bacterial communities in high arsenic sediments of Hetao Plain, Inner Mongolia. *Ecotoxicology* **2014**, *23*, 1890–1899. [CrossRef] [PubMed]
35. Lowers, H.A.; Breit, G.N.; Foster, A.L.; Whitney, J.; Yount, J.; Uddin, M.N.; Muneem, A.A. Arsenic incorporation into authigenic pyrite, Bengal Basin sediment, Bangladesh. *Geochim. Cosmochim. Acta* **2007**, *71*, 2699–2717. [CrossRef]
36. Wang, Y.; Li, P.; Jiang, Z.; Liu, H.; Wei, D.; Wang, H.; Wang, Y. Diversity and abundance of arsenic methylating microorganisms in high arsenic groundwater from Hetao Plain of Inner Mongolia, China. *Ecotoxicology* **2018**, *27*, 1047–1057. [CrossRef] [PubMed]
37. Li, P.; Jiang, Z.; Wang, Y.; Deng, Y.; Van Nostrand, J.D.; Yuan, T.; Liu, H.; Wei, D.; Zhou, J. Analysis of the functional gene structure and metabolic potential of microbial community in high arsenic groundwater. *Water Res.* **2017**, *123*, 268–276. [CrossRef]
38. Zhou, J.; Li, P.; Wang, Y.; Liu, H.; Wei, D.; Yuan, C.; Wang, H. Arsenic mobilization in a high arsenic groundwater revealed by metagenomic and Geochip analyses. *Sci. Rep.* **2019**, *9*, 12972. [CrossRef]
39. Liu, E.; Yang, Y.; Xie, Z.; Wang, J.; Chen, M. Influence of Sulfate Reduction on Arsenic Migration and Transformation in Groundwater Environment. *Water* **2022**, *14*, 942. [CrossRef]

40. Lu, S.; Li, S.; Liu, Z.; Gao, X.; Zhang, L.; Sun, C. Hydrochemical evolution of pore water in riverbed sedimentation zone during riverbank infiltration. *J. Water Supply Res. Technol.* **2021**, *70*, 696–709. [CrossRef]
41. Lu, S. Biogeochemical Process of Arsenic in Groundwater and Its Simulation Affected by Groundwater Exploitation in Riverside. Ph.D. Thesis, Jilin University, Changchun, China, 2018. (In Chinese)
42. Chander, K.; Brookes, P.C.; Harding, S.A. Microbial biomass dynamics following addition of metal-enriched sewage sludges to a sandy loam. *Soil Biol. Biochem.* **1995**, *27*, 1409–1421. [CrossRef]
43. Chen, Y.; Jiang, Y.; Huang, H.; Mou, L.; Ru, J.; Zhao, J.; Xiao, S. Long-term and high-concentration heavy-metal contamination strongly influences the microbiome and functional genes in Yellow River sediments. *Sci. Total Environ.* **2018**, *637–638*, 1400–1412. [CrossRef] [PubMed]
44. Li, H.; Zeng, X.-C.; He, Z.; Chen, X.; Guo-ji, E.; Han, Y.; Wang, Y. Long-term performance of rapid oxidation of arsenite in simulated groundwater using a population of arsenite-oxidizing microorganisms in a bioreactor. *Water Res.* **2016**, *101*, 393–401. [CrossRef] [PubMed]
45. Jiang, X.; Peng, X.; Deng, G.; Sheng, H.; Wang, Y.; Zhou, H.; Yee Tam, N.F. Illumina Sequencing of 16S rRNA Tag Revealed Spatial Variations of Bacterial Communities in a Mangrove Wetland. *Microb. Ecol.* **2013**, *66*, 96–104. [CrossRef] [PubMed]
46. Sheik, C.S.; Mitchell, T.W.; Rizvi, F.Z.; Rehman, Y.; Faisal, M.; Hasnain, S.; Mclnerney, M.J.; Krumholz, L.R. Exposure of Soil Microbial Communities to Chromium and Arsenic Alters Their Diversity and Structure. *PLoS ONE* **2012**, *7*, e40059. [CrossRef]
47. Peng, Q.; Shaaban, M.; Wu, Y.; Hu, R.; Wang, B.; Wang, J. The diversity of iron reducing bacteria communities in subtropical paddy soils of China. *Appl. Soil Ecol.* **2016**, *101*, 20–27. [CrossRef]
48. Champ, D.R.; Gulens, J.; Jackson, R.E. Oxidation–reduction sequences in ground water flow systems. *Can. J. Earth Sci.* **1979**, *16*, 12–23. [CrossRef]
49. Li, P.; Wang, Y.; Jiang, Z.; Jiang, H.; Li, B.; Dong, H.; Wang, Y. Microbial Diversity in High Arsenic Groundwater in Hetao Basin of Inner Mongolia, China. *Geomicrobiol. J.* **2013**, *30*, 897–909. [CrossRef]
50. Pan, Y.; Yang, X.; Xu, M.; Sun, G. The Role of Enriched Microbial Consortium on Iron-Reducing Bioaugmentation in Sediments. *Front. Microbiol.* **2017**, *8*, 462. [CrossRef]
51. Fang, Y.; Du, Y.; Hu, L.; Xu, J.; Long, Y.; Shen, D. Effects of sulfur-metabolizing bacterial community diversity on H_2S emission behavior in landfills with different operation modes. *Biodegradation* **2016**, *27*, 237–246. [CrossRef]
52. Chung, J.; Kim, Y.; Lee, D.; Shim, H.; Kim, J. Characteristics of Denitrifying Phosphate Accumulating Organisms in an Anaerobic-Intermittently Aerobic Process. *Environ. Eng. Sci.* **2006**, *23*, 981–993. [CrossRef]
53. Ghosh, S.; Sar, P. Identification and characterization of metabolic properties of bacterial populations recovered from arsenic contaminated ground water of North East India (Assam). *Water Res.* **2013**, *47*, 6992–7005. [CrossRef] [PubMed]
54. Li, P.; Jiang, D.; Li, B.; Dai, X.; Wang, Y.; Jiang, Z.; Wang, Y. Comparative survey of bacterial and archaeal communities in high arsenic shallow aquifers using 454 pyrosequencing and traditional methods. *Ecotoxicology* **2014**, *23*, 1878–1889. [CrossRef] [PubMed]
55. Drewniak, L.; Matlakowska, R.; Rewerski, B.; Sklodowska, A. Arsenic release from gold mine rocks mediated by the activity of indigenous bacteria. *Hydrometallurgy* **2010**, *104*, 437–442. [CrossRef]
56. Drewniak, L.; Styczek, A.; Majder-Lopatka, M.; Sklodowska, A. Bacteria, hypertolerant to arsenic in the rocks of an ancient gold mine, and their potential role in dissemination of arsenic pollution. *Environ. Pollut.* **2008**, *156*, 1069–1074. [CrossRef]
57. Oremland, R.S.; Stolz, J.F. The Ecology of Arsenic. *Science* **2003**, *300*, 939–944. [CrossRef]
58. Niggemyer, A.; Spring, S.; Stackebrandt, E.; Rosenzweig, R.F. Isolation and characterization of a novel As(V)-reducing bacterium: Implications for arsenic mobilization and the genus Desulfitobacterium. *Appl. Environ. Microb.* **2001**, *67*, 5568–5580. [CrossRef]
59. Oremland, R.S.; Stolz, J.F. Arsenic, microbes and contaminated aquifers. *Trends Microbiol.* **2005**, *13*, 45–49. [CrossRef]
60. Harrington, J.M.; Fendorf, S.E.; Rosenzweig, R.F. Biotic generation of arsenic(III) in metal(loid)-contaminated freshwater lake sediments. *Environ. Sci. Technol.* **1998**, *32*, 2425–2430. [CrossRef]
61. Oremland, R.S.; Dowdle, P.R.; Hoeft, S.; Sharp, J.O.; Schaefer, J.K.; Miller, L.G.; Blum, J.S.; Smith, R.L.; Bloom, N.S.; Wallschlaeger, D. Bacterial dissimilatory reduction of arsenate and sulfate in meromictic Mono Lake, California. *Geochim. Cosmochim. Acta* **2000**, *64*, 3073–3084. [CrossRef]
62. Malasarn, D.; Saltikov, C.W.; Campbell, K.M.; Santini, J.M.; Hering, J.G.; Newman, D.K. arrA Is a Reliable Marker for As(V) Respiration. *Science* **2004**, *306*, 455. [CrossRef] [PubMed]
63. Cummings, D.E.; Caccavo, F.; Fendorf, S.; Rosenzweig, R.F. Arsenic Mobilization by the Dissimilatory Fe(III)-Reducing Bacterium Shewanella alga BrY. *Environ. Sci. Technol.* **1999**, *33*, 723–729. [CrossRef]
64. Ahmann, D.; Krumholz, L.R.; Hemond, H.F.; Lovley, D.R.; Morel, F.M.M. Microbial Mobilization of Arsenic from Sediments of the Aberjona Watershed. *Environ. Sci. Technol.* **1997**, *31*, 2923–2930. [CrossRef]
65. Zobrist, J.; Dowdle, P.R.; Davis, J.A.; Oremland, R.S. Mobilization of Arsenite by Dissimilatory Reduction of Adsorbed Arsenate. *Environ. Sci. Technol.* **2000**, *34*, 4747–4753. [CrossRef]
66. Jia, Y.; Huang, H.; Chen, Z.; Zhu, Y. Arsenic Uptake by Rice Is Influenced by Microbe-Mediated Arsenic Redox Changes in the Rhizosphere. *Environ. Sci. Technol.* **2014**, *48*, 1001–1007. [CrossRef] [PubMed]
67. Zhang, S.; Zhao, F.; Sun, G.; Su, J.; Yang, X.; Li, H.; Zhu, Y. Diversity and Abundance of Arsenic Biotransformation Genes in Paddy Soils from Southern China. *Environ. Sci. Technol.* **2015**, *49*, 4138–4146. [CrossRef]

68. Salmassi, T.M.; Venkateswaren, K.; Satomi, M.; Newman, D.K.; Hering, J.G. Oxidation of Arsenite by Agrobacterium albertimagni, AOL15, sp. nov., Isolated from Hot Creek, California. *Geomicrobiol. J.* **2002**, *19*, 53–66. [CrossRef]
69. Santini, J.M.; vanden Hoven, R.N. Molybdenum-Containing Arsenite Oxidase of the Chemolithoautotrophic Arsenite Oxidizer NT-26. *J. Bacteriol.* **2004**, *186*, 1614–1619. [CrossRef]
70. Stolz, J.F.; Basu, P.; Oremland, R.S. Microbial Arsenic Metabolism: New Twists on an Old Poison. *Microbe* **2010**, *5*, 53–59. [CrossRef]
71. Richey, C.; Chovanec, P.; Hoeft, S.E.; Oremland, R.S.; Basu, P.; Stolz, J.F. Respiratory arsenate reductase as a bidirectional enzyme. *Biochem. Biophys. Res. Commun.* **2009**, *382*, 298–302. [CrossRef]
72. Zargar, K.; Hoeft, S.; Oremland, R.; Saltikov, C.W. Identification of a Novel Arsenite Oxidase Gene, arxA, in the Haloalkaliphilic, Arsenite-Oxidizing Bacterium Alkalilimnicola ehrlichii Strain MLHE-1. *J. Bacteriol.* **2010**, *192*, 3755–3762. [CrossRef] [PubMed]
73. Zargar, K.; Conrad, A.; Bernick, D.L.; Lowe, T.M.; Stolc, V.; Hoeft, S.; Oremland, R.S.; Stolz, J.; Saltikov, C.W. ArxA, a new clade of arsenite oxidase within the DMSO reductase family of molybdenum oxidoreductases. *Environ. Microbiol.* **2012**, *14*, 1635–1645. [CrossRef] [PubMed]

water

MDPI

Article

Impacts of River Bank Filtration on Groundwater Hydrogeochemistry in the Upper of Hutuo River Alluvial Plain, North China

Baoyun Zhang [1,2], Lining Chen [2], Yasong Li [2,3], Yaci Liu [2,3], Chao Li [4], Xiangke Kong [2,3] and Yuanjing Zhang [2,3,*]

1 Institute of Geophysical and Geochemical Exploration, Chinese Academy of Geological Sciences, Langfang 065000, China
2 Fujian Provincial Key Laboratory of Water Cycling and Eco-Geological Processes, Xiamen 361000, China
3 Institute of Hydrogeology and Environmental Geology, Chinese Academy of Geological Sciences, Shijiazhuang 050061, China
4 Shijin Irrigation District Affairs Center of Hebei Province, Shijiazhuang 050051, China
* Correspondence: mirrorzyj@163.com

Abstract: River bank filtration (RBF) under human supervision has been applied for groundwater recharge. This study clarified the characteristics of water composition and its origins during the reservoir water recharge RBF. The groundwater samples were collected during four periods: pre-recharge (June 2018), early recharge (November 2018), intermediate recharge (May 2019), and late recharge (October 2019). Hydrogeochemical methods (Piper diagram, chlor-alkali index, and ion correlation) and principal component analysis (PCA) were used to analyze the chemical evolution of groundwater in the aforementioned periods. TDS concentration tended to increase in the later stage due to the aggravation of carbonate rock dissolution and cation exchange adsorption. Results demonstrated a small, temporary influence of reservoir water on groundwater, characterized as the Ca–Mg–HCO_3–SO_4 and Ca–Mg–SO_4–HCO_3 types, both before and after the recharge. The research on water chemistry changes under different mixing ratios depicts that the continuation of the recharge process promotes mineral dissolution. Rock dissolution was the primary environmental control factor of groundwater components during the recharge period. This strengthened the water–rock reaction and caused potential risk impacts such as the increase of nitrate.

Keywords: river bank filtration (RBF); management aquifer recharge (MAR); hydrochemistry; water–rock reaction; Hutuo River

Citation: Zhang, B.; Chen, L.; Li, Y.; Liu, Y.; Li, C.; Kong, X.; Zhang, Y. Impacts of River Bank Filtration on Groundwater Hydrogeochemistry in the Upper of Hutuo River Alluvial Plain, North China. *Water* **2023**, *15*, 1343. https://doi.org/10.3390/w15071343

Academic Editors: Yuanzheng Zhai, Jin Wu and Aizhong Ye

Received: 10 February 2023
Revised: 9 March 2023
Accepted: 28 March 2023
Published: 30 March 2023

1. Introduction

The over-exploitation of groundwater is a global challenge at present because it is jointly affected by increased water demand and the instability brought by climate change [1]. The uncertainty of water resources availability demands a solvent that can function both as the water storage area and an energy-efficient treatment process [2]. River bank filtration (RBF), which is also known as Bank filtrate or filtration (BF), has been a natural process for managed aquifer recharge (MAR) with low regulation cost and high efficiency in the removal of contaminants [3–5]. As reviewed by Dillon et al. [6], the aquifer recharge started at the end of the 18th century, when a few human intervention measures to improve the efficiency of groundwater recharge were adopted in some European countries; however, those processes were generally incidental and not well-managed. In addition to increased scientific research and practical experience, more MAR technics have been developed within the last 60 years [6]. MAR can be categorized into three types by the way of recharge: (i) infiltration method, (ii) direct injection method, and (iii) filtration method [6–9]. RBF is an induced recharge method and is considered as filtration in the abovementioned categorization [10].

When river water is replenished into the aquifer, it undergoes hydrogeochemical processes such as water–rock interaction, mixing, and cation exchange adsorption. These processes can lead to changes in groundwater chemistry by altering its components [10–12]. Studies in India and Finland have discovered the effect of ion exchange adsorption and water–rock interaction on groundwater chemistry during RBF replenishment [13,14]. In a specific location in Mexico, replenishing groundwater through river infiltration caused a change in its acid–base balance, resulting in an altered water–rock interaction [15]. During ecological water replenishment in Ejina Banner, the water chemistry type remained relatively stable, but with an increase in the concentration of major ions [16]. In an experimental study of infiltrating south-to-north water to replenish groundwater in the Yongding River, Beijing, researchers found that the replenishment process can change water–rock interaction [17].

It is a current research hotspot in RBF by using unique water sources. The infiltration of river water into the aquifer is affected by factors such as temperature and redox potential, resulting in changes in the concentrations of major ions and TDS in groundwater [18,19]. When using reclaimed water to replenish groundwater through RBF, mixing, water–rock interaction, and cation exchange are enhanced [20,21]. In addition, colloidal particles present in the soil can adsorb organic pollutants in reclaimed water [22,23]. The use of flood or rainwater to infiltrate and replenish groundwater through rivers can easily alter the redox conditions of groundwater, and thus, lead to further changes in hydrogeochemical processes [24,25]. The replenishment of reservoir water through rivers is highly dependent on the amount and timing of replenishment, resulting in a phased change in groundwater volume and chemical composition [26–28]. Although research on reservoir water replenishment tends to focus on the replenishment amount [29], there has been relatively less attention on studying the changes in groundwater chemistry. Therefore, it is crucial to investigate the changes in groundwater chemistry resulting from the river bank filtration of reservoir water to replenish groundwater.

The Hutuo River alluvial fan of the North China Plain is a critical local economic development zone, but ecological and environmental problems such as spring water cutoff and river drying are severe [30], which have attracted considerable attention from hydrogeologists [31]. To improve the situation, the Huangbizhuang Reservoir in the upper Hutuo River intermittently released water to the dry Hutuo River channel to recharge the groundwater by RBF. The purpose of this study was to analyze the characteristics and impacts of the intrinsic changes for using reservoir water to recharge groundwater by RBF. A conclusion on the possible environmental problems caused by groundwater recharge was reached to support water management strategies and to further increase both the quantity and quality of local groundwater resources.

2. Study Area

The Hutuo RBF area is located in the North China Plain. The river began to dry up after 1980, and the rainy season did not contribute much water. The investigation site covers the area from Huangbizhuang Town (in southwestern Hebei Province, China), to Zhengding County in the middle and lower reaches (38.15°–38.31° N, 114.21°–114.49° E), with an approximate area of 236.88 km^2 (Figure 1). The temperate semi-humid and semi-arid continental monsoon climate is the dominant climate type in this region. The upstream average temperature for many years has been 6.4 °C. The average annual precipitation is 484.0 mm, whereas the spatial and temporal distribution of precipitation is uneven [32,33]. The rainy season (from June to August) accounts for 70–85% of the total annual rainfall [34]. The study area is located in the upper of the Hutuo River alluvial fan. The eastern part of this region is the groundwater divide, and the Huangbizhuang Reservoir is situated northwest of this zone. The groundwater in the study area flows from the northwest to the southeast, and the flow direction changes a little before and after water replenishment.

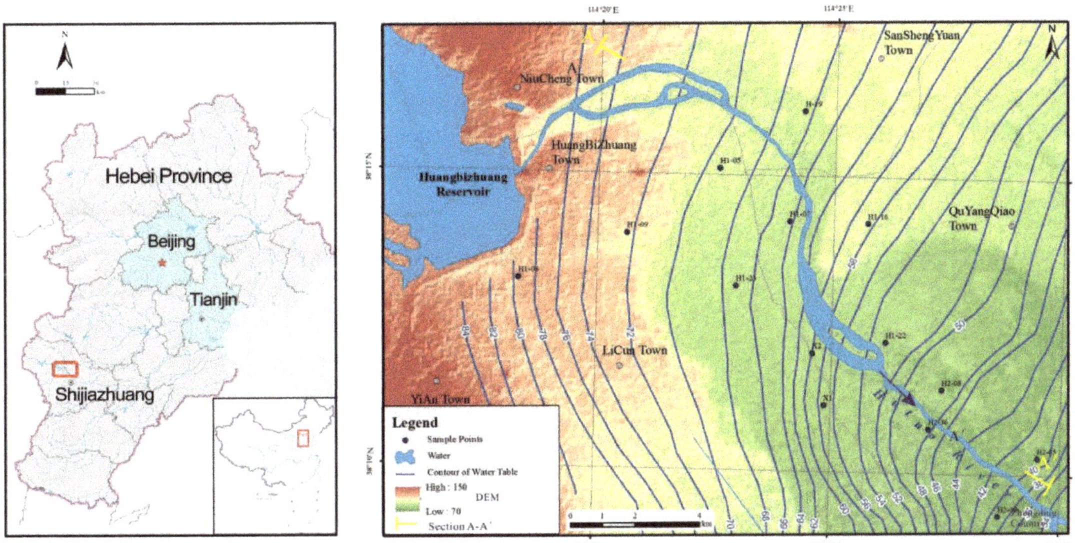

Figure 1. Scope of the study area.

The groundwater in this section is a phreatic, slightly confined water aquifer group composed of the Holocene (Q_4) and Upper Pleistocene ($Q_3{}^3$) sediments. The aquifer primarily consists of quaternary cobbles, gravel, sand, laminated or lensed loam, and sandy loam. The aquifer is mainly composed of pebbles and gravel sand, and its upper part is dry. The southeast of the region has a thinning interlaced sedimentary zone, where the vadose zone is mainly composed of cohesive soil. In accordance with the drilling core test results and previous studies, the aquifer contains calcite, dolomite, rock salt, gypsum, and mirabilite [35–38]. The water table depth in this area is mainly 8.2–40.1 m. The bottom interface depth is generally 40–60 m. It is a single and two-layer aquifer structure, with good water permeability, and the water conductivity is more than 5000 m^2/d. The sources of groundwater recharge mainly include atmospheric precipitation, lateral runoff, reservoir channel leakage, and channel infiltration recharge. In accordance with the local government policy, groundwater has not been exploited in recent years for level restoration [39]. Near the river bed, there are continuous sand, gravel, pebble layers, and partially viscous clay lens. The upper part is mainly fine sand and medium-coarse sand, with a thickness of 5–22 m and large permeability, which provides a great infiltration condition. In addition, the near-mountain plains are connected with the water bodies of mountain valleys and plains, and the lateral runoff conditions are good. The salinity of groundwater is less than 1 g/L. With the gradual transition of the southeast alluvial fans to the inter-fan and front-fan zones, the thickness of the aquifer decreases, the medium particles become finer, the clay layers of varying thickness are sandwiched between the layers, and the water permeability and water conductivity are significantly reduced (Figure 2).

With a severe drought condition in the Hutuo river, the study area is mainly recharged by the discharge of the Huangbizhuang Reservoir that implements a non-sustainable water drainage measure. The water drainage operation is carried out within a few months before the rainy season. Most of the water released leaks into the groundwater layer through the river channel, while the rest continues to flow to the downstream area. Hence, the study area is considered as the infiltration site. From 2018 to 2020, water had been intermittently recharged in the upper of the Hutuo River. The amount of monthly reservoir water discharge is shown in Figure 3. In June 2018, the amount of reservoir discharge was 352×10^4 m^3. From October 2018 to February 2019, the discharge of reservoir water reached its peak, of which the maximum discharge was 0.478×10^8 m^3 in December 2018. The total amount of water discharge in 2018 was 1.21×10^8 m^3. Draining was not observed in March, April, November, and December of 2019. The maximum amount of water discharge and the total water discharge were 0.323×10^8 m^3 and 1.21×10^8 m^3, respectively, in 2019. The reservoir water release in 2019 was 900×10^4 m^3 more than that in 2018. From June 2018 to December 2019, the total accumulated water release of the reservoir was 2.51×10^8 m^3, with an average monthly replenishment of 0.13×10^8 m^3.

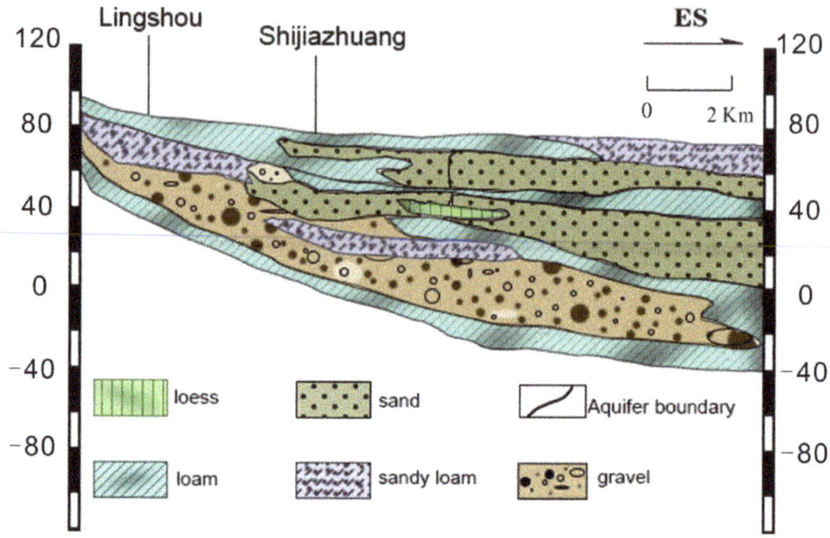

Figure 2. Hydrogeological profile of Section A-A′.

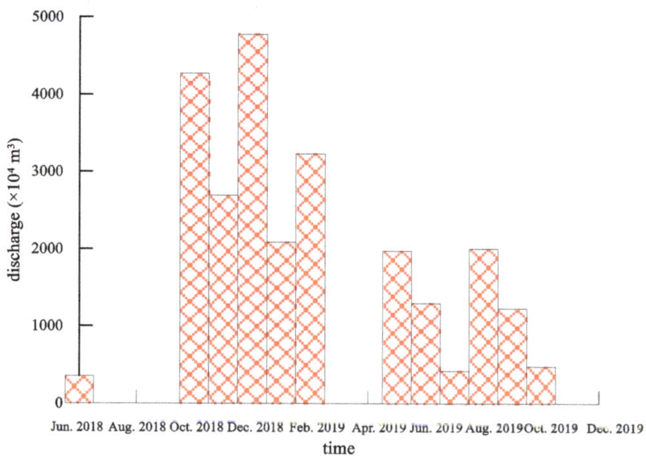

Figure 3. Amount of monthly reservoir discharge.

3. Materials and Methods

3.1. Water Sample Collection

To research the changes in the chemical characteristics of groundwater caused by BRF of reservoir water, the reservoir water samples and groundwater samples at four different periods were collected. Their locations are shown in Figure 1. The reservoir water samples were collected in November 2018. The groundwater sampling time was June 2018 (pre-recharge), November 2018 (early recharge), May 2019 (intermediate recharge), and October 2019 (late recharge), and a total of 51 samples were collected. The groundwater samples were taken from wells for domestic and agricultural purposes, with a depth range of 35–70 m. The length of screen pipes in all sampling wells ranged from 1 to 10 m, and each sampling well had only one screen pipe rather than multiple screens. The distance between the bottom of the screen pipe and the total well depth ranged from 10 to 20 m in the study area. The samples were collected using pumps installed in these wells. Prior to sampling, several water volumes were removed. The polyethylene sampling bottles were purged with ultrapure water and subsequently rinsed thrice with local groundwater samples. For routine anion and cation determination, 1.5 L groundwater was collected. For trace element detection, 500 mL groundwater samples were collected, and to adjust pH < 2, 1:1 HNO_3 was added.

3.2. Analysis and Testing

On-site water chemistry parameters were monitored using calibrated instruments. Total dissolved solids (TDS), temperature, DO, and pH of the water samples were measured on-site. The stopping criterion was the steady state of the measured values.

The cations K^+, Na^+, Ca^{2+}, and Mg^{2+} were analyzed using inductively coupled plasma (ICP-900, Thermo, Waltham, MA, USA). The anions Cl^-, SO_4^{2-} and NO_3^- were analyzed using ion chromatography (ICS-900, Dionex, Sunnyvale, CA, USA), and the anions HCO_3^- and CO_3^{2-} were analyzed using on-site titration. HCO_3^- concentrations in all groundwater samples were determined by the titration method using 0.0048 M H_2SO_4; methyl orange endpoint titration was adopted with a final pH of 4.2–4.4. CO_3^{2-} concentrations were also analyzed by titration; phenolphthalein was used as an indicator of endpoint titration. The charge balance errors for all the samples were within 10%. The samples taken were sent to the Groundwater Mineral Water and Environmental Supervising and Testing Center of the Ministry of Natural Resources for testing.

4. Results

4.1. Hydrochemical Characteristics of Groundwater

The on-site water quality parameters such as pH and TDS of groundwater samples, as well as the analysis data of major ion compositions, are shown in Table 1. The average pH value of groundwater decreased from 7.82 to 7.69 (June 2018–October 2019), the maximum decreased from 8.10 to 7.85, and the medium value decreased from 7.86 to 7.67. The average TDS value of groundwater decreased from 737 mg/L to 710 mg/L (June 2018–October 2019), and the maximum decreased from 1183 mg/L to 1009 mg/L. The average concentration of TDS (mg/L) in June 2018 (pre-recharge) was 737 mg/L, which was reduced to 632 mg/L in November 2018 (early recharge), and then rebounded to a higher level of 753 mg/L in November 2019 (late recharge). DO and temperature changed slightly, and the average temperature and DO are approximately 14 °C and 7.3.

The overall distribution of the chemical types of groundwater in the study area is given by the average values of ion contents. Bicarbonate and sulfate water is the main type and their content is much higher than that of the other ions. The calcium and magnesium ion content is substantially larger than that of the other cations. Irrespective of the recharge, the general order of abundance of the major cations was Ca^{2+}, Mg^{2+}, Na^+, and K^+, whereas the abundance of major anions was that HCO_3^- and SO_4^{2-} far exceeded Cl^-. The June 2018 (pre-recharge) concentrations of major ions were the highest, whereas those of November 2018 (the early recharge) were the lowest. From November 2018 (early recharge) to October 2019 (late recharge), the concentrations of major ions increased gradually. From June 2018 (pre-recharge) to October 2019 (late recharge), the mean concentrations of HCO_3^- varied from 270.48 to 298.3 mg/L, and those of Mg^{2+} varied from 39.65 to 45.01 mg/L. The concentration of Mg^{2+} and HCO_3^- in the late recharge period is higher than that in the pre-recharge period, whereas it is the opposite for the other ions.

Table 1. Concentrations of Inorganic Ions of Samples (mg/L).

Time	June 2018				November 2018				May 2019				October 2019			
Number of Sample Points	11				11				14				14			
Inorganic Ions	Max	Min	Mean	Median	Max	Min	Mean	Median	Max	Min	Mean	Median	Max	Min	Mean	Median
K^+	3.67	1.72	2.65	2.70	4.09	1.51	2.65	2.54	3.19	1.10	2.70	2.82	3.18	0.99	2.71	2.87
Na^+	76.15	28.27	39.87	35.63	51.92	22.39	34.12	31.82	52.26	28.72	37.98	37.33	52.62	30.12	38.62	37.32
Ca^{2+}	226.00	119.20	148.15	142.20	195.20	69.10	126.68	124.80	207.60	95.07	152.17	156.05	217.51	99.90	148.58	147.02
Mg^{2+}	66.79	31.67	39.65	36.81	59.80	23.43	36.43	32.19	56.69	24.62	39.12	40.61	70.84	29.31	45.01	39.82
Cl^-	210.60	39.66	77.12	54.75	106.80	37.11	61.81	61.27	110.00	38.42	67.78	67.59	109.51	40.71	68.99	68.79
SO_4^-	350.80	157.90	224.47	221.90	231.10	144.00	182.00	172.80	263.00	143.90	218.41	223.05	255.70	140.60	207.10	210.55
HCO_3-	361.90	186.60	270.48	284.30	347.50	193.70	254.45	233.70	441.00	206.00	302.46	296.00	480.38	196.93	317.13	298.37
CO_3^{2-}	ND	ND	ND	ND	ND	ND	ND	ND	ND	ND	ND	ND	ND	ND	ND	ND
NO_3^-	99.02	21.81	54.19	56.53	103.80	10.51	45.45	35.68	149.80	4.35	66.34	66.00	152.90	3.94	66.66	69.93
DO	7.98	3.35	7.32	7.39	8.0	3.34	7.23	7.35	8.0	4.04	7.28	7.06	7.99	3.56	7.31	7.38
TDS	1183	610	737	703	895	433	632	630	1045	505	745	738	1088	547	753	714
pH	8.10	7.61	7.82	7.86	7.94	7.51	7.72	7.75	7.73	7.31	7.50	7.51	7.93	7.45	7.69	7.67
temperature		15.8			11.9	13.8	14.3	14.2	15.5	12.1	14.1	14.5	15.9	11.4	13.9	14.3

Note: ND—not detected.

To minimize the seasonal variation of nitrate, the comparison was between the same seasons of the years. After recharging, the average concentration of nitrate in groundwater increased by 12.15 mg/L from 54.19 mg/L (June 2018, pre-recharge) to 66.34 mg/L (May 2019, intermediate recharge). Except for H1-07 and H1-19, the nitrate content of other wells decreased from June 2018 to May 2019. H1-19 was located in the farmer's yard, and there was a sudden increase of nitrate caused by human pollution. Therefore, anthropogenic factors led to an increase in nitrate content.

4.2. Hydrochemical Facies

The Piper diagram has not been disturbed by human subjective factors [40], and is a common method to identify the main ion composition of water chemistry [41,42]. Piper plots are created using the main ion data for groundwater in the study area (Figure 4). Two samples in October 2019 (late recharge) and one sample in May 2019 (intermediate recharge) are scattered in the zone of the left lower triangle predominated by the bicarbonate-type. Two samples in June 2018 (pre-recharge) are scattered in the zone of the upper triangle where the sulfate-type predominates. The central diamond plot provides important information to discriminate between separate groups of samples. Most samples are scattered in the mixed-type zone, and such samples are the most common in groundwater chemistry in the recharge period. From June 2018 (pre-recharge) to October 2019 (late recharge), groundwater types in the study area were mainly Ca-Mg-HCO_3-SO_4 and Ca-Mg-SO_4-HCO_3, with no obvious change in chemistry. The results of the Piper plot reveal Ca^{2+} and Mg^{2+} as the main groundwater cations in the study area, and the Ca^{2+} content is the highest. Cation contents follow the order $Ca^{2+} > Mg^{2+} > Na^+ + K^+$. The anions are mainly HCO_3^- and SO_4^{2-}.

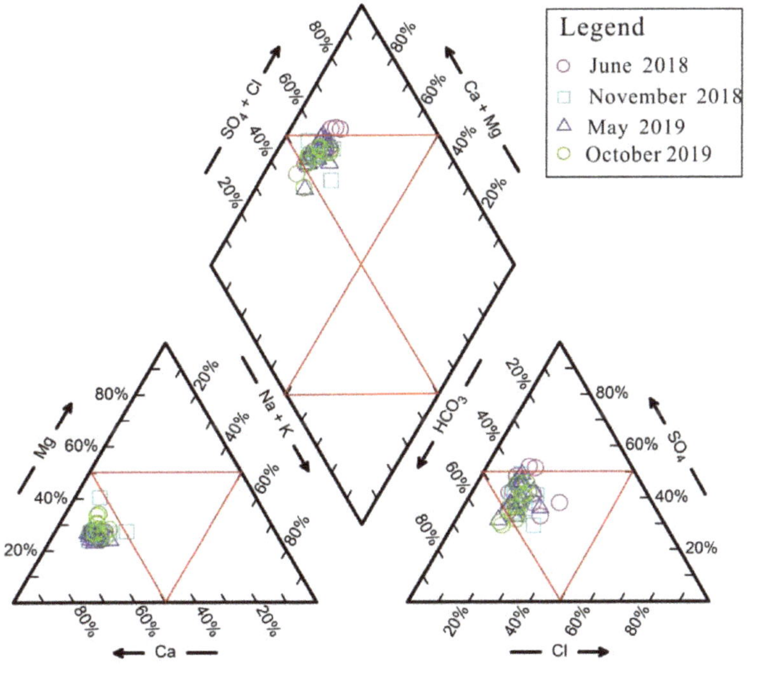

Figure 4. Piper diagram of groundwater in four stages of the recharge area.

The above analysis detected a slight change in the ion concentration of the study area after recharge. The follow-up of this study focuses on the above indicators to analyze the hydrochemical effects of groundwater after recharge. A small change was noticed in groundwater composition after recharge; therefore, the average recharge period (the average concentration of November 2018, May 2019, and October 2019) and June 2018 (pre-recharge) data were selected for principal component analysis to identify the control factors and scientifically deduce the water chemistry reaction caused by the recharge.

5. Discussion

Water–rock interactions and mixed interaction are the most important factors affecting the hydrogeochemical processes [43,44]. The test results of hydrogeochemistry are important data for studying its evolution mechanisms [45–47]. During the reservoir water recharge by RBF, groundwater is influenced by the mixed dilution of recharge water and water–rock interaction [48–51]. This study of groundwater through hydrogeochemistry can clarify the formation mechanisms and evolution of groundwater hydrochemical characteristics.

5.1. Hydrogeochemical Processes

The correlation between anions and cations and changes of the correlation coefficient can be used to reasonably speculate geochemical processes such as rock–water interaction [52,53]. In accordance with the drilling core test results and previous studies, the aquifer contains calcite, dolomite, rock salt, gypsum, and mirabilite in the study area [35–38]. The scale relationship of the main ions is plotted (Figure 5), and the correlation coefficients between different ions for each group of water samples are obtained. The scatter plot of Ca^{2+} and HCO_3^- (Figure 5a) shows that the two groups of groundwater samples, pre-recharge ($R = 0.88$) and recharge ($R = 0.88$), are strongly correlated. These two groups of water samples may have experienced weathering and dissolution of calcite, and this hydrogeochemical process can be expressed as R1. The scatter plots for Ca^{2+} and SO_4^{2-} (Figure 5b) exhibit well-correlated groundwater samples for the pre-recharge ($R = 0.77$) group. Weathering and dissolution of gypsum during the formation of groundwater in this group is a highly important hydrogeochemical process, and the reaction is expressed as R2. The scatter plot of Mg^{2+} and HCO_3^- (Figure 5c) demonstrates two well-correlated groups of groundwater samples, pre-recharge ($R = 0.78$) and recharge ($R = 0.79$). The sample may have undergone dolomite dissolution, and the expression for this reaction is R3. The scatter plot for Na^+ and HCO_3^- (Figure 5f) shows that pre-recharge ($R = 0.65$) is slightly correlated, whereas the pre-recharge groundwater ($R = 0.91$) is well correlated between Ca^{2+} and Na^+ (Figure 5g). The pre-recharge groundwater group experiences calcite dissolution and cation exchange. That is, Ca^{2+} in groundwater exchanges the adsorbed Na^+ during rock formation. The correlation between Na^+ and SO_4^{2-} in the pre-recharge groundwater group ($R = 0.78$) is good (Figure 5e). The dissolution of mirabilite may affect the distribution of chemical elements in the pre-recharge groundwater group. The dissolution reaction equation for mirabilite is shown in R4. The pre-recharge groundwater group ($R = 0.95$) has strong correlations in the scatter plots for Na^+ and Cl^- (Figure 5d). The dissolution of salt rock also occurs in this group and is expressed as Equation R5.

$$CaCO_3 + CO_2 + H_2O = Ca^{2+} + 2HCO_3^- \tag{1}$$

$$CaSO_4 = Ca^{2+} + SO_4^{2-} \tag{2}$$

$$CaMg(CO_3)_2 + 2CO_2 + 2H_2O = Ca^{2+} + Mg^{2+} + 4HCO_3^- \tag{3}$$

$$Na_2SO_4 \cdot 10H_2O = 2Na^+ + SO_4^{2-} + 10H_2O \tag{4}$$

$$NaCl = Na^+ + Cl^- \tag{5}$$

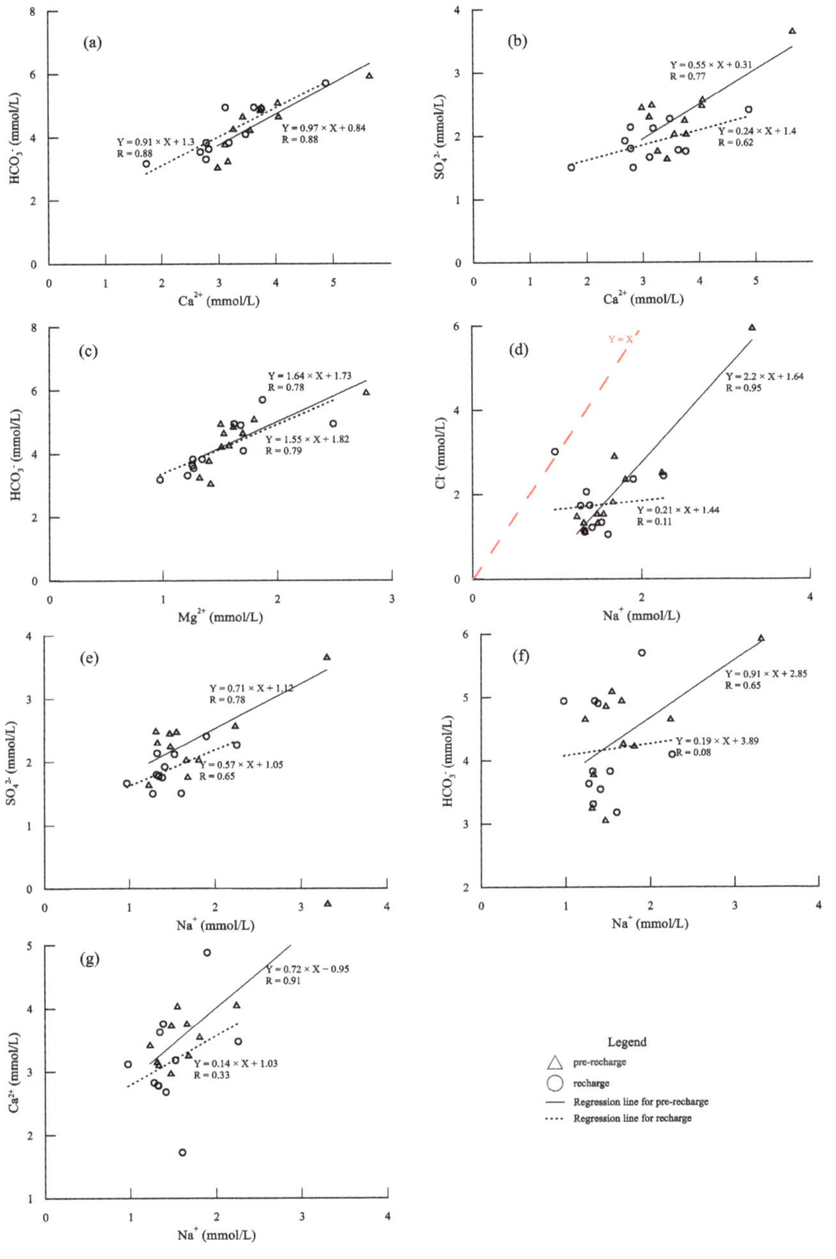

Figure 5. Relationship between the main ion pairs in groundwater. (**a**) Relationship between Ca^{2+} and HCO_3^-. (**b**) Relationship between Ca^{2+} and SO_4^{2-}. (**c**) Relationship between Mg^{2+} and HCO_3^-. (**d**) Relationship betweenNa$^+$ and Cl^-. (**e**) Relationship betweenNa$^+$ and SO_4^{2-}. (**f**) Relationship between Na^+ and HCO_3^-. (**g**) Relationship between Na^+ and Ca^{2+}.

There is a strong correlation between Na^+ and Cl^-, which indicates the main source of Na^+ and Cl^- to be the dissolution of halite. In theory, the ratio of Na^+ and Cl^- contents thus obtained should be 1:1, but most of the groundwater sampling points in this area are located in the lower right of the line y = x (Figure 5d). This means Na^+ content in the groundwater samples is greater than the Cl^- content. This is because Ca^{2+} and Mg^{2+} in the water are exchanged with Na^+ adsorbed in the rock formation, and the content of Na^+ in the water is more than that of Cl^-. The equation for this reaction is R6.

$$Ca^{2+} \text{ (in water)} + 2Na^+ \text{ (clay bound)} = 2\,Na^+ \text{ (in water)} + Ca^{2+} \text{ (clay bound)} \quad (6)$$

Ratio plots were used to further investigate the origins of the ions and the major hydrogeochemical processes. If the ions are controlled only by the dissolution of gypsum and carbonate rocks, (SO_4^{2-} + HCO_3^-) and (Ca^{2+} + Mg^{2+}) are equal [54,55]. The (SO_4^{2-} + HCO_3^-) and (Ca^{2+} + Mg^{2+}) ion proportional relationship should be located at the 1:1 line in the scatter plot (Figure 6). If y = x, the curve represents the dissolution line of carbonatite and gypsum. The water sample points in Figure 6 tend to be distributed downward, and almost all the points fall into the area with a relatively high content of (SO_4^{2-} + HCO_3^-). This indicates the occurrence of pronounced cation exchange in this area, that is, the ion exchange of Ca^{2+} + Mg^{2+} dissolved in water and Na^+ is adsorbed by clay.

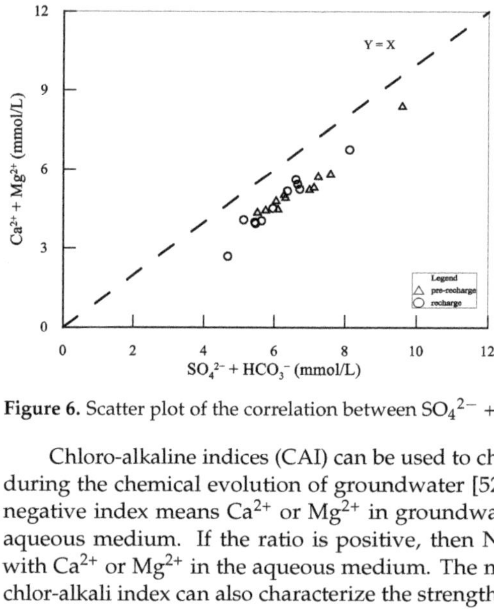

Figure 6. Scatter plot of the correlation between SO_4^{2-} + HCO_3^- and Ca^{2+} + Mg^{2+}.

Chloro-alkaline indices (CAI) can be used to characterize the strength of ion exchange during the chemical evolution of groundwater [52], with the expressions R7 and R8. A negative index means Ca^{2+} or Mg^{2+} in groundwater is ion-exchanged with Na^+ in the aqueous medium. If the ratio is positive, then Na^+ in groundwater is ion-exchanged with Ca^{2+} or Mg^{2+} in the aqueous medium. The magnitude of the absolute value of the chlor-alkali index can also characterize the strength of the ion exchange.

$$\text{CAI\_1} = \frac{Cl^- - (Na^+ + K^+)}{Cl^-} \quad (7)$$

$$\text{CAI\_2} = \frac{Cl^- - (Na^+ + K^+)}{\left(SO_4^{2-} + HCO_3^- + CO_3^{2-} + NO_3^-\right)} \quad (8)$$

Figure 7 shows the changes in CAI-1 and CAI-2 indices for the two groups of water sampling points. Most of the water sample points are greater than 0 from the perspective of the CAI index, except for a few points where indices are less than 0. Some pre-recharge and few recharge points are ion exchange between Ca^{2+} or Mg^{2+} in the groundwater and Na^+ in the aqueous medium, and most of the points are ion exchange between Na^+ in the groundwater and Ca^{2+} or Mg^{2+} in the aqueous medium. The average values of CAI-1 and CAI-2 of the water sampling points reveal that the pre-recharge samples are slightly lower than that of the recharge group, and the intensity of ion exchange in the recharge stage is higher than that of the pre-recharge stage.

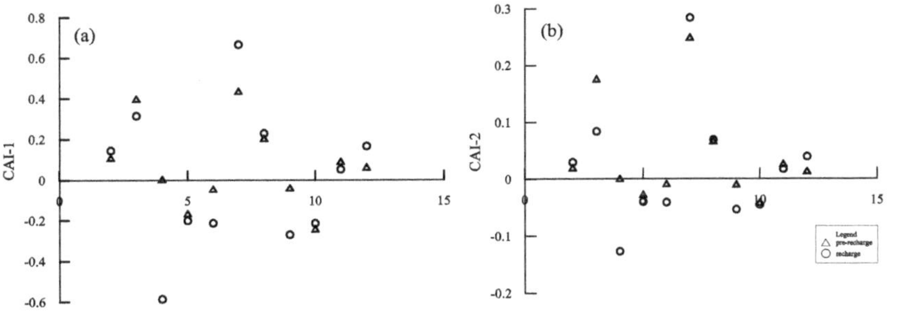

Figure 7. Changes in CAI-1 and CAI-2 of water samples. (**a**) CAI-1 in the two groups of water samples. (**b**) CAI-2 in the two groups of water samples.

Samples close to the 1:1 equilibrium line in (Ca^{2+} + Mg^{2+})/(HCO_3^-) and (Ca^{2+} + Mg^{2+})/(SO_4^{2-}) are obvious evidence to conclude that Ca^{2+} and Mg^{2+} in the groundwater mainly come from sulfate dissolution [55–57]. The samples are mostly distributed above the straight line 1:1 (Figure 8). This indicates the occurrence of short-term dilution after the recharge of the reservoir, and also an increase in the salt leaching and filtration of groundwater, in addition to the cation exchange effect.

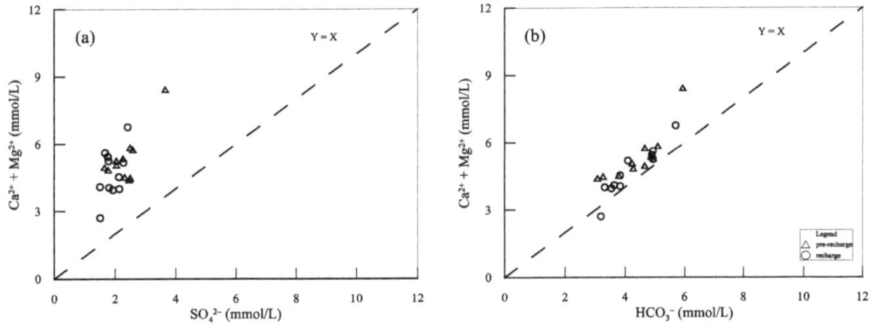

Figure 8. Correlation comparison of (Ca^{2+} + Mg^{2+})/HCO_3^- and (Ca^{2+} + Mg^{2+})/SO_4^{2-}. (**a**) Relationship between SO_4^{2-} and (Ca^{2+} + Mg^{2+}). (**b**) Relationship between HCO_3^- and (Ca^{2+} + Mg^{2+}).

Gibbs [58] divided the chemical effects of groundwater into three categories by comparing the relationship between Na^+/(Ca^{2+} + Na^+) and TDS and between Cl^-/(Cl^- + HCO_3^-) and TDS: "evaporation control type," "water-rock interaction type," and "precipitation control type." Many researchers have used Gibbs diagrams to study the sources of water chemical components and to analyze the formation mechanisms of water chemistry [59–61]. The Gibbs diagram for the study area was drawn (Figure 9), and the main controlling factors of the groundwater hydrochemical composition were analyzed. All points are in the area controlled by water–rock interaction. The formation of regional hydrochemical types is predominantly controlled by water–rock interactions, and evaporation has little impact on water regional hydrochemical types.

Principal component analysis was performed on the pre-recharge and recharge data, and the correlation matrix was found using the Kaiser–Meyer–Olkin (KMO) test. The test results reveal the suitability of the two groups of data for principal component analysis (Table 2). Table 2 illustrates that in pre-recharge, F1 accounts for 65.58% of the total variance and is characterized by the association with TDS, Na^+, Ca^{2+}, Mg^{2+}, SO_4^{2-}, and Cl^-. The correlation coefficients were all above 0.85, but the correlation with HCO_3^- was weak (the correlation coefficient is as low as 0.32). Hence, F1 can be a direct reason for the increase in TDS concentration in groundwater. F1 appears as the dissolution of gypsum and halite and the ion exchange process which plays a dominant role in groundwater pre-recharge. F2 contributes to 23.32% of the total variance, which depicts a significant positive correlation with HCO_3^-. F2 has a low correlation with other components that characterize the groundwater hydrochemical environment and indicates only the dissolution of carbonate. In the recharge period, F1 had a direct positive correlation with Na+, Mg^{2+}, TDS, Ca^{2+}, Cl^-, and HCO_3^-. The correlation coefficient with HCO_3^- increased from 0.32 to 0.92, and the variance contribution of F1 is 61.75%, which is 3.83% lower than pre-recharge. This indicates the dissolution of carbonates such as calcite and dolomite. If other factors such as groundwater pressure extraction or human activities are excluded, the presumed reason for the reduced correlation coefficient can be the short-term dilution effect of the reservoir's supplementary on groundwater.

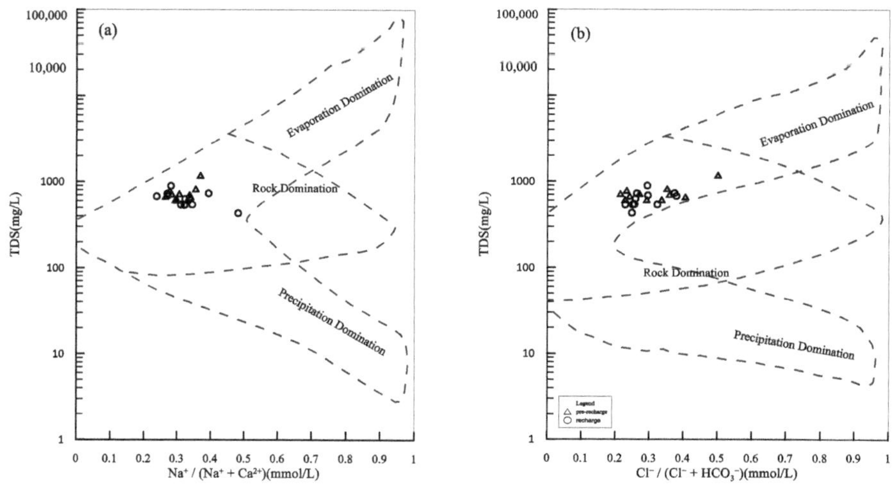

Figure 9. Gibbs diagram of the research area. (**a**) The relationship between $Na^+/(Na^+ + Ca^{2+})$ and TDS. (**b**) The relationship between $Cl^-/(Cl^- + HCO_3^-)$ and TDS.

Table 2. Results of Principal Component Analysis.

Time	Pre-Recharge		Recharge	
Factors	Fac_1	Fac_2	Fac_1	Fac_2
Na^+	0.90	0.33	0.94	0.38
Ca^{2+}	0.89	0.35	0.85	0.55
Mg^{2+}	0.87	0.28	0.95	−0.14
Cl^-	0.88	0.11	0.88	0.09
SO_4^{2-}	0.85	−0.18	0.24	0.82
HCO_3^-	0.32	0.81	0.92	0.33
TDS	0.92	0.39	0.89	0.58
Kaiser–Meyer–Olkin Measure of Sampling Adequacy	0.71		0.68	
Cumulative variance (%)	65.58	88.90	61.75	83.02

For F1, the correlation coefficient of HCO_3^- increased from 0.32 (pre-recharge) to 0.92 (recharge), whereas the SO_4^{2-} correlation coefficient decreased from 0.85 (pre-recharge) to 0.24. This indicates the dissolution of dolomite and calcite and the mixed dilution of recharge water. For F2, the SO_4^{2-} correlation coefficient increased from −0.18 to 0.82. This is because the groundwater chemical reaction was affected during recharge, resulting in the dissolution of mirabilite.

5.2. The Change of Groundwater Quality during Water Recharge

A mixing model with the help of the MIX module in the PHREEQC software was established. Research on water chemistry changes under mixing ratios was studied. By setting different mixing ratio schemes, the software's own phreeqc.dat database is used to calculate the saturation index and the amount of mineral phase transfer. The scenario at the beginning of the simulation is set to a 9:1 ratio of groundwater to recharge water, and the ratio of recharge water is gradually increased until the final simulation. Set June 2018 (pre-recharge) groundwater samples as the groundwater sample and reservoir water as the recharge sample. The specific mixing plan is shown in Table 3.

Table 3. Recharge Schemes with Different Mixing Ratios.

Water Type	Mixing Ratio								
Groundwater	0.9	0.8	0.7	0.6	0.5	0.4	0.3	0.2	0.1
Recharge Water	0.1	0.2	0.3	0.4	0.5	0.6	0.7	0.8	0.9

Cation exchange adsorption in the recharge area is weakened, and mixing also promotes Na^+ and Cl^- entering groundwater in recharge with an increase in the proportion of recharge water. The concentration change trend of the same ion components in different ratios is compared (Table 4). This reveals a gradual decrease in the ion concentration of each component in the mixed groundwater and the dilution effect as the main factor during the recharge process.

Table 4. Simulation Results of Main Ion Components in Different Ratios (mg/L).

G:R	9:1	8:2	7:3	6:4	5:5	4:6	3:7	2:8	1:9
Na^+	38.34	37.03	35.7	34.39	33.05	31.74	30.41	29.11	27.76
HCO_3^-	248.27	236.62	224.97	213.26	201.48	189.71	177.82	165.81	153.65
Mg^{2+}	32.712	31.2	29.688	28.176	26.664	25.152	23.6232	22.0968	20.568
Ca^{2+}	122.36	115.4	108.44	101.48	94.52	87.48	80.48	73.4	66.32
SO_4^{2-}	216	161.37	155.71	150.05	144.29	138.62	132.86	127.11	121.34

Note: G:R—the ratio of Groundwater (G): Recharge water (R).

The saturation index of mineral phases in each mixed solution calculated by different mixing schemes can be used to analyze the changing trend of the water–rock reaction after recharge. The saturation index of all minerals decreases gradually with the increase of water recharge proportion (Table 5). This results in the increase of mixing proportion in the unsaturated dissolution of all minerals, in which the saturation index (negative value) of dolomite and halite decreases significantly. When the proportion of recharge water between calcite and dolomite is 10–30%, the saturation index is more than 0 (positive value); therefore, the mixed solution is still saturated and insoluble for calcite and dolomite. When the proportion of recharge water increases to 40%, the saturation index suddenly changes to negative. As the proportion of recharge water continues to increase, calcite and dolomite become unsaturated. Consequently, the continuation of the recharge process promotes mineral dissolution.

Table 5. Simulation Results of SI in Different Ratios.

G:R	9:1	8:2	7:3	6:4	5:5	4:6	3:7	2:8	1:9
calcite	0.03	0.03	0.02	−0.02	−0.1	−0.21	−0.36	−0.59	−0.96
dolomite	0.05	0.06	0.02	−0.07	−0.23	−0.47	−0.8	−1.27	−2.04
gypsum	−0.08	−0.16	−0.25	−0.35	−0.47	−0.6	−0.75	−0.95	−1.2
halite	−0.23	−0.48	−0.76	−1.05	−1.42	−1.85	−2.37	−3.07	−5.23
mirabilite	−2.04	−2.12	−2.21	−2.34	−2.49	−2.65	−2.79	−2.92	−3.21

Note: G:R—the ratio of Groundwater (G): Recharge water (R).

6. Conclusions

This study focused on the river bank filtration (RBF site at the upper of the Hutuo River Alluvial Plain, North China, and reasonably analyzed the hydrochemical changes of groundwater during the period of bank infiltration recharge. From June 2018 (pre-recharge) to October 2019 (late recharge), groundwater in the study area was mainly of type Ca–Mg–HCO_3–SO_4 and Ca–Mg–SO_4–HCO_3, with no obvious change in chemistry. As the concentration of various ions in groundwater decreased, the short-term dilution effect of groundwater recharge temporarily improved its quality. Anthropogenic factors lead to the increase of nitrate content in some wells. Reservoir water recharge will further promote groundwater–rock reaction and filtration by RBF. The ion proportional relationship diagram and principal component analysis demonstrate the dissolution of calcite, dolomite, gypsum, mirabilite, and halite in the pre-recharge stage. The recharge stage mainly experienced the dissolution of calcite and dolomite. The ion-pair scatter plot of Na^+ and Cl^- and the chlor-alkali index reveal cation exchange as one of the main hydrogeochemical processes in the study area. Some pre-recharge and few recharge points are ion exchange between Ca^{2+} or Mg^{2+} in the groundwater and Na^+ in the aqueous medium, and most of the points are ion exchange between Na^+ in the groundwater and Ca^{2+} or Mg^{2+} in the aqueous medium. The intensity of ion exchange in the recharge stage was higher than that of the pre-recharge stage. The research of water chemistry changes under mixing ratios discloses the dilution effect as the main factor during the recharge process. Calcite and dolomite become unsaturated as the proportion of recharge water continues to increase, which in turn, promotes mineral dissolution.

The results of this study are promising and suggest a successful application of the combination of these techniques to decipher the characteristics and impacts of intrinsic changes for using reservoir water to recharge groundwater by RBF. The quantity and quality of local groundwater resources can be increased by the reservoir water supplement in RBF.

Author Contributions: Wrote the manuscript, B.Z., L.C. and Y.L. (Yasong Li); designed the research, B.Z. and Y.Z.; processed the data, L.C. and Y.L. (Yaci Liu); contributed to the theoretical interpretation of the results, B.Z., X.K. and Y.Z.; supervised the project, Y.L. (Yasong Li), Y.L. (Yaci Liu), C.L., X.K. and Y.Z.; All authors have read and agreed to the published version of the manuscript.

Funding: The research work was financially supported by the National Water Pollution Control and Treatment Science and Technology Major Project (No. 2018ZX07109-004), with contributions by the China Geological Survey project, Grant/Award Number DD201903034, and National Natural Science Foundation of China, Grant/Award Number: 41907175. The authors gratefully acknowledge the Institute of Geophysical and Geochemical Exploration, Chinese Academy of Geological Sciences (AS2020Y04), for financial support.

Data Availability Statement: The data that support the findings of this study are available within the article.

Acknowledgments: The authors gratefully acknowledge Peng-Fei Han and Dong Zhang for their guidance and revision of the paper.

Conflicts of Interest: The authors declare no competing interest.

References

1. Regnery, J.; Li, D.; Lee, J.; Smits, K.M.; Sharp, J.O. Hydrogeochemical and microbiological effects of simulated recharge and drying within a 2D meso-scale aquifer. *Chemosphere* **2020**, *241*, 9. [CrossRef] [PubMed]
2. Regnery, J.; Gerba, C.P.; Dickenson, E.R.V.; Drewes, J.E. The importance of key attenuation factors for microbial and chemical contaminants during managed aquifer recharge: A review. *Crit. Rev. Environ. Sci. Technol.* **2017**, *47*, 1409–1452. [CrossRef]
3. Dillon, P.; Stuyfzand, P.; Grischek, T.; Lluria, M.; Pyne, R.D.G.; Jain, R.C.; Bear, J.; Schwarz, J.; Wang, W.; Fernandez, E.; et al. Sixty years of global progress in managed aquifer recharge. *Hydrogeol. J.* **2019**, *27*, 1–30. [CrossRef]

4. Zhu, Y.; Zhai, Y.; Du, Q.; Teng, Y.; Wang, J.; Yang, G. The impact of well drawdowns on the mixing process of river water and groundwater and water quality in a riverside well field, Northeast China. *Hydrol. Process.* **2019**, *33*, 945–961. [CrossRef]
5. Cao, Y.; Wei, Y.; Fan, W.; Peng, M.; Bao, L. Experimental study of land subsidence in response to groundwater withdrawal and recharge in Changping District of Beijing. *PLoS ONE* **2020**, *15*, e0232828. [CrossRef] [PubMed]
6. Dillon, P.; Toze, S.; Page, D.; Vanderzalm, J.; Bekele, E.; Sidhu, J.; Rinck-Pfeiffer, S. Managed aquifer recharge: Rediscovering nature as a leading edge technology. *Water Sci. Technol.* **2010**, *62*, 2338–2345. [CrossRef]
7. Hu, B.; Teng, Y.; Zhai, Y.; Zuo, R.; Li, J.; Chen, H. Riverbank filtration in China: A review and perspective. *J. Hydrol.* **2016**, *541*, 914–927. [CrossRef]
8. Massmann, G.; Nogeitzig, A.; Taute, T.; Pekdeger, A. Seasonal and spatial distribution of redox zones during lake bank filtration in Berlin, Germany. *Environ. Geol.* **2007**, *54*, 53–65. [CrossRef]
9. Yuan, J.; Dyke, M.I.V.; Huck, P.M. Water reuse through managed aquifer recharge (MAR): Assessment of regulations/guidelines and case studies. *Water Pollut. Res. J. Can.* **2016**, *51*, 357–376. [CrossRef]
10. Hiscock, K.M.; Grischek, T. Attenuation of groundwater pollution by bank filtration. *J. Hydrol.* **2002**, *266*, 139–144. [CrossRef]
11. Grischek, T.; Bartak, R. Riverbed clogging and sustainability of riverbank filtration. *Water* **2016**, *8*, 604. [CrossRef]
12. Ganot, Y.; Holtzman, R.; Weisbrod, N.; Russak, A.; Katz, Y.; Kwtzman, D. Geochemical processes during managed aquifer recharge with desalinated seawater. *Water Resour. Res.* **2018**, *54*, 978–994. [CrossRef]
13. Kurki, V.; Lipponen, A.; Katko, T. Managed aquifer recharge in community water supply: The Finnish experience and some international comparisons. *Water Int.* **2013**, *38*, 774–789. [CrossRef]
14. Bartak, R.; Page, D.; Sandhu, C.; Grischek, T.; Saini, B.; Mehrotra, I.; Chakresh, K.J.; Narayan, C. Ghosh Application of risk-based assessment and management to riverbank filtration sites in India. *J. Water Health* **2015**, *13*, 174–189. [CrossRef] [PubMed]
15. Daess, L.W.; Andrade-Tafoya, P.D.; Lafarga-Moreao, J.; Mahlknecht, J.; van Geldern, R.; Beramendi-Orosco, L.E.; Barth, J.A.C. Groundwater recharge sites and pollution sources in the wine-producing Guadalupe Valley (Mexico): Restrictions and mixing prior to transfer of reclaimed water from the US-México border. *Sci. Total Environ.* **2020**, *713*, 136715. [CrossRef]
16. Wang, D.; Yu, J.; Wang, P.; Zhu, B. Shallow Groundwater Chemistry Characteristics and Their Controlling Factors in the Ejina Delta. South-to-North Water Diversion. *Water Sci. Technol.* **2013**, *11*, 51–55. (In Chinese)
17. Zhu, Y.; Zhai, Y.; Teng, Y.; Wang, G.; Du, Q.; Wang, J.; Yang, G. Water supply safety of riverbank filtration wells under the impact of surface water-groundwater interaction: Evidence from long-term field pumping tests. *Sci. Total Environ.* **2020**, *711*, 135141. [CrossRef]
18. AlBassam, M.A.; Awad, H.S.; Al Alawi, J.A. DurovPlot: A computer program for processing and plotting hydrochemical data. *Ground Water* **1997**, *35*, 362–367. [CrossRef]
19. Boving, T.B.; Patil, K.; D'Souza, F.; Barker, S.F.; McGuinness, S.L.; O'Toole, J.; Sinclair, M.; Forbes, A.B.; Leder, K. Performance of riverbank filtration under hydrogeologic conditions along the upper Krishna River in Southern India. *Water* **2019**, *11*, 12. [CrossRef]
20. Li, C.; Li, B.; Bi, E. Characteristics of hydrochemistry and nitrogen behavior under long-term managed aquifer recharge with reclaimed water: A case study in north China. *Sci. Total Environ.* **2019**, *668*, 1030–1037. [CrossRef]
21. Yu, Y.; Song, X.; Zhang, Y.; Zheng, F.; Liu, L. Impact of reclaimed water in the watercourse of Huai River on groundwater from Chaobai River basin, Northern China. *Front. Earth Sci.* **2016**, *11*, 643–659. [CrossRef]
22. Pan, W.; Huang, Q.; Huang, G. Nitrogen and organics removal during riverbank filtration along a reclaimed water restored river in Beijing, China. *Water* **2018**, *10*, 491. [CrossRef]
23. Horriche, F.J.; Benabdallah, S. Assessing aquifer water level and salinity for a managed artificial recharge site using reclaimed water. *Water* **2020**, *12*, 341. [CrossRef]
24. Sprenger, C.; Lorenzen, G.; Hulshoff, I.; Grutzmacher, G.; Ronghang, M.; Pekdeger, A. Vulnerability of bank filtration systems to climate change. *Sci. Total Environ.* **2011**, *409*, 655–663. [CrossRef] [PubMed]
25. Munz, M.; Oswald, S.E.; Schaefferling, R.; Lensing, H.-J. Temperature-dependent redox zonation, nitrate removal and attenuation of organic micropollutants during bank filtration. *Water Res.* **2019**, *162*, 225–235. [CrossRef]
26. Guyennon, N.; Salerno, F.; Portoghese, I.; Romano, E. Climate Change Adaptation in a Mediterranean Semi-Arid Catchment: Testing Managed Aquifer Recharge and Increased Surface Reservoir Capacity. *Water* **2017**, *9*, 689. [CrossRef]
27. Xanke, J.; Liesch, T.; Goeppert, N.; Klinger, J.; Gassen, N.; Goldscheider, N. Contamination risk and drinking water protection for a large-scale managed aquifer recharge site in a semi-arid karst region, Jordan. *Hydrogeol. J.* **2017**, *25*, 1795–1809. [CrossRef]
28. Zammouri, M.; Brini, N. Efficiency of artificial groundwater recharge, quantification through conceptual modelling. *Water Resour. Manag.* **2020**, *34*, 3345–3361. [CrossRef]
29. Groeschke, M.; Frommen, T.; Taute, T.; Schneider, M. The impact of sewage-contaminated river water on groundwater ammonium and arsenic concentrations at a riverbank filtration site in central Delhi, India. *Hydrogeol. J.* **2017**, *25*, 2185–2197. [CrossRef]
30. Li, Y.; Zhang, Z.; Fei, Y.; Chen, H.; Qian, Y.; Dun, Y. Investigation of quality and pollution characteristics of groundwater in the Hutuo River Alluvial Plain, North China Plain. *Environ. Earth Sci.* **2016**, *75*, 581. [CrossRef]
31. Zhang, X.; He, J.; He, B.; Sun, J. Assessment, formation mechanism, and different source contributions of dissolved salt pollution in the shallow groundwater of Hutuo River alluvial-pluvial fan in the North China Plain. *Environ. Sci. Pollut. Res.* **2019**, *26*, 35742–35756. [CrossRef] [PubMed]

32. Zhang, Q.; Wang, H.; Wang, L. Tracing nitrate pollution sources and transformations in the over-exploited groundwater region of north China using stable isotopes. *J. Contam. Hydrol.* **2018**, *218*, 1–9. [CrossRef]
33. Zhang, Q.; Wang, H.; Wang, Y.; Yang, M.; Zhu, L. Groundwater quality assessment and pollution source apportionment in an intensely exploited region of northern China. *Environ. Sci. Pollut. Res.* **2017**, *24*, 16639–16650. [CrossRef] [PubMed]
34. *GB/T 14848-2017*; Quality Standard for Ground Water. Standardization Administration of the PRC: Beijing, China, 2017; p. 20. (In Chinese)
35. Su, X.; Xu, W.; Du, S. Responses of groundwater vulnerability to artificial recharge under extreme weather conditions in Shijiazhuang City, China. *J. Water. Supply Res. T.* **2014**, *63*, 224–238. [CrossRef]
36. Tian, X.; Meng, S.; Cui, X.; Zhang, X.; Zhang, Z.; Fei, Y. Hydrochemical Effect of Groundwater Recharge in Over-Exploited Area of Hutuo River Basin. *Res. Environ. Sci.* **2021**, *34*, 629–636. (In Chinese)
37. He, P. Primary Analysis of Groundwater Circumstances Influence of Huangbizhuang Village Reservoir Waterproof Project to the Hutuo River Alluvium. *Ground Water* **2009**, *31*, 121–123. (In Chinese)
38. Men, G. Design flood analysis of Huangbizhuang Reservoir downstream river channel. *Water Sci. Eng. Technol.* **2015**, *4*, 25–28. (In Chinese)
39. Zheng, Y.; Vanderzalm, J.; Hartog, N.; Escalante, E.F.; Stefan, C. The 21st century water quality challenges for managed aquifer recharge: Towards a risk-based regulatory approach. *Hydrogeol. J.* **2022**, *31*, 189. [CrossRef]
40. Piper, A.M. *A Graphic Procedure in the Geochemical Interpretation of Water Analysis*; United States Department of the Interior, Geological Survey, Water Resources Division, Ground Water Branch: Washington, DC, USA; CRC Press: Boca Raton, FL, USA, 1953; p. 63.
41. Zhai, Y.; Zheng, F.; Zhao, X.; Xia, X.; Teng, Y. Identification of hydrochemical genesis and screening of typical groundwater pollutants impacting human health: A case study in Northeast China. *Environ. Pollut.* **2019**, *252*, 1202–1215. [CrossRef]
42. Kang, X.; Niu, Y.; Yu, H.; Gou, P.; Hou, Q.; Lu, X.; Wu, Y. Effect of rainfall-runoff process on sources and transformations of nitrate using a combined approach of dual isotopes, hydrochemical and Bayesian model in the Dagang River basin. *Sci. Total Environ.* **2022**, *837*, 1–11. [CrossRef]
43. Appelo, C.A.J.; Willemsen, A. Geochemical calculations and observations on salt water intrusions, I. A combined geochemical/minxing cell model. *J. Hydrol.* **1987**, *94*, 313–330. [CrossRef]
44. Duan, R.; Li, P.; Wang, L.; He, X.; Zhang, L. Hydrochemical characteristics, hydrochemical processes and recharge sources of the geothermal systems in Lanzhou City, northwestern China. *Urban Clim.* **2022**, *43*, 1–15. [CrossRef]
45. Su, Y.; Zhu, G.; Feng, Q.; Li, Z.; Zhang, F. Environmental isotopic and hydrochemical study of groundwater in the Ejina Basin, northwest China. *Environ. Geol.* **2009**, *58*, 601–614. [CrossRef]
46. Bekele, E.; Zhang, Y.; Donn, M.; McFarlane, D. Inferring groundwater dynamics in a coastal aquifer near wastewater infiltration ponds and shallow wetlands (Kwinana, Western Australia) using combined hydrochemical, isotopic and statistical approaches. *J. Hydrol.* **2019**, *568*, 1055–1070. [CrossRef]
47. Noble, J.; Ansari, M.A. Isotope hydrology and geophysical techniques for reviving a part of the drought prone areas of Vidarbha, Maharashtra, India. *J. Hydrol.* **2019**, *570*, 495–507. [CrossRef]
48. Wang, Z.; Yin, J.; Pu, J.; Wang, P.; Liang, X.; Yang, P.; He, Q.; Gou, P.; Yuan, D. Integrated understanding of the Critical Zone processes in a subtropical karst watershed (Qingmuguan, Southwestern China): Hydrochemical and isotopic constraints. *Sci. Total Environ.* **2020**, *749*, 141257. [CrossRef]
49. Dugga, P.; Pervez, S.; Tripathi, M.; Siddiqui, M.N. Spatiotemporal variability and source apportionment of the ionic components of groundwater of a mineral-rich tribal belt in Bastar, India. *Groundw. Sustain. Dev.* **2020**, *10*, 100356. [CrossRef]
50. Hou, G.; Zhao, M.; Wang, Y. *Groundwater Investigation in the Ordos Basin*; China Geological Survey: Beijing, China, 2006. (In Chinese)
51. Kong, X.; Wang, S.; Liu, B.; Sun, H.; Sheng, Z. Impact of water transfer on interaction between surface water and groundwater in the lowland area of North China Plain. *Hydrol. Process.* **2018**, *32*, 2044–2057. [CrossRef]
52. Guo, X.; Zuo, R.; Wang, J.; Meng, L.; Teng, Y.; Shi, R.; Gao, X.; Ding, F. Hydrogeochemical Evolution of Interaction Between Surface Water and Groundwater Affected by Exploitation. *Groundwater* **2019**, *57*, 430–442. [CrossRef]
53. Liu, J.; Gao, Z.; Wang, Z.; Xu, X.; Su, Q.; Wang, S.; Qu, W.; Xing, T. Hydrogeochemical processes and suitability assessment of groundwater in the Jiaodong Peninsula, China. *Environ. Monit. Assess.* **2020**, *192*, 17. [CrossRef]
54. Lasaga, A.C. Chemical kinetics of water–rock interactions. *J. Geophys. Res.* **1984**, *89*, 4009–4025. [CrossRef]
55. Wu, J.; Li, P.; Qian, H.; Duan, Z.; Zhang, X. Using correlation and multivariate statistical analysis to identify hydrogeochemical processes affecting the major ion chemistry of waters: A case study in Laoheba phosphorite mine in Sichuan, China. *Arab. J. Geosci.* **2014**, *7*, 3973–3982. [CrossRef]
56. Liu, Y.; Fei, Y.; Meng, S.; Cui, X. Hydrochemical evolution of groundwater and soils in the water-level-fluctuation zone. *Environ. Earth Sci.* **2019**, *78*, 12. [CrossRef]
57. Han, Y.; Zhai, Y.; Guo, M.; Cao, X.; Lu, H.; Li, J.; Wang, S.; Yue, W. Hydrochemical and isotopic characterization of the impact of water diversion on water in drainage channels, groundwater, and Lake Ulansuhai in China. *Water* **2021**, *13*, 3033. [CrossRef]
58. Gibbs, R.J. Mechanisms controlling world water chemistry. *Science* **1970**, *170*, 1088–1090. [CrossRef] [PubMed]
59. Marghade, D.; Malpe, D.B.; Zade, A.B. Major ion chemistry of shallow groundwater of a fast growing city of Central India. *Environ. Monit. Assess.* **2012**, *184*, 2405–2418. [CrossRef]

60. Xing, L.; Guo, H.; Zhan, Y. Groundwater hydrochemical characteristics and processes along flow paths in the North China Plain. *J. Asian Earth Sci.* **2013**, *70–71*, 250–264. [CrossRef]
61. Pant, R.R.; Zhang, F.; Rehman, F.U.; Wang, G.; Ye, M.; Zeng, C.; Tang, H. Spatiotemporal variations of hydrogeochemistry and its controlling factors in the Gandaki River Basin, Central Himalaya Nepal. *Sci. Total Environ.* **2018**, *622–623*, 770–782. [CrossRef]

water

MDPI

Article

Groundwater Pollution Model and Diffusion Law in Ordovician Limestone Aquifer Owe to Abandoned Red Mud Tailing Pit

Yueming Qi, Pei Zhou, Junping Wang *, Yipeng Ma, Jiaxing Wu and Chengzhi Su

School of Resources and Geosciences, China University of Mining and Technology, Xuzhou 221116, China; ym_qi@cumt.edu.cn (Y.Q.); ts21010106a31@cumt.edu.cn (P.Z.); ts19010098a31@cumt.edu.cn (Y.M.); ts21010105a31@cumt.edu.cn (J.W.); ts21010092a31@cumt.edu.cn (C.S.)
* Correspondence: ts19010031a31@cumt.edu.cn

Abstract: Red mud is a strong alkaline solid waste pollutant produced in the process of aluminum smelting, which causes great pollution to the regional groundwater environment due to its high content of fluorine and aluminum and high concentration of strong alkali. In this study, fluoride ion was selected as the model contaminant, and a numerical model of the groundwater flow field and solute transport was developed using GMS software to simulate and analyze the migration patterns of fluoride contaminants caused by the red mud pit for the fractured karst geohydrological conditions. The results demonstrated that the groundwater model and flow pattern were mainly controlled by atmospheric precipitation recharge, given flow boundary conditions and leakage of rivers and drains. When the concentration of fluorine pollutants in the red mud yard was 60.0 mg/L, the maximum migration distance of F^- in the groundwater of the ordovician limestone aquifer was 473, 1160, 1595 and 1750 m after 1, 5, 10 and 15 years of bottom leakage, and the additional transport distances were 687, 435 and 155 m every 5 years, respectively. The range of F^- pollution plume was 0.37 km^2, 1.15 km^2, 1.95 km^2 and 2.14 km^2, respectively and the range of newly added pollution plume was 0.78 km^2, 0.80 km^2 and 0.19 km^2, respectively, every five years. Both indicated that with the extension of time, the migration and diffusion rate of pollutants slow down, and the diffusion volume increased first and then decreased. The F^- pollution plume spread from the red mud pit to the northeast, which was consistent with the flow of groundwater. The high-concentration pollution plume was mainly distributed in the Ordovician limestone fractured aquifer in the northeast. This study revealed the migration law of red mud pollutants, and provided a scientific decision-making basis for the prevention and control of red mud groundwater pollution in the future.

Keywords: red mud; groundwater; fluorine pollution; migration model; diffusion law

check for updates

Citation: Qi, Y.; Zhou, P.; Wang, J.; Ma, Y.; Wu, J.; Su, C. Groundwater Pollution Model and Diffusion Law in Ordovician Limestone Aquifer Owe to Abandoned Red Mud Tailing Pit. *Water* **2022**, *14*, 1472. https://doi.org/10.3390/w14091472

Academic Editor: Domenico Cicchella

Received: 30 March 2022
Accepted: 30 April 2022
Published: 4 May 2022

Publisher's Note: MDPI stays neutral with regard to jurisdictional claims in published maps and institutional affiliations.

1. Introduction

In recent years, with the rapid development of economy and society, China's demand for aluminum has been increasing. Because the bauxite horizon in North China is stable and easy to mine, the aluminum smelting industry has developed rapidly and become a local pillar industry. At the same time, the red mud produced by aluminum smelting has also increased greatly, with incomplete statistics demonstrated that over 70 million tonnes of red mud are discharged annually [1]. The discharge and storage of red mud not only occupied a lot of land, but also under the leaching effect of natural rainfall [2], the harmful components in red mud will enter the surrounding soil and groundwater, causing serious pollution [3]. He (2020) et al. found that the fluoride concentration in 89% of crops and soil exceeded the standard by detecting the fluoride in vegetables and shallow soil within 10 km of the waste aluminum plant [4]. Fluoride has a serious impact on human health [5], and when daily fluoride intake exceeds 6 mg, it can lead to fluorosis [6]. The problem of groundwater fluorine pollution is imminent.

Many scholars have used numerical simulation methods to simulate the movement and reaction of water flow and solute in confined aquifer and phreatic aquifer. Yang et al. (2014) used the HYDRUS-1D model to simulate the migration and transformation law of ammonia nitrogen in landfill leachate in the aeration zone. In this study, the predicted source intensity value is used to evaluate the degree of site pollution, and finally predicted the concentration value of ammonia nitrogen when it reached the groundwater, so as to provide data reference for site pollution air defense [7]. Zhu et al. (2001) used MODFLOW and MT3D software to simulate the migration of petroleum pollutants in limestone karst fissure water in the Dawu water source area, and found that the permeability coefficient and effective porosity have a great impact on the results of solute migration [8]. Sathe et al. (2019) studied the diffusion of arsenic pollution in eastern India through soil column experiment and GMS software, and found that it has the risk of further polluting deep groundwater [9].

In the past experiments on red mud, they usually focused on the resource utilization of red mud or the hydro-geochemistry of fluorine in groundwater. Meanwhile, there is a lack of relevant research on the non-point source pollution caused by the red mud pit after bauxite mining. Therefore, this paper investigates the migration law of red mud pollutant leachate infiltrating into the karst fissure aquifer under the strong alkaline condition and red mud tailing pit contact with fissured karst aquifers. A multi-theoretical and multi-method crossover was adopted for the study of contaminant transport in fractured karst aquifers. The research used hydro-geological methods such as borehole data, groundwater tracing experiments and pumping test methods, and adopted multi-modules such as SOILD module, MODFLOW groundwater flow calculation module and MT3DMS solute transport module to investigate the spatial and temporal distribution and migration patterns of fluorine contamination in the red mud tailing pit. On this basis, the fluorine pollution control scheme was put forward and provides a scientific basis for fluorine pollution control and groundwater remediation on a large scale.

2. Overview of the Study Area

The non-standard red mud tailing pit is situated in northern China (Figure 1). The study area comprises in between 118°02′15″ N–118°07′09″ N latitudes and 36°43′11″ E–36°39′57″ E longitudes, covering a total area of about 12.692 km^2. The average annual precipitation is 671.7 mm, mainly from June to September in the wet season. The average annual evaporation is 1 345.7 mm. The main river in the study area is the Mansi River, located in the west of the study area. Generally, there is surface flow only from September to October in the wet season, and the recharge to groundwater is weak. The main exposed strata (Figure 2) in the study area are Quaternary (Q), Carboniferous-Permian (C-P) and Ordovician (O). Among them, the Ordovician strata are mainly carbonate sedimentary strata, located in the northern part of the study area, trending northeast, dipping southwest, with a dip angle of about 10°. The Ordovician lithology is limestone, dolomitic limestone and dolomite (Figures 3 and 4). With a large variation in production near the fracture zone, its karst fissure are extremely developed, which is the main water recharge source of the main aquifer and local groundwater in the study area. The Carboniferous and Permian strata are distributed in the piedmont of the study area, and strip exposed in the northwest of the Ordovician strata. The bauxite mudstone and shale of Benxi Formation are mainly exposed in the study area. The bauxite mudstone of Benxi Formation is the main source of local bauxite. The red mud tailing pit after bauxite mining by the aluminum refinery is located here. The average thickness of Quaternary stratum is 25 m, mainly composed of gravel, sand and a small amount of clay, which is discontinuous in some areas of the study area.

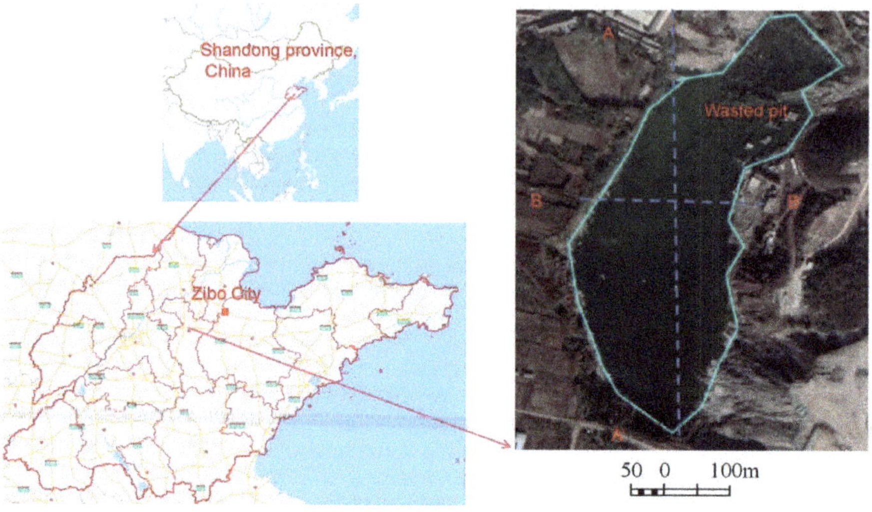

Figure 1. Map of wasted pit in the study area.

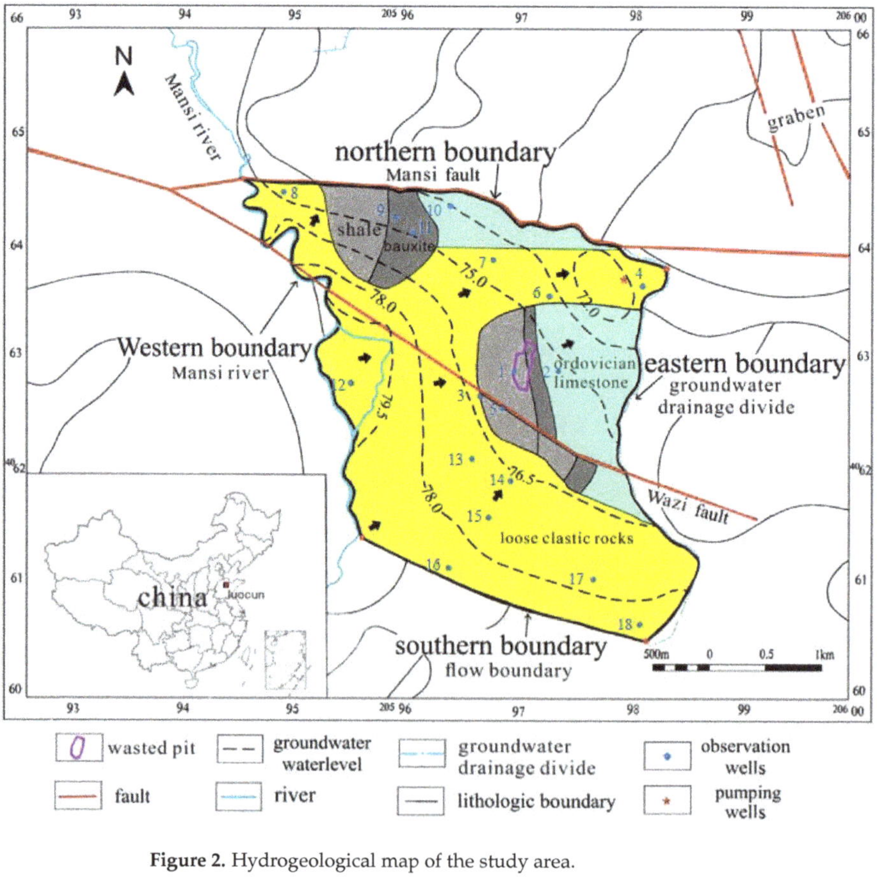

Figure 2. Hydrogeological map of the study area.

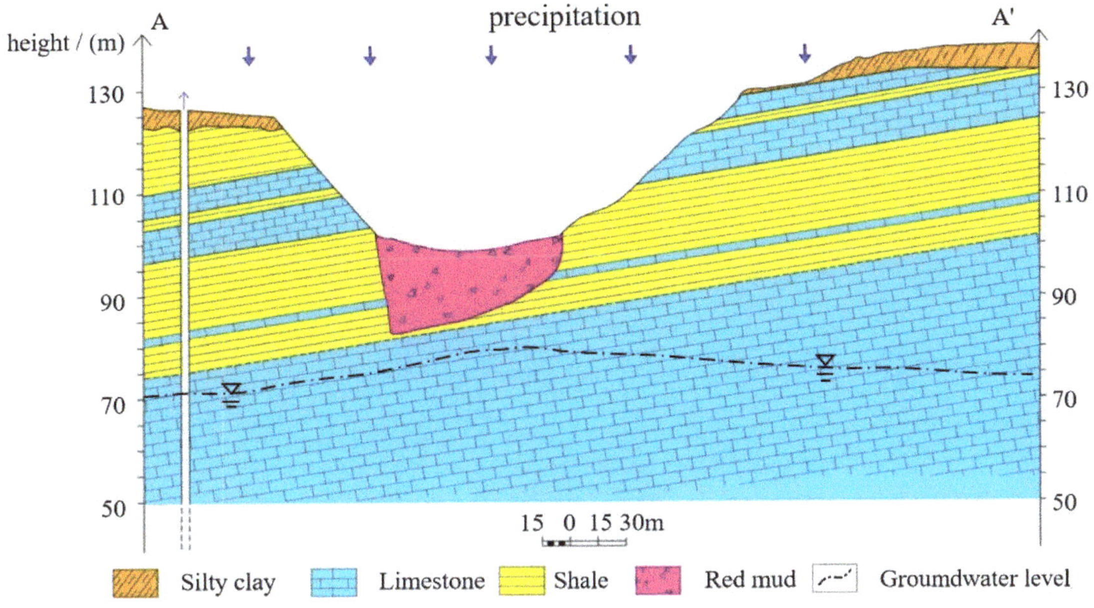

Figure 3. A-A′ Section.

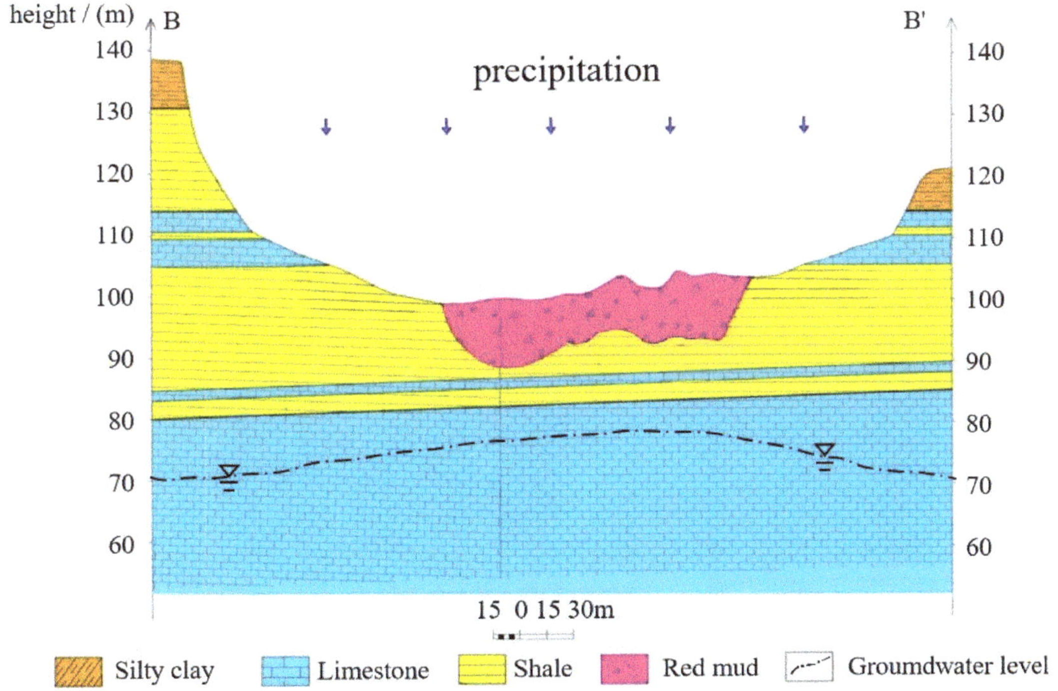

Figure 4. B-B′ section. (the location of the sections was described in Figure 1).

The study area is located in the plate uplift zone, affected by multi-stage tectonic activities, with many folds and faults in the area, and the terrain is mostly mountainous and hilly. Based on the distribution characteristics of faults in the whole region, the faults basically strike west-east, and the stratum integrity is damaged. Under the action of long-term water flow, the karst phenomenon of karst pipelines or fissures is developed, and groundwater is mainly stored in the corrosion fissure network of faults, joints and bedding fissures [10].

The water recharge of Ordovician limestone aquifer under the red mud tailing pit mainly comes from atmospheric precipitation recharge, river infiltration recharge and lateral recharge of the south flow boundary. The upper phreatic aquifer flows along the terrain and discharges to the northeast. The lower Ordovician limestone aquifer is controlled by regional structure and topography, and the runoff direction is basically consistent with the upper phreatic water.

3. Groundwater Pollution Sources

After the bauxite deposit in the study area was mined, the red mud was directly stacked in the tailing pit (Figure 1), which is in direct contact with the broken Ordovician limestone fissure karst aquifer in the form of a "skylight + water-conducting fissure zone" combination. The red mud leachate generated by rainfall could directly enter the limestone fissure aquifer, which contains alkali, fluoride and other pollutants [11]. According to the satellite image survey, the red mud tailing pit have been leaching the surrounding areas of the pit and groundwater after soaking and leaching for at least 15 years [12].

Water samples were taken from the observation wells of the contaminated site (As shown in Figure 1) and tested with hash DR2800 spectrophotometer. The results are shown in Table 1. Compared with the Groundwater Quality Standard (GB14848-2017) in China for class III quality standard, F^-, SO_4^{2-}, Al^{3+} and other ions exceed the standard, of which F^- exceeds the standard with the highest frequency (100%). The distribution of F^- shows obvious gradient distribution characteristics, with many exceeding standard sites, with wide coverage that are harmful to human health. Therefore, this paper took the red mud tailing pit as non-point sources of pollution and selected F^- as the main characteristic pollutant.

Table 1. Test results of groundwater pollutants.

Scheme 42.	F^- (mg/L)	SO_4^{2-} (mg/L)	Al^{3+} (mg/L)	TDS (mg/L)
1	52.40	784.34	37.96	2491.24
2	18.22	663.39	0.01	1590.86
3	6.41	205.36	0.01	891.13
4	5.84	216.36	0.01	782.10
5	5.57	275.76	0.01	964.06
6	4.53	408.72	0.01	920.20
7	2.00	1529.03	0.15	3279.13
8	1.20	378.64	0.01	1034.02

4. Construction of Numerical Simulation Model

4.1. Hydro-Geological Conceptual Model

According to the stratum lithology, thickness and hydro-geological parameters revealed by the drilling data and surface elevation data, the whole stratum can be generalized into two layers vertically, which are the overlying stratum and Ordovician limestone aquifer. According to the drilling data, the top and bottom plate elevations of each layer were obtained by the kriging interpolation method. The conceptual model of groundwater flow was established by using the MODFLOW module in GMS, with a 50 m spacing grid with two layers and three surfaces (Slice) in the vertical direction, for a total of 177,400 effective cells. The simulation period was from 1 January 2003 to 31 December 2019, with three months as a stress cycle. Finally, the aquifer in the study area was generalized as a heterogeneous anisotropic three-dimensional unsteady flow groundwater system.

4.2. Mathematical Model

Through the analysis of hydro-geological conceptual model in the study area, the above heterogeneous anisotropic karst fissure aquifer can be described by the groundwater flow continuity equation and its definite solution conditions [13]. The Formulas (1)–(4) are as follows:

$$\frac{\partial}{\partial x}\left(K_{xx}\frac{\partial H}{\partial x}\right)+\frac{\partial}{\partial y}\left(K_{yy}\frac{\partial H}{\partial y}\right)+\frac{\partial}{\partial z}\left(K_{zz}\frac{\partial H}{\partial z}\right)+W=\mu_s\frac{\partial H}{\partial t}(x,y,z)\in\Omega,t\geq 0 \quad (1)$$

$$\text{Initial heads}: H(x,y,z,t)|_{t=0}=H_0(x,y,z) \quad (x,y,z)\in\Omega \quad (2)$$

$$\text{Constant head boundary}: H(x,y,z,t)|_{\Gamma_1}=H_1 \quad (x,y,z)\in\Gamma_1,\ t\geq 0 \quad (3)$$

$$\text{Zero flow boundary}: K_n\frac{\partial H}{\partial n}\Big|_{\Gamma 2}=0 \quad (x,y,z)\in\Gamma_2,t\geq 0 \quad . \quad (4)$$

where: H is the aquifer head (m); K_{xx}, K_{yy} and K_{zz} are the permeability coefficient in the x, y and z directions ($m\cdot d^{-1}$); H_0 is the initial water level of the flow field (m); H_1 is the river water level (m); K_n is the permeability coefficient in the normal direction of the Γ_2 boundary ($m\cdot d^{-1}$); μs is the unit storage coefficient (L^{-1}); W is the source and sink (d^{-1}); Ω is the analog range; Γ_1 is the constant head boundary (Dirichlet boundary); Γ_2 is the Zero flow boundary or hydraulic barrier boundary (Neumann boundary); n is the outer normal direction of the Γ_2 boundary.

The pollutant transport was simulated with the MT3DMS module in GMS. According to the characteristics of the groundwater system, the groundwater solute transport in the study area was analyzed, and the corresponding groundwater solute transport Equation (5) [14] and the initial concentration condition (6) was established as follows:

$$\frac{\partial c}{\partial t}=\frac{\partial}{\partial x}\left(D_{xx}\frac{\partial c}{\partial X}\right)+\frac{\partial}{\partial y}\left(D_{yy}\frac{\partial c}{\partial y}\right)+\frac{\partial}{\partial z}\left(D_{zz}\frac{\partial c}{\partial z}\right)-\frac{\partial(u_x c)}{\partial x}-\frac{\partial(u_y c)}{\partial y}-\frac{\partial(u_z c)}{\partial z}(x,y,z\epsilon\Omega,t\geq 0) \quad (5)$$

$$c(x,y,z,t)|_{t=0}=c_0(c,y,z,t)\ (x,y,z\epsilon\Omega,t=0) \quad (6)$$

where: D_{xx}, D_{yy}, and D_{zz} are the dispersion coefficient in x, y and Z directions (m^2/d); u_x, u_y, and u_z are the actual flow velocity in three directions (m/d); c is the solute concentration (mg/L); c_0 is the initial concentration (mg/L).

4.3. Boundary Conditions

According to the measured flow field and regional hydrogeological conditions, the eastern, southern and northern boundaries of the study area are generalized as constant flow boundaries, including some zero flow separation boundaries. The eastern boundary is the groundwater ridge, which is set as the zero flow boundary; the western boundary is the Mansi River, which is set as the constant head boundary. According to the measured results of the river, it is set as 80 m from south to north; the southern and northern boundaries are set as constant flow boundaries, which are determined to be 2.5×10^4 m^3/d according to the hydrogeological survey results. The upper boundary selects atmospheric rainfall as the infiltration boundary and rainfall as the main recharge source. According to the monthly rainfall data provided by the local meteorological bureau and the multi-year average rainfall infiltration coefficient of 0.326, the daily rainfall infiltration is calculated as the surface recharge for the assignment. The Ordovician limestone with a buried depth of 500 m is selected as the bottom boundary of the model.

Generally, the pH value has an important effect on the leaching of fluorine [15]. By collecting fresh red mud samples and soaking them for 30 days, the maximum leaching concentration of F^- was detected to be 60.0 mg/L after 30 days of leaching at a pH of 10.0 or so. This value is used as the initial maximum concentration of the pollution source in the simulation. Considering that the red mud tailing pit covers 25,342 m^2, it is regarded as a surface pollution source in the model simulation. The red mud tailing pit is set as the specified concentration recharge area, and the pollutant concentration in other areas including the boundary is set to zero.

4.4. Parameter Partition and Value

Hydro-geological parameters are very important in groundwater numerical simulation. Their rationality and correctness will directly affect the accuracy and reliability of the groundwater model. Hydro-geological parameters are determined and assigned according to regional pumping test and tracer experiments (Figure 5). The tracer experiment used ammonium molybdate as the tracer and the JP-2 polarograph combined with the standard curve comparison method to determine the concentration of Mo^{6+}. Draw the time and concentration curve according to the detection results of each receiving point, and calculate the permeability coefficient (K) and Longitudinal dispersivity (α) of groundwater aquifer [16]. The calculation formula is as follows:

$$\alpha = \frac{(t_1 - t_2)\left(x^2 - U^2 t_1 t_2\right)}{4Ut_1 t_2 \text{In}\left(\frac{C_1 t_1}{C_2 t_2}\right)} \tag{7}$$

where: C_1, C_2 are the concentrations of tracers detected at t_1 and t_2, respectively (mg/L); x is the distance from the detection hole to the source hole (m); U is the groundwater velocity (m/d).

$$K = \frac{U}{J} \tag{8}$$

where: J is the hydraulic gradient (m/m).

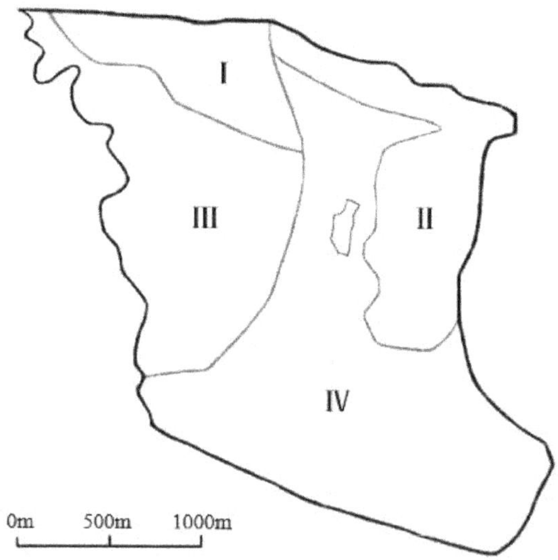

Figure 5. Parameter partition diagram of aquifer permeability coefficient.

The initial values of other parameters are the empirical values of specific yield, storage rate, effective porosity, river infiltration coefficient and other parameters of corresponding lithology [17], and the vertical permeability coefficient was set as 1/10 of the horizontal permeability coefficient. Table 2 were final parameter values.

Table 2. Initial and calibrated values of hydrogeological parameters in different zones.

Parameter	Zone	Initial Value	Calibrated
Horizontal permeability coefficient(m/d)	I	15.0	17.0
	II	10.0	20.0
	III	0.8	1.0
	IV	5.0	8.0
Porosity (%)	I	0.1	0.15
	II	0.1	0.25
	III	0.1	0.05
	IV	0.1	0.10
Specific yield (%)	-	0.01	0.02
Specific storage(1/m)		1×10^{-4}	9.6×10^{-4}
River conductance (m/d)	-	3.56	3.56
Longitudinal dispersivity (m)	-	50	66.7

5. Model Calibration

In this unsteady flow model, the simulated flow field calculated by the MODFLOW module was used as the initial flow field, and the parameter estimation module in GMS software was used to automatically identify and correct the three-dimensional groundwater numerical model to obtain the hydrogeological parameters of the flow model. The root mean square (RMSE) and correlation coefficient (R) were selected as the identification conditions. When the RMSE of the calculated value and the simulated value was the smallest and R approached 1, the most accurate model parameters were obtained.

$$\mathrm{RMSE} = \sqrt{\frac{1}{N}\sum_{i=1}^{N}\left(h_i - h_i^{obs}\right)^2} \tag{9}$$

$$R = \frac{\Sigma\left[\left(h_i - \bar{h}_i\right)\left(h_i^{obs} - \bar{h}_i^{obs}\right)\right]}{\sqrt{\Sigma\left[\left(h_i - \bar{h}_i\right)^2\left(h_i^{obs} - \bar{h}_i^{obs}\right)^2\right]}} \tag{10}$$

where: N is the number of groundwater head data (N = 82); h_i is the simulated value of the groundwater head; $h_i{}^{obs}$ is the measured value of groundwater head; $\bar{h}_i$ and $\bar{h}_i^{obs}$ are the average of the simulated value of groundwater head and the measured value of the groundwater head.

5.1. Groundwater Flow Field Simulation

The observed groundwater table in the study region was observed within 70~80 m, which well corroborate with the Zhu X(2000) results for the same study area. The MODFLOW groundwater simulation results are shown in Figure 6. The color error bars were used to show a MODFLOW calibration, and the center of the target bar signifies the observed head value whereas, the top of the error bar corresponds to the observed value plus the confidence interval and bottom of the error bar represents the observed value minus the confidence interval. If the simulated values lie entirely within the target, the color bar is drawn in green. If the simulated value is outside the target but the error is less than 200%, the bar is drawn in yellow. Whereas if the error greater than 200%, the bar is drawn in red.

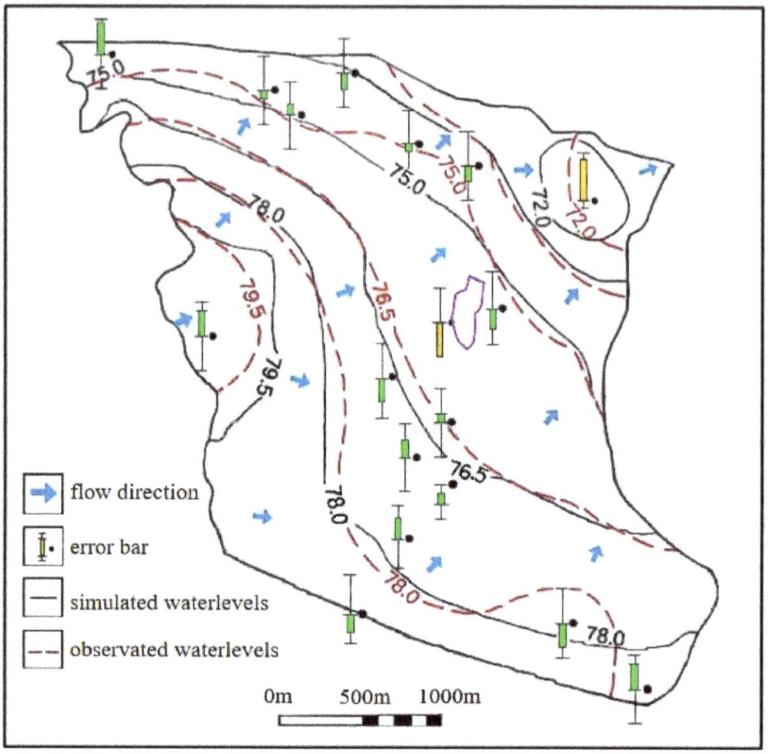

Figure 6. Fitting result of simulated and actual groundwater flow field.

The confidence interval of the instantaneous head of observation well was set as 1 m, and the confidence of each observation well was 95%. The groundwater heads of 18 observation wells were recorded in the study area. In Figure 6, the RMSE between the simulated and measured values of all observation wells is 0.95 m; the overall correlation coefficient reaches 0.970. In this study area, 89% of the error bars are green, and only the water level error bars at the discharge outlet in the northeast corner and the red mud tailing pit are yellow. On the whole, it can be considered that the simulation results of the groundwater flow field are consistent with the current groundwater flow field in the study area.

5.2. Multi-Year Water Level Calibration

From the multi-year water level data of the groundwater observation well, the period from January 2003 to December 2014 (144 months) was selected as the simulated stress period, and the groundwater flow model was run to obtain the observation groundwater level fitting results in different periods (Figure 7). It has been noticed between a plot of observation heads vs. computed groundwater heads that at the middle of each stress period (i.e., July–August) the limestone fissure aquifer were replenished by rainfall and surface water bodies. Figure 7 shows that 92% of the water level fall within the confidence interval of the simulated value. According to the Groundwater Resources Management Model Work Requirements (China, GB/T14497-93), nodes with a relative error of less than 10% must account for more than 70% of the water level node; the accuracy of this calculation can meet the requirements, indicating that the model can accurately reflect the multi-year variation characteristics of the groundwater level in the study area.

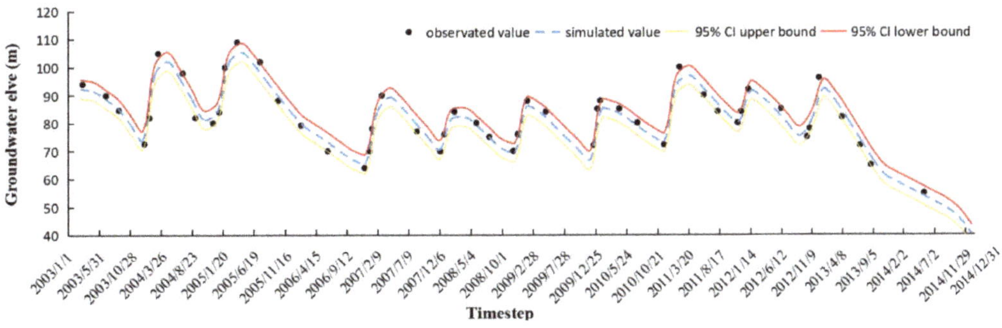

Figure 7. Prediction groundwater level at observation well under 95% confidence interval.

5.3. Observation Water Level Calibration

It can be observed from the simulation results that during the simulation period, the groundwater level demonstrated a downward trend due to the dryness of the local climate and the impact of man-made mining. With the increase in the scope of human activities, the forest coverage area decreases year by year, the soil and water conservation capacity is degraded, and the surface water and the moisture in the unsaturated zone is lost, reducing the recharge of the groundwater.

After running the model, Figure 8 shows the correlation diagram between the simulated value and the measured value. A total of 82 measured values were monitored in all observation wells. In addition, the root mean square error (RMSE) and correlation coefficient were calculated to be 1.30 m and 0.991, respectively, during the simulation time. The fitting result indicated that the simulation model was correct and the parameters were selected reasonably.

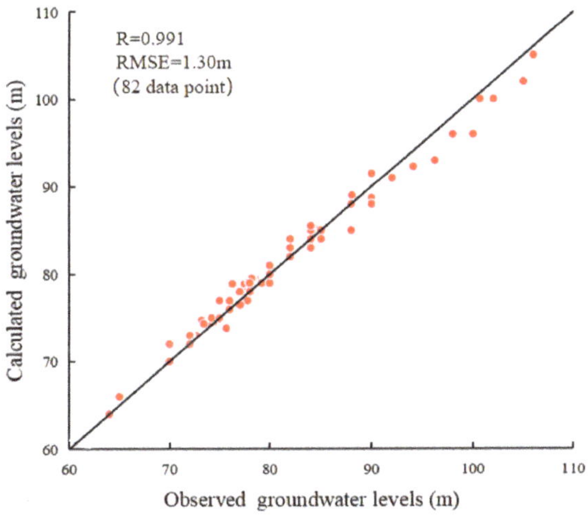

Figure 8. Scatter plot for observed groundwater level vs. simulated groundwater level.

5.4. Water Balance Analysis

Further analysis of water balance in the study area: the results (Table 3) show that 65.9% of the groundwater was recharged by the constant flow boundary on the south side, 30.9% was recharged from atmospheric rainfall and 3.1% of groundwater was recharged by rivers. The recharge of the constant head inflow in this area accounts for the least amount of total recharge, indicating that the river's recharge of groundwater in the region is weak. It is due to the natural surface flow of Mansi River, which has basically been lost because of the pumping project in the study area since the 1950s, with surface flow occurring only during the wet season. A total of 64.2% of the groundwater is discharged from the flow boundary on the north side, 20.7% is discharged from the pumping well, 12.9% of the groundwater is discharged by evaporation and 2.0% of the groundwater is discharged from the river. Combining the recharge volume on the north side and the discharge volume on the south side, it can be found that the constant flow boundary is the main source of groundwater recharge and the main hydrodynamic condition for the formation of the flow field.

Table 3. Summary of total MODFLOW simulation groundwater flow budget.

Sources/Sinks	Flow In ($\times 10^4$ m^3/d)	Flow Out ($\times 10^4$ m^3/d)	Annual Flow In ($\times 10^4$ m^3/a)	Annual Flow Out ($\times 10^4$ m^3/a)
Constant flow boundary	2.5432	−2.4751	964.70	866.90
Wells	0.0000	−0.8000	0.00	292.00
River leakage	0.1213	−0.0815	44.27	29.75
Recharge	1.1925	0.0000	398.76	0.00
Evapotranspiration	0.0000	−0.5045	0.00	220.64
Total source/sink	3.8570	−3.8575	1407.80	1408.00
Summary	In-Out	% difference	In-Out	% difference
Sources/sinks	−0.0005	0.0000	−0.20	0.00
Cell-to-cell	0.0000	0.0000	0.00	0.00
Total	−0.0005	0.0000	−0.20	0.00

6. Simulation Result

This numerical simulation takes into account the maximum initial contaminant concentration and convective dispersion effect of groundwater on pollutants. The research set contaminant concentration observation point at the place was 200 m, 500 m and 1 000 m away from the pollution source along the diffusion direction of the pollutants, respectively (Figure 9), and the concentration duration curve was calculated (Figure 10). The standard value of F^- is 1 mg/L in accordance with the groundwater quality standard for class III (China, GB14848-2017). When the concentration of F^- exceeds this value, that means it has exceeded the Chinese national standard.

6.1. Distribution of Pollution Plumes

The period from December 2004 to December 2019 (180 months) was selected as the simulated stress period. The simulation results of MT3DMS demonstrate that the pollutant migration was mainly controlled by atmospheric precipitation replenishment, constant flow boundary conditions and leakage of rivers and drainage ditches, which are the main catchment parameters in the study area model at the time of these three factors. The corresponding distribution of pollution plumes at 1, 5, 10 and 15 years after the occurrence of a pollutant spill is shown in Figure 9. As shown in Figure 9d, the results of the transport modeling are successfully validated by F^- concentration observation data. It can be observed that when the pollution duration was 1, 5, 10 and 15 years, the maximum horizontal migration distances were 473 m, 1160 m, 1595 m and 1750 m, respectively. The pollution center concentration was 60 mg/L, 53.2 mg/L, 45.2 mg/L and 42.3 mg/L, the area of F^- pollution plumes was 0.37 km^2, 1.15 km^2, 1.95 km^2 and 2.14 km^2, respectively. It shows that the vertical pollution of F^- migration was affected by the hydraulic connection of the aquifer. The maximum vertical influence distance was 492 m, and the pollution of deep aquifers is seriously affected.

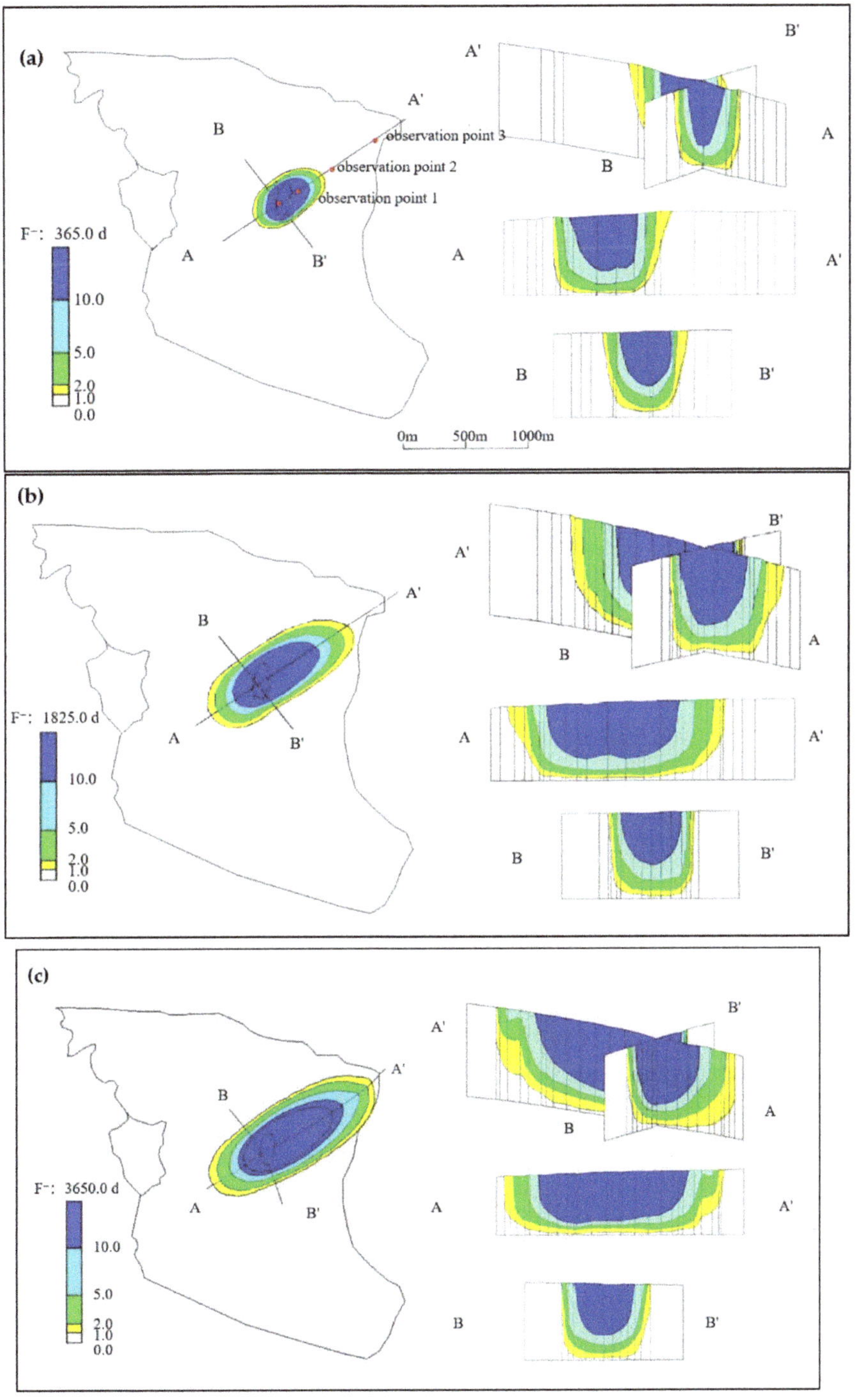

Figure 9. *Cont.*

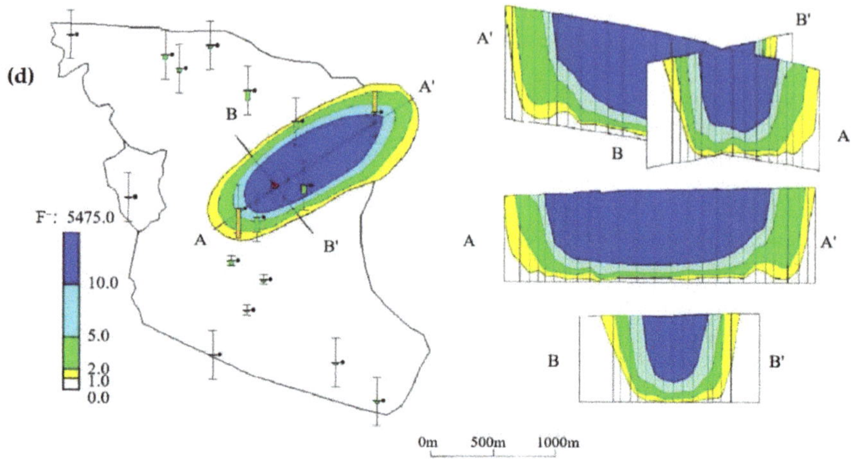

Figure 9. Distribution of pollution plumes in different pollution situations: (**a**) one year later, (**b**) five years later, (**c**) ten years later and (**d**) fifteen years later.

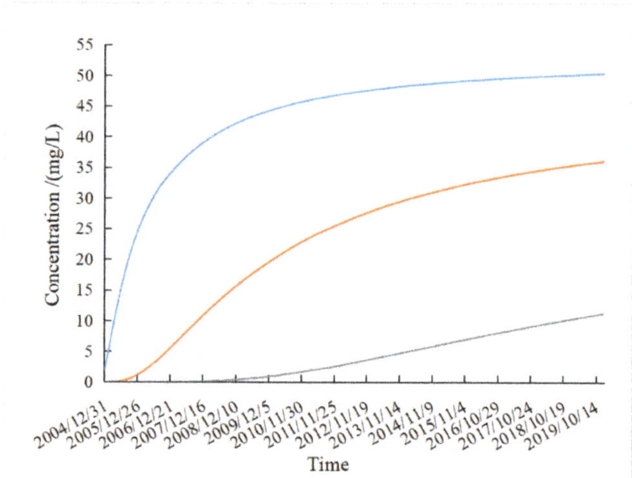

Figure 10. Variation of pollutant concentration in three observation wells.

6.2. The Concentration of Pollutants in Different Locations

According to the geological conditions of the red mud pit, after the F^- pollutant in the red mud pit is leached out from the red mud, it seeps down from the bottom of the pit until it flows to the Ordovician limestone aquifer with a high permeability coefficient.

Figure 10 showed the changes of pollutant concentrations in each observation hole from 31 December 2004 to 31 December 2019. It can be observed from Figure 10 that due to the close distance and continuous leakage at the No. 1 hole, the F^- concentration increased rapidly and then remained at a high level; no obvious downward trend was observed during the simulation period. In the evaluated period, the maximum concentration of the No. 1 observation well is 50.1 mg/L, No. 2 observation well is 36.2 mg/L, and No. 3 point is 11.4 mg/L, which reflects that the farther away from the pollution source, the lower the pollutant concentration and the concentration gradient, and the smaller the rule.

7. Discussion

7.1. Fitting Error and Parameter Error

Figure 6 shows that the simulated water level of the observation well at the red mud tailing pit is lower than the measured value, which may be due to the complex geological conditions here, where red mud is stacked on the Ordovician limestone aquifer, and the permeability coefficient of the red mud is usually less than 4.5×10^{-6} m/d [18]. At the red mud tailing pit, the rainwater is difficult to infiltrate, resulting in high simulation value in the pit. The simulated water level of the observation well at the discharge of the groundwater flow field is higher than the measured value, while the Ordovician limestone fissure aquifer in the northeast corner is connected with Mansi fault and directly exposed to the surface to receive atmospheric rainfall recharge, resulting in the simulated value being larger than the actual value.

In Figure 7, there are four observation well data that do not fall within the confidence interval, which are 10 March 2004, 2 April 2005, 1 February 2011 and 15 February 2012, respectively. These four time points are the peak in the annual variation of groundwater level and the four years are all wet years. Through data investigation and analysis, this may be due to the rapid rise of groundwater level caused by heavy rainfall in the wet season of wet years whereas the model rainfall data adopts the average of monthly rainfall data, as the daily rainfall due to the lack of daily rainfall data; thus, an averaging effect occurs, resulting in the low peak value of the simulation value [19].

After model validation and calibrated parameter, the results are shown in Table 2. Table 2 shows that the values of permeability coefficient in different areas of Ordovician limestone aquifer vary greatly. The permeability coefficient of Ordovician limestone aquifer obtained by model inversion is generally large, with a variation range of 1 ~ 20 m/d and a size difference of 20 times, indicating that the karst fissure aquifer has strong heterogeneity.

Generally, the karst is not developed below the depth of 100~150 m of the karst fissure aquifer, and storage coefficient is usually less than 10^{-4} m^{-1}. However, the storage coefficient of Ordovician limestone aquifer is as high as 9.6×10^{-4} m^{-1}. According to the rock core and the borehole records, the development depth of karst fissure and dissolved pores in the study area can reach 500~700 m. The total storage capacity of karst fissure aquifer is large and has strong regulation and storage capacity; thus, the water storage coefficient is relatively large [20].

7.2. Precipitation of Fluorine

Water quality monitoring results demonstrated that the maximum migration distance of F^- was only 1750 m, and the migration distance of pollutants usually reached tens or even tens of kilometers under the long-term convection and diffusion of groundwater, indicating that precipitation occurred in the migration process, and the following reactions could also be existed in the analysis:

$$Ca^{2+} + 2F^- \Leftrightarrow CaF_2 \downarrow \tag{11}$$

$$Ca^{2+} + 2OH^- \Leftrightarrow Ca(OH)_2 \downarrow \tag{12}$$

$$Mg^{2+} + 2F^- \Leftrightarrow MgF_2 \downarrow \tag{13}$$

$$Mg^{2+} + 2OH^- \Leftrightarrow Mg(OH)_2 \downarrow \tag{14}$$

The relative abundance of main cations in the groundwater of Ordovician limestone is $Ca^{2+} > Mg^{2+} > Na^+ > K^+$, because there are a lot of limestone ($CaCO_3$), dolomite (CaMg $(CO_3)_2$) and fluorite (CaF_2) in Ordovician limestone. The high level of Ca^{2+} and Mg^{2+} in groundwater is caused by hydrolysis, and the reactions (1) and (3) mainly occur in the initial stage of F^- entering the groundwater. According to chemical Equation (1) and Equation (3), it can be introduced that at higher initial F^- concentrations, the increase of Ca^{2+} and Mg^{2+} concentration promotes Equations (1) and (3) to move to the right, generating more CaF_2 and MgF_2 precipitation, and thus reducing the F^- concentration in the groundwater. Therefore, the concentration of F^- is negatively correlated with the concentration of groundwater Ca^{2+}+Mg^{2+} (Figure 11a). The leachate of red mud contains a large amount of Al^{3+}, which is prone to hydrolysis reaction and generates positively charged hydrolysate $Al_2 (OH)_2{}^{4+}$, which generates electrostatic adsorption with negatively charged F^-. Under the combined action of both, the concentration of F^- in water decreases rapidly. After that, with the migration of pollutants in the groundwater, the content of OH^- decreased gradually. Under the action of electrostatic adsorption, the F^- was accelerated to be closer to the adsorption site, resulting in an increase in the adsorption amount of SOIL colloid for F^-, which resulted in a shorter migration distance of F^-.

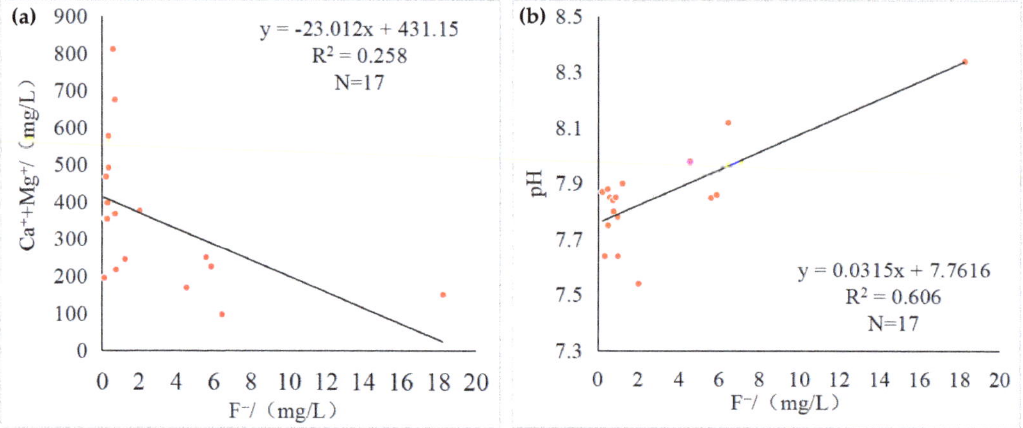

Figure 11. (**a**) Ca^{2+}+Mg^{2+} and F^- correlation coefficient diagram; (**b**) pH and F^- correlation coefficient diagram.

At the same time, previous studies have demonstrated that the acid-base environment of groundwater has a greater impact on F^- concentration [21], and F^- concentration is positively correlated with groundwater pH (Figure 11b). The higher the pH value, the excess OH^- in the solution will combine with the preferentially dissociated Ca^{2+} to form $Ca(OH)_2$ precipitation, which reduces the concentration of Ca^{2+}, thereby inhibiting the formation of CaF_2 in the water, which thus facilitates the enrichment of F^- in the groundwater. With the flow of groundwater in the aquifer, the pH value of the water gradually decreases, and the abundant Ca^{2+} in the groundwater will preferentially combine with F^- to form CaF_2 precipitation, resulting in the rapid reduction of F^- in the groundwater.

8. Conclusions

In this study, the method of combining pumping experiments, tracing experiments and numerical simulations were used to study the fluorine pollution process and diffusion law of red mud pit-polluted Ordovician limestone aquifer groundwater, and the following conclusions were obtained:

(1) A three-dimensional groundwater mathematical model was established for the special red mud pit storage conditions and complex boundary changes in the study area, and the GMS software was used to simulate the migration law of fluorine pollutants in the aquifer. The anti-seepage layer was damaged, and the groundwater pollution plume caused by the red mud yard moved to the northeast as a whole. After 1 year of red mud leachate leakage, the maximum migration distance of pollutant F^- in groundwater was about 473 m, and the pollution plume range was 0.37 km^2. When the leakage occurred for 5 years, the maximum migration distance of F^- was about 1160 m, and the pollution plume range was 1.15 km^2 after 10 years; the maximum migration distance of pollutants in groundwater F^- was about 1595 m, and the pollution plume range was 1.95 km^2. When

the seepage occurred 15 years ago, the maximum migration distance of F^- was about 1750 m, and the pollution plume range was 2.14 km^2. The maximum vertical influence distance is 498 m.

(2) The migration of F^- pollutants in red mud into groundwater was mainly controlled by atmospheric precipitation recharge, flow boundary conditions and seepage in rivers and drains. The simulation results of MT3DMS demonstrated that after 15 years of simulation, the diffusion area of the pollution plume reached 2.14 km^2 and then gradually stabilized, indicating that red mud F^- pollutants have less and less impact on groundwater farther than 1750 m away.

There were some errors in the process of the model simulation, which may be mainly caused by complex hydrogeological conditions and hydrogeochemical reaction factors. The former was mainly due to the influence of structure; the karst aquifer has strong heterogeneity, and there may be strong runoff zones, resulting in high local permeability coefficient and water storage coefficient. Meanwhile the latter, during the migration process of F^- pollutants, may be associated with complex hydrolysis and precipitation, which leads to some errors in the simulation results.

Author Contributions: Data curation, J.W. (Junping Wang); conceptualization, P.Z. and J.W. (Jiaxin Wu); formal analysis, P.Z.; funding acquisition, Y.Q.; investigation, Y.Q.; methodology, Y.Q.; resources, C.S.; software, J.W. (Junping Wang); writing—original draft, Y.Q.; writing—review and editing, P.Z. and Y.M. All authors have read and agreed to the published version of the manuscript.

Funding: The research was financially supported by the National Natural Science Foundation of China (Nos. 41741020 and 41572218), a project funded by the Priority Academic Program Development of Jiangsu Higher Education Institutions, and the Fundamental Research Funds for the Central Universities (No. 2014QNB51).

Institutional Review Board Statement: Not applicable.

Informed Consent Statement: Not applicable.

Data Availability Statement: Not applicable.

Acknowledgments: This work was financially supported by the National Natural Science Foundation of China (Nos. 41741020 and 41572218), a project funded by the Priority Academic Program Development of Jiangsu Higher Education Institutions, and the Fundamental Research Funds for the Central Universities (No. 2014QNB51). The authors also thank the anonymous reviewers and editors for their constructive comments and suggestions on improving this manuscript.

Conflicts of Interest: The authors declare no conflict of interest.

References

1. Zhu, Q.S.; Xu, G.Q. The Current Situation and Research Progress of GroundWater Fluorine Pollution in China. *Environ. Sci. Manag.* **2009**, *34*, 34–42. [CrossRef]
2. Yi, C.Y.; Wang, B.G.; Jin, M.G. Research Progress of Migration and Transformation Laws of Fluoride in Groundwater-soil-plant System. *Saf. Environ. Eng.* **2013**, *20*, 59–64. [CrossRef]
3. Aschonitis, V.G.; Mastrocicco, M.; Colombani, N.; Salemi, E.; Kazakis, N.; Voudouris, K.; Castaldelli, G. Assessment of the Intrinsic Vulnerability of Agricultural Land to Water and Nitrogen Losses via Deterministic Approach and Regression Analysis. *Water Air Soil Pollut.* **2012**, *223*, 1605–1614. [CrossRef]
4. He, L.; Tu, C.; He, S.; Long, J.; Sun, Y.; Lin, C. Fluorine enrichment of vegetables and soil around an abandoned aluminium plant and its risk to human health. *Environ. Geochem. Health* **2021**, *43*, 1137–1154. [CrossRef] [PubMed]
5. Wang, J.; Zheng, N.; Liu, H.; Cao, X.; Teng, Y.; Zhai, Y. Distribution, Formation and Human Health Risk of Fluorine in Groundwater in Songnen Plain, NE China. *Water* **2021**, *13*, 3236. [CrossRef]
6. Yang, J.Y.; Gou, M. The Research status of fluorine cintamination in soils of China. *Ecol. Environ. Sci.* **2017**, *26*, 506–513. [CrossRef]
7. Yang, Y.; Li, J.; Li, M.X.; Li, X.; Bai, S.; Xi, B.; Lyn, N.; Yang, Y. Application of HYDRUS-1D model in quantitative assessment of groundwater pollution resource intensity. *Chin. J. Environ. Eng.* **2014**, *8*, 5293–5298. [CrossRef]
8. Zhu, X.Y.; Liu, J.L.; Zhu, J.J.; Chen, Y.D. Numerical Study of contaminants transport in fracture-karst water in dawu well field. *Chin. Sci. (D)* **2000**, *30*, 479–485. [CrossRef]
9. Sathe, S.S.; Mahanta, C. Groundwater flow and arsenic contamination transport modeling for a multi aquifer terrain: Assessment and mitigation strategies. *J. Environ. Manag.* **2019**, *231*, 166–181. [CrossRef]

10. Liu, J.L.; Zhu, X.Y.; Qian, X.X. Study of Some problems on the development and protection of fracture-karst water resources in North China. *Acta Geol. Sin.* **2000**, *74*, 344–352. [CrossRef]
11. Hallett, B.M.; Dharmagunawardhane, H.A.; Atal, S.; Valsami-Jones, E.; Ahmed, S.; Burgess, W. Mineralogical sources of groundwater fluoride in Archaen bedrock/regolith aquifers: Mass balances from southern India and north-central Sri Lanka. *J. Hydrol. Reg. Stud.* **2015**, *4*, 111–130. [CrossRef]
12. Guo, S.H.; Gao, P.; Wu, B.; Zhang, L.Y. Fluorine emission list of China's key industries and soil fluorine concentration estimation. *Chin. J. Appl. Ecol.* **2019**, *30*, 1–9. [CrossRef]
13. Tong, X.X.; Ning, L.B.; Dong, S.G. GMS Model for Assessment and Prediction of Groundwater Pollution of a Garbage Dumpling Site in Luoyang. *Environ. Sci. Technol.* **2012**, *35*, 197–201. [CrossRef]
14. Zhang, Y. Numerical Simulation and Prediction of Contaminated Groundwater in an Animal Protein Production Site by GMS. *Geotech. Eng. Tech.* **2017**, *31*, 258–262. [CrossRef]
15. Rong, S.R.; Peng, D.P.; Chen, J.N. Effect of PH Value on Leaching if Rare Mentals from Red Mud in Simulated Natural Storage Conditions. *Env. Eng.* **2020**, *38*, 155–159. [CrossRef]
16. Nielsen, D.R.; Van Genuchten, M.T.; Biggar, J.W. Water flow and solute transport processes in the unsaturated zone. *Water Resour. Res.* **1986**, *22*, 89S–108S. [CrossRef]
17. Wang, Q.K.; Liang, L.; Chen, C.; Yang, J.Y.; Li, W.; Gong, H.Z.; Zhao, J. Study on Simulations and Control Measures of Leaking Oil Pollution Based on GMS. *Environ. Prot. Oil Gas Fields* **2018**, *28*, 12–16. [CrossRef]
18. Wang, G.H.; Zhang, B.J.; Chen, J.H.; Liu, S.C.; Wang, Q.J.; Li, Y.D. Permeation and Migration of Red Mud in Porous Media. *J. Water Resour. Archit. Eng.* **2020**, *18*, 1–5. [CrossRef]
19. Yang, Y.; Yin, G.X.; Zhu, L.X. Fluorine Pollution and its Formation Analysis of the Shallow Groundwater in Jiaozuo City. *Environ. Sci. Manag.* **2009**, *34*, 68–72. [CrossRef]
20. Li, Q.; Liao, C.N.; Liao, M.X.; Peng, D.P.; Huang, T. Analysis of infiltration mechanism of red mud leachate into sodium-bentonite clay pad. *Env. Eng.* **2021**, *39*, 148–153. [CrossRef]
21. Chen, W.; Hao, C.M.; Ma, Z.Y.; Wang, Y.; Zhang, L.; Xu, R.; Chen, P. Geochemical behavior of fluoride in residual coal in groundwater reservoirs. *J. North China Inst. Sci. Technol.* **2021**, *18*, 67–73. [CrossRef]

water

MDPI

Review

Geo-Environmental Models of In-Situ Leaching Sandstone-Type Uranium Deposits in North China: A Review and Perspective

Fuxin Zheng, Yanguo Teng, Yuanzheng Zhai *, Jingdan Hu, Junfeng Dou and Rui Zuo

Engineering Research Center for Groundwater Pollution Control and Remediation of Ministry of Education of China, College of Water Sciences, Beijing Normal University, Beijing 100875, China
* Correspondence: diszyz@vip.163.com

Abstract: Since the 1990s, sandstone-type uranium in the northern basin of China has become the main target for mining. Uranium mining can cause a series of impacts on the environment. A conceptual model of the geo-environment for sandstone-type uranium in northern China was described, which covers the changes in the geo-environmental characteristics in the natural state, in the mining process, during decommissioning and after treatment. Sandstone-type uranium is mainly distributed in the Songliao, Erlian, Ordos, Turpan–Hami and Ili Basins, which have arid climates and poor stratum permeability. Pitchblende is the main uranium-bearing mineral and is associated with iron, copper, coal, organic matter and other minerals. The mineral often has a low ore grade (0.01–1.0%) and high carbonate content (2–25%). Uranyl carbonate accounts for more than 90% of the total uranium in groundwater. The uranyl content is closely related to the TDS. The TDS of groundwater in the eastern and central ore belts is usually lower than 2 g/L, while in the western region, such as Xinjiang, it can exceed 10 g/L. In situ leaching (ISL) is the main mining method that results in groundwater pollution. Acid leaching leads to a pH decrease (<3), and heavy metals represented by U and Fe exceed the background values by hundreds of times, resulting in groundwater pollution. CO_2 leaching is more environmentally friendly, and the excess ions are usually Ca^{2+}, Mg^{2+}, NO_3^- and HCO_3^-. Soil chemical anomalies originate mostly from wind erosion and precipitation leaching of decommissioned tailings. Uranium pollution is mainly concentrated within 20 cm of the surface, and the exceedance generally varies from two to 40 times. During ISL, a series of environmental measures will be taken to prevent pollution from being exposed to the surface. After treatment, the decommissioned uranium mines will likely have no impact on the surrounding environment. In the future, the protection of groundwater should be strengthened during production, and remediation methods based on electrokinetic, microbial and permeable reactive barrier (PRB) technology should be further researched.

Keywords: sandstone-type uranium; geo-environmental model; environmental properties; geochemical baseline; environmental impact; northern China

check for updates

Citation: Zheng, F.; Teng, Y.; Zhai, Y.; Hu, J.; Dou, J.; Zuo, R. Geo-Environmental Models of In-Situ Leaching Sandstone-Type Uranium Deposits in North China: A Review and Perspective. *Water* **2023**, *15*, 1244. https://doi.org/10.3390/w15061244

Academic Editor: Domenico Cicchella

Received: 5 February 2023
Revised: 8 March 2023
Accepted: 20 March 2023
Published: 22 March 2023

1. Introduction

As the world's largest carbon-emitting economy, the Chinese government has pledged to achieve carbon neutrality by 2060. To achieve this goal, the number of nuclear power plants in the next 30 years will be increased five times on the current basis [1]. Uranium is the main raw material for nuclear power generation, and the development of nuclear power is bound to significantly increase the mining of uranium [2,3]. The largest reserve uranium deposits in China are sandstone-type deposits, which account for approximately 43% of the total [4]. Since the 1990s, with the introduction of the in situ leaching (ISL) mining method, sandstone-type uranium in northern China basins has gradually replaced southern granite-type and volcanic-type uranium and has become the main mining body [5,6]. Compared with open pit mines, the ISL of uranium is mainly conducted underground, which is relatively friendly to the environment, especially in the atmosphere and soil. However, ISL often pollutes groundwater due to the use of geological drilling and chemical solvents [7,8]. In addition, heap leaching of tailings poses potential environmental threats [9–11].

A geo-environmental model is a collection of multiple conceptual models that contain information on geology, hydrogeology, geochemistry, engineering, etc. [12]. In 1998, the U.S. The Department of the Interior and the Federal Geological Survey first systematically described the concepts on their web pages and used the host deposit types as the main index to classify environmental models [12,13]. The construction of geo-environment models can provide reference targets and ideas for mine environment management and remediation [12–14]. Uranium deposits usually show some special hydrogeological and geochemical characteristics [15–17], which usually change during and after mining [18,19]. How to repair the geo-environment according to these changes presents new challenges for researchers.

Based on the above, this article describes geo-environmental models of sandstone-type uranium in northern China, which covers the changes in the Earth's environment before, during and after mining, and discusses the prospects of geo-environmental protection and restoration methods in the future. Section 2 describes the distribution of sandstone-type uranium in northern China and summarizes the Earth's environmental characteristics and baselines of these deposits. Section 3 concentrates on the soil, rocks and groundwater chemical anomalies, as well as radiation and vegetation environment changes. The environmental measures during and after mining are addressed in Section 4. Finally, Section 5 provides suggestions for the geo-environmental protection and remediation of sandstone-type uranium mine development.

2. Environmental Properties and Geochemical Baselines

2.1. Environmental Properties

Most of the sandstone-type uranium deposits in northern China are located in arid or semiarid areas [20]. The main ore belts are the Songliao Basin, Erlian Basin, Ordos Basin, Turpan-Hami Basin and Ili Basin from east to west (Figure 1) [21–23]. The largest deposit is Nuheting, located in the Erlian Basin, with a reserve of more than 50,000 t. There are more than ten medium-scale deposits of more than 3000 t. Limited by technical conditions and economic factors, scaled mining is currently mainly concentrated in the Mengqiguer area of the Ili Basin and Qianjiadian area of the Songliao Basin. Uranium ore is usually low-consolidated sandstone composed of detritus of different sizes. The main components of the detritus are quartz and feldspar, and there are also some cemented substances (such as calcareous cement) [24–28]. The ore has a high carbonate content and is rich in organic matter and other reducing substances (such as organic carbon, H_2S and CH_4), which have good reducing abilities [29].

The stratigraphic chronology of uranium deposits varies by region (Figure 2a). Affected by interlayer oxidation, the distribution of ore belts can be divided into oxidation zones, transition zones and primary zones, among which the transition zone is the main metallogenic zone (Figure 2b) [30]. The oxidation zone is mainly associated with pyrite, limonite or siderite. Therefore, the ore is dark yellow or brown, and some ore may even contain "red layers," whose main component is iron oxide [31–33]. In the reduction zone and transition zone, ores are generally gray due to iron reduction (Figure 2) [4,34–36]. In the Ordos and Ili basins, uranium deposits are often associated with coal or copper mines. Paragenesis of uranium and coal is also one of the main resource development methods in these areas [37–40]. Other minerals, such as trace elements and rare earth elements, are usually characterized by abnormal abundance and have higher contents than other ores [41–43]. The main component of uranium minerals is pitchblende [27,28,44,45], while in the Turpan-Hami Basin, the U-bearing minerals are brannerite and coffinite [46]. Uranium minerals are usually found in the crevices of ore and cementitious material and are often associated with hematite and pyrite [45].

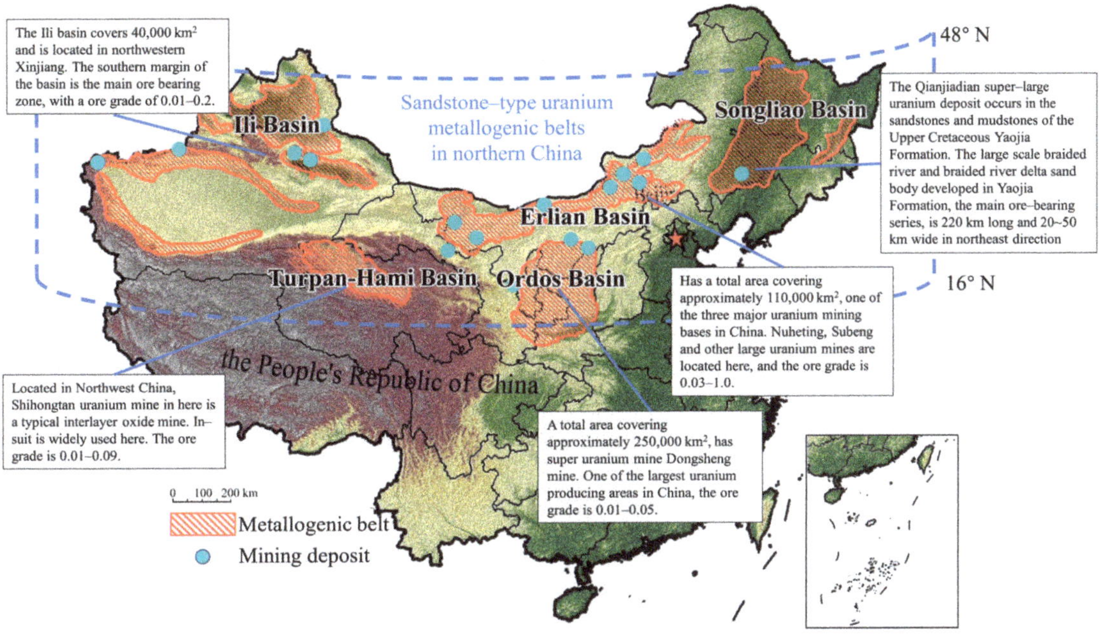

Figure 1. Distribution of sandstone-type uranium deposits in northern China.

Multilayer aquifer systems are universal in all these deposits [47–50]. For example, the Qianjiadian uranium deposit in the Songliao Plain has four aquifers with an average thickness of 55 m [48], and the Shihongtan deposit in the Turpan-Hami Basin has seven aquifers with an average thickness of approximately 23 m [49]. The aquifer is often composed of light gray or reddish gravelly sandstone mixed with argillaceous cementation. Dense and continuous groundwater barriers are usually composed of mudstone with a thickness of 1–6 m [47–50]. These aquifers determine the ore occurrence characteristics through the control of redox conditions and then affect the geo-environment of the deposits (Figure 3). Similar to the metallogenic differences between the eastern and western Shihongtan ore belts [51], the uranium occurrence differences in the northern and southern Ili Basin [50] are all related to the interlayer oxidation of groundwater.

Precipitation is the main source of groundwater [36,47]. In Northwest China (Ili and Turpan-Hami Basins), there is also Quaternary phreatic recharge from mountains in the basin margins [50,52]. Controlled by the tectonic system and geological conditions, the groundwater flows from the basin edge to the central depression. The hydraulic gradient and flow velocity vary from different basins. In open basins (Songliao, Erlian or Ordos Basin), the hydraulic gradient is generally less than 2‰, and the groundwater flow velocity is low. Similar to the Erlian Basin, the hydraulic gradient is 0.3–1.4‰, and the groundwater flow velocity is 0.7–2.9 m/a. In intermountain basins, the hydraulic gradient is generally 4–5‰, and the groundwater velocity can reach approximately 4 m/a [53]. This value will be higher in areas with large local topographic fluctuations, such as the No. 513 mine in Sudongbulake and the No. 510 mine in Honghaigou, where the hydraulic gradient reaches 7–9‰ and 23–35‰, respectively. The groundwater flow velocity exceeds 17 m/a [53]. Due to the arid climate, groundwater discharge is usually dominated by evaporation [47,54,55]. Except for the Shihongtan area in the Turpan-Hami Basin, the groundwater in most deposits is of good quality, with low salinity and moderate to weakly alkaline pH. Ca^{2+}, Na^{+}, HCO_3^{-}, SO_4^{2-} and Cl^{-} are the main ions [38,47,53,56]. These ions come from rock dissolution (such as gypsum and calcite) under water-rock interactions. In addition to conventional ions, groundwater also contains sulfides (such as H_2S, HS^{-},

S^{2-}, etc.), reducing gas (such as CH_4), Fe^{3+}/Fe^{2+}, and some trace elements [57–59]. Uranium in groundwater often exists in the form of uranyl. U(VI) is dominant in the oxidation environment of shallow aquifers, while U(IV) is dominant in deep reduction aquifers [47,52,60]. The formation of the Ili uranium deposit is due to uranium valence changes in different aquifers. Uranium in the shallow aquifer is continuously oxidized and enters groundwater in the form of uranyl [such as $UO_2(CO_3)_2{}^{2-}$ and $UO_2(CO_3)_3{}^{4-}$] (Table 1). The uranyl enters the deep aquifer with the recharge and flow of groundwater, where uranyl is restored to U (IV) precipitation and mineralization [50,55].

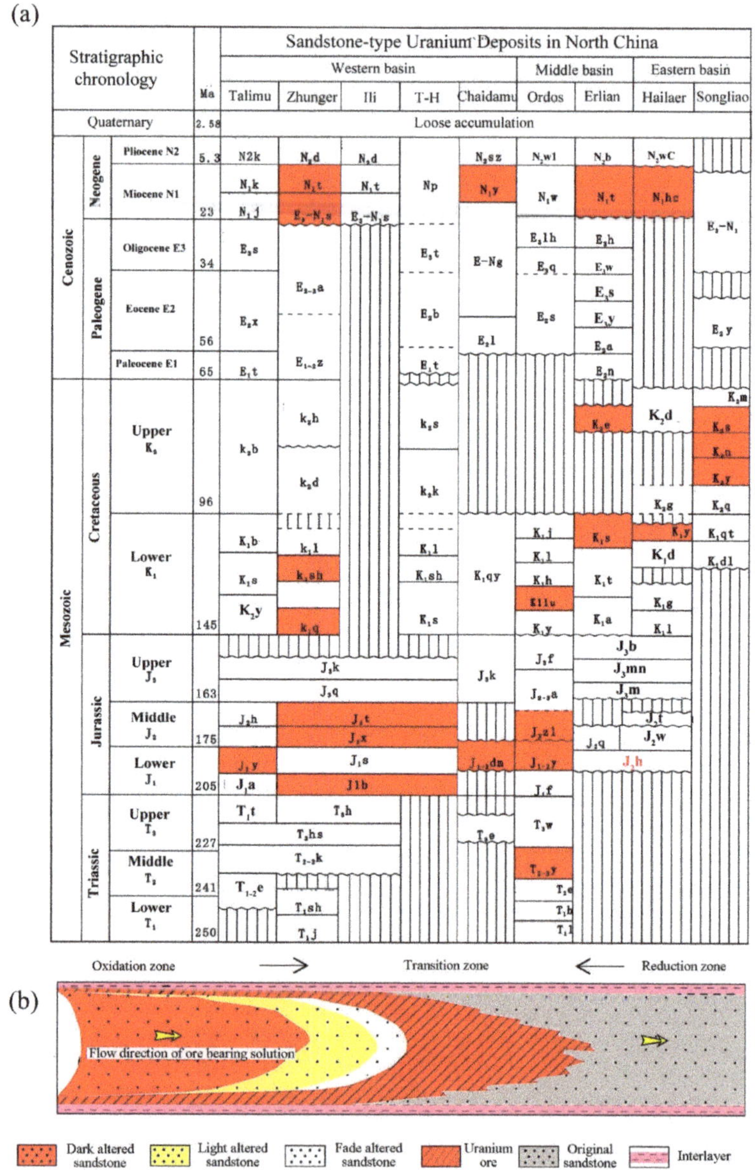

Figure 2. Stratigraphic chronology of stone-type uranium deposits in northern China (**a**) and a sketch of the oxidation-reduction zone (**b**) in the metallogenic belt. T-H means Turpan-Hami Basin.

Table 1. Hydrochemical baseline of the representative ore belt in each region.

Area	Location	Water Type	Hydrochemical Type	pH	TDS (mg/L)	Uranyl Types	Concentration of U (μg/L)	References
Songnen Basin	Qianjiadian	Confined groundwater of Yaojia Fm	HCO_3· Cl-Na, Cl·HCO_3-Na	7.15–7.79	3.64–4.12	$UO_2(CO_3)_3^{4-}$, $UO_2(CO_3)_2^{2-}$	130–312	[48,61]
Erlian Basin	Yihegaole	Bedrock fissure water	HCO_3·SO_4-Na·Mg, HCO_3·SO_4-Na·Ca	7.12–10.87	<2000	$UO_2(CO_3)_3^{4-}$, $UO_2(CO_3)_2^{2-}$	0.02–244	[62]
Ordos Basin	Northern basin	Surface water	/	9.31–9.32	412–3710	/	1.14–1.18	[47]
		Jurassic confined water	SO_4·Cl-Na	8.58–12.34	157–1060	$UO_2(OH)_3^{-}$, $UO_2(CO_3)_3^{4-}$, $UO_2(OH)_2$	0.44–1120	
Turpan-Hami Basin	Southern Shihongtan	Bedrock fissure water	Cl·SO_4-Na, Cl-Na, SO_4-Ca·Mg	8.02–8.25	8250–12,160	$UO_2(CO_3)_3^{4-}$, $UO_2(CO_3)_2^{2-}$	44.5–4237	[51,52]
Ili Basin	South edge of the basin	Jurassic interlayer porous confined water	HCO_3-Ca, HCO_3-Ca·Na, HCO_3·SO_4-Na·Ca, SO_4·HCO_3-Ca·Na, SO_4·Cl-Na·Ca	7.8–8.5	50–2790	$UO_2(CO_3)_3^{4-}$, $UO_2(CO_3)_2^{2-}$	10–1200	[50,55]

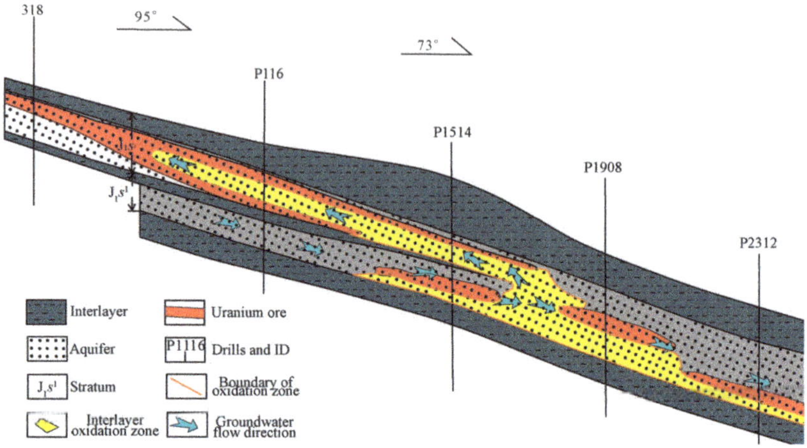

Figure 3. Aquifer and interlayer oxidation in the metallogenic belt.

2.2. Geochemical Baseline

The geochemical baselines in different regions are shown in Table 2. The proportion of detritus in the ore is generally more than 80%, and the detritus is composed of more than 60% quartz [26,63,64]. Approximately 90% of the ores have a carbonate content of 1–3% (carbonate-containing ores), and a few can reach more than 3% (carbonate ores) [65–68]. Some ore carbonate contents in the Ili Basin can reach up to 5–25% [64,69,70]. The uranium grade of the ore is generally between 0.01–0.05%. The Ili Basin has the highest average level in the phreatic oxidation zone of the southern basin margin (0.1–0.2%) and in the coal seam (88.7 μg/g) [27]. Approximately 70–90% of uranium minerals are UO_2. The pH of the metallogenic belt (transition zone) is nearly neutral to weakly alkaline, and the organic matter is generally 0.1–1.5% [65]. The Songliao Basin has the lowest organic matter content, with an average of only approximately 0.2%, while the Ordos Basin has the highest organic matter content, and some ores can reach 5–10% [28,68].

Table 2. Geochemical baselines of representative ore belts in each region.

Area	Location	Ore Types	Grade (%)	Lithic Fragments Ratio (%)	Carbonate Content (%)	Organic Matter Content (%)	Associated Minerals	Enriched Trace Elements	References
Songnen Basin	Qianjiadian	Coarse brownish red feldspathic sandstone, fine–coarse gray sandstone, red mudstone	0.01–0.03	90	2	Ave: 0.2 Max: 1.0	Pyrite, siderite	V, Cr, Co, Ni, Zn	[26,36,66,68]
Erlian Basin	Nuheting	Dark gray silty mudstones, siltstone	0.03–1.0	60–90	1.9	0.1–0.82	Pyrit	V, Mo, Cu, Pb, Zn, Se, Ni	[34,63]
Ordos Basin	Nalinggou	Red, green and gray mudstones, siltstones and sandstones	0.01–0.05	80–90	2.67	Ave: 0.5 Max: 5–10	Coal, pyrite, chalcopyrite	Rb, Ba, Th, K, Pb	[28,65,70]
Turpan-Hami Basin	Shihongtan	Gray white, light gray, light brown, yellow sandstone	0.01–0.09	20–50	2–3	0.1–1.5	Pyrite	V, Se, Mo, Re	[56,67,71]
Ili Basin	South edge of the basin	Gritstone, gravel sandstone, sandstone, siltstone, mudstone, and coalstone	0.01–0.2	46–79	5–25	0.1–2.8	Pyrite, coal	As, Ag, Sb, Se, Fe, S, Mo	[64,69,72]

The associated trace elements in ores of different regions have similarities and differences in element types and enrichment (Figure 4). For example, the main associated trace elements in the Songliao and Erlian Basins are Re, Th, Ba, Se, Co, Ni, Pb, V, Zn, etc., but the Songliao Basin has a stronger enrichment effect [43,73]. The ores in the Dongsheng mine (Ordos Basin) and No. 511 mine (Ili Basin) are both associated with and enriched with Se, Mo and Ge [24,25]. However, the No. 511 mine is additionally associated and enriched with Re and V. The contents of Se and Re in the No. 511 mine are higher, and the ore has a comprehensive utilization value, which is not available in the Dongsheng mine [42]. In addition, rare earth elements (REEs) are usually high in the Erlian Basin and the Ordos Basin. The ∑REE contents in the Dongsheng mine are 94.29×10^{-6}~203.53×10^{-6}, while those in the Bayintala mine are 105.01×10^{-6}~169.80×10^{-6} [74].

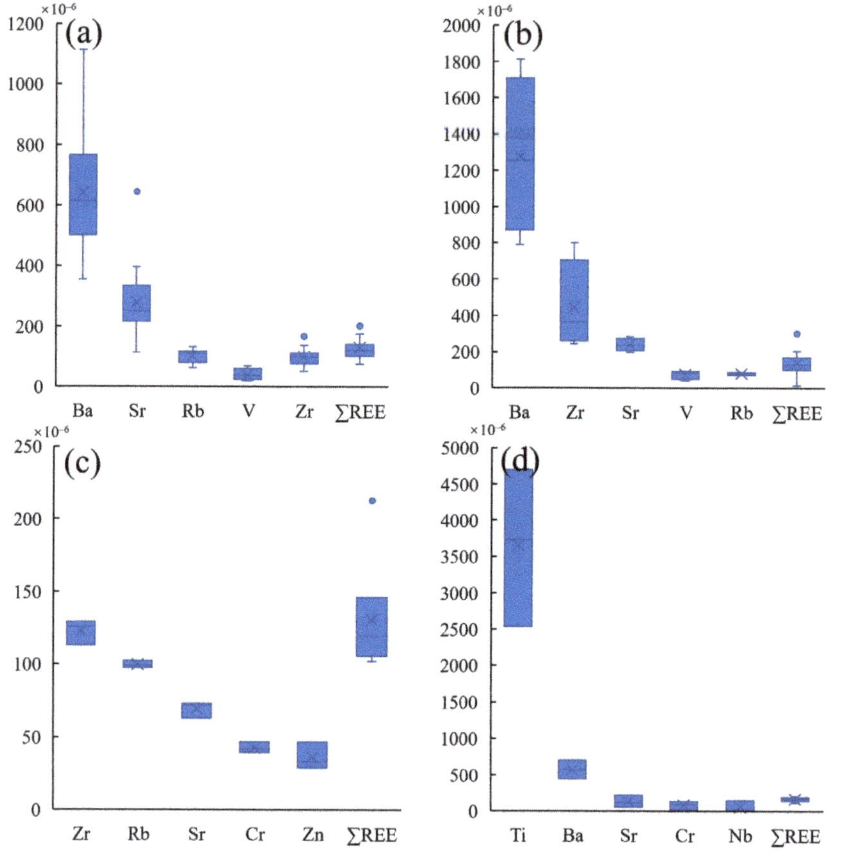

Figure 4. Statistics of the main trace elements and ∑REEs in each ore zone (10^{-6}) (**a**) for Qianjiadian in the Songliao Basin; (**b**) for Dongsheng in the Ordos Basin; (**c**) for Shihongtan in the Turpan-Hami Basin; and (**d**) for the southern edge of the Ili Basin [38,41,54,75–78].

The location of uranium deposits usually shows some hydrochemical anomalies (Figure 5). These chemical differences can be used as prospecting indicators for uranium deposits in the area and provide a reference for subsequent mining. The hydrochemical baselines of the same area show differences due to the different groundwater occurrences and types (Table 2). In the Songliao Basin, the groundwater chemical type in the upper aquifer is HCO_3·Cl-Na, the TDS is approximately 580 mg/L, and the pH is 7.65. The groundwater in the lower aquifer has better quality than that of the upper aquifer, and it can reach class III for the "standard for groundwater quality (GB/T 14848-2017). The groundwater chemical type is HCO_3·Cl-Na and Cl·HCO_3-Na, the TDS is approximately 3.64–4.89 mg/L, and the pH is 7.15–7.79 [48]. In the Erlian Basin, the bedrock fissure water in the low mountain area is mostly HCO_3-Ca·Na type water, and the TDS is less than 2000 mg/L. The bedrock fissure water in the hilly area is the HCO_3·SO_4-Na·Mg type or HCO_3·SO_4-Na·Ca type, the TDS is generally greater than 1000 mg/L, and some parts are as high as 2000–4000 mg/L. The pore water quality is mainly affected by evaporation and concentration, which increases the TDS of groundwater. From upstream to downstream, evaporation and concentration become stronger, and the TDS changes from low to high. The water type downstream is Cl-Na, and the TDS significantly exceeds 4000 mg/L [62]. The shallow groundwater (<20 m) in the northern Ordos Basin is of the HCO_3-Ca and HCO_3-Ca·Mg types, the pH is 7.53–8.06, and the TDS is 126–497 mg/L. Deep Cretaceous confined water with good quality ranges from 28 m to 300 m, and the TDS is 877–968 mg/L, which supplies local people and livestock drinking water [47]. The shallow groundwater in the Ili Basin is mainly the SO_4·Cl-Na·Ca type, and the TDS is low (<500 mg/L) or medium (500–5000 mg/L) [50,55]. The groundwater in the Turpan-Hami Basin has a high TDS, the deposit aquifers are mainly Cl SO_4-Na type water (TDS: 8–12 g/L), and the anticline area of the southern deposit is Cl-Na type water (TDS > 12 g/L) [52].

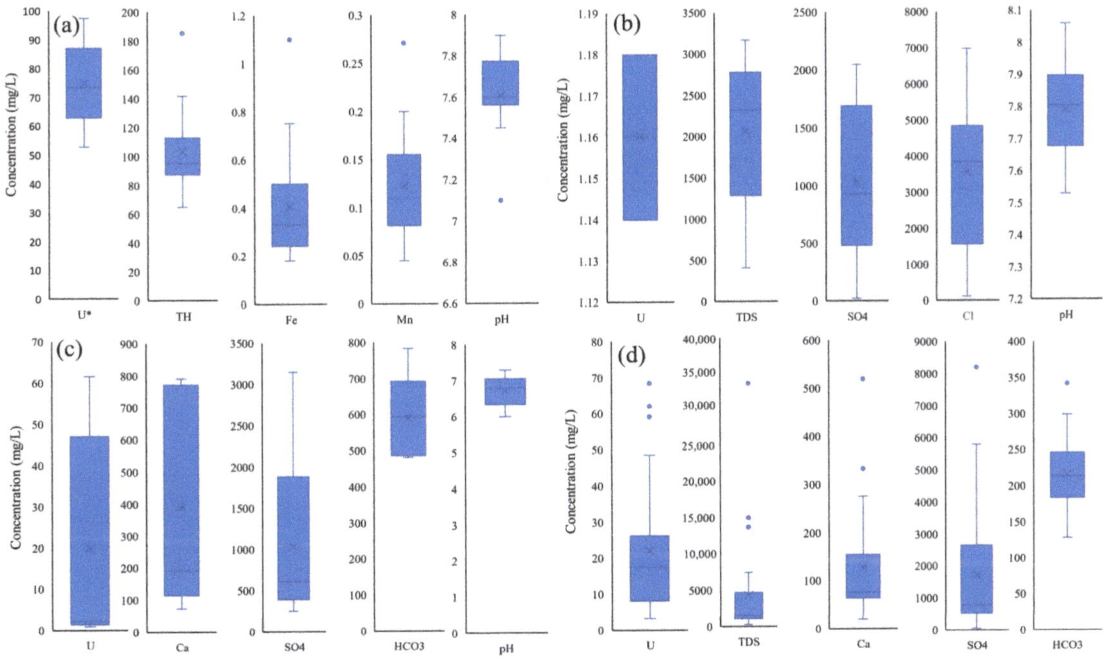

Figure 5. Representative constant elements of groundwater in the ore belt (**a**) for the Songliao Basin; (**b**) for the Ordos Basin; (**c**) for the Turpan-Hami Basin; and (**d**) for the Ili Basin. The pH is unitless, and * represents μg/L [14,75,79–91].

More than 90% of the uranium in groundwater is uranyl carbonate, followed by uranyl hydroxide [47,60]. The U content in groundwater is positively correlated with the U content in rocks [47,60]. In Shihongtan, Turpan-Hami Basin, the U content in the groundwater of the southern deposit (uranium-rich ore) is 44.5–4237 μg/L, while in the northern deposit (uranium-poor ore), the highest U content in the groundwater is only 65.1 μg/L, with an average of 29.3 μg/L [52]. In addition, the U content in groundwater is negatively correlated with TDS. In the Songliao Basin, the average U concentration in the upper aquifer with high TDS is approximately 4.6 μg/L and that in the lower aquifer with low TDS is 130.0–312.0 μg/L [48]. In the southeastern Alanor area, Erlian Basin, the groundwater discharge has high TDS, and the uranium concentration is 0.021–244 μg/L, while in the Nuheting area, which has low TDS, the concentration of U can reach 453 μg/L [92]. In the Ili Basin, the uranium concentration in $SO_4{\cdot}Cl$-Na-type water with high TDS is generally less than 10 μg/L, while in $HCO_3{\cdot}SO_4$-Ca-type water with low TDS, it is more than 30 μg/L [50,55] (Table 1).

Surface water in sandstone-type uranium deposits is often of poor quality, and the water-rock interaction with uranium ore is small. For example, studies on streams and lakes in northern Ordos [47] found that the local surface water is weakly alkaline, with a pH of approximately 9.3. The TDS is 412–3710 mg/L, approximately 6.4 times that of groundwater at the same location. The main ions in water are Na^+, Mg^{2+} and HCO_3^- and Cl^-. The U concentration is 1.14–1.18 μg/L, which is only approximately 1/24 of the groundwater. There are no significant differences between uranium ore belts and other places for surface water. Compared with groundwater, uranium ore does not significantly affect the quality of surface water.

3. Environmental Impact of Uranium Mining and Extraction

3.1. Overall Characteristics of Environmental Impact

In situ leaching (ISL) of uranium directly injects the solvent into the underground mining layer and extracts uranium minerals through the reaction between the solvent and ore [93,94]. The most common ISL method is acid leaching, which dissolves and mines uranium by injecting H_2SO_4 + H_2O_2 [94,95]. Secondly, in the Songliao, Ordos and Ili Basins, the alkaline method (or CO_2 method, CO_2 + O_2, weakly alkaline) has also been adopted [28,96]. Compared with open-pit mining, ISL has limited surface impact but a noticeable effect at depth [96]. Therefore, the mining process and tailings have a certain impact on the water and soil environment in and around the mine.

The pH of the deposits and groundwater are usually near-neutral. Acid leaching leads to pH reduction because of H_2SO_4 pollution. Meanwhile, leaching leads to the dissolution and release of other minerals, such as pyrite, which releases Fe^{3+} into the groundwater. Fe^{3+} as an oxidant further increases the release of other metal or alkali metal ions [94,95]. In alkaline leaching, CO_2 reacts with carbonates, resulting in an increase in Ca^{2+} and Mg^{2+}. Other heavy metals are also released by mineral dissolution. However, due to the near-neutral environment, insoluble metal sediment is formed quickly, so the content of heavy metal ions in groundwater is low [28,96].

3.2. Environmental Impacts on Soil

Soil chemical anomalies usually originate from eroding and leaching tailings rather than production (Figure 6). Wind-borne pollution and water leaching are the main sources of pollution [79]. Wind propagation affects a wide range, but pollution is generally only distributed on the soil surface. For example, in a U-Th tailing in northern China [83], the radioactive elements U and Th were abnormally high in the southeast of the tailing due to the prevailing northwest wind. The content of U in the downwind soil was 2.65 times the background value and 1.61 times the upwind value. The soil pollution from tailings spread to approximately 4 km but was basically distributed within 10 cm of the surface. The water leaching range is small, but the vertical effect is deeper. For example, in a tailing pond [97], the U content reached its peak at 8.5 m from the surface due to perennial precipitation leaching. The maximum migration depth reached 14 m.

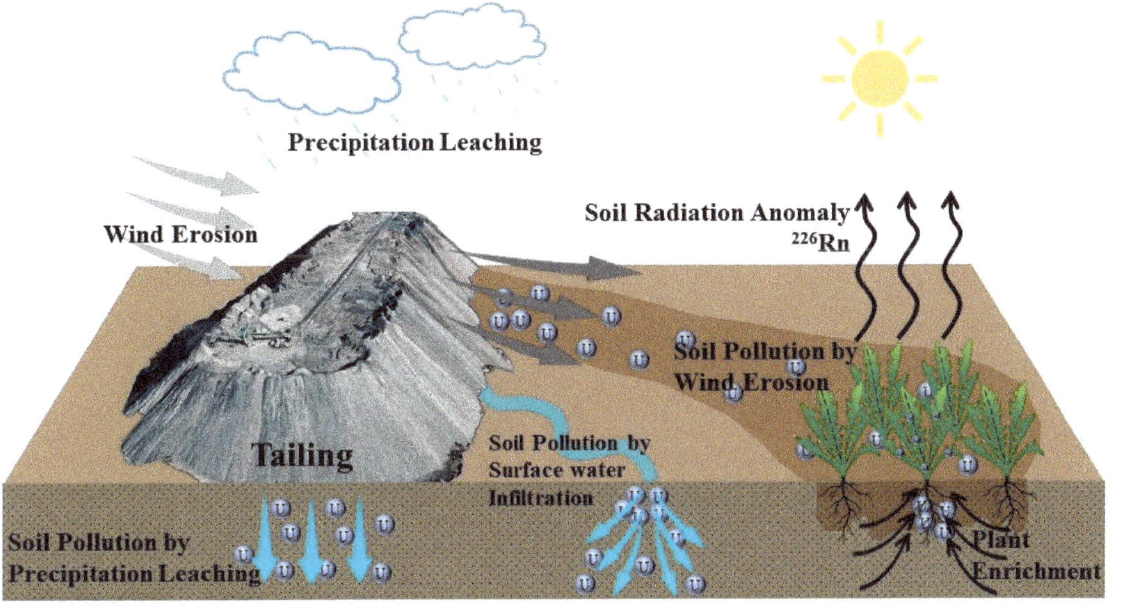

Figure 6. Soil pollution under wind erosion and precipitation leaching.

Open wastewater leakage can also cause soil pollution. For example, in a mine in the northern basin [98], due to the discharge of wastewater along the surface, heavy metals were carried downstream for more than 450 m but were mainly concentrated within 20 cm of the surface. The U content in the soil was 74.78–79.30 mg/kg, which was more than 27 times higher than the background value (2.72 mg/kg). In addition, other heavy metals have different degrees of pollution. The average exceedances of Cd, Cr, Cu, Mn, Ni, Pb and Zn were approximately 946.0, 5.7, 4.2, 1.7, 1.5, 4.7 and 4.4 times, respectively. Among them, Cd pollution is the most serious, and it is even worse than U pollution. Therefore, special attention should be given during production and remediation.

3.3. Environmental Impacts on the Groundwater Environment

Compared with soil, uranium ISL has a greater impact on groundwater. Similar to uranium deposits No. 737, No. 738 and No. 739 in Xinjiang [99,100], soil indicators remained within limits from mining to decommissioning. However, for groundwater, indicators were still in an abnormal state after years of decommissioning. Among them, ∑Fe (275 mg/L) and Mg^{2+} (252 mg/L) were more than 200 times higher than the background value, and $SO_4{}^{2-}$, U, Ca^{2+}, and $NO_3{}^{-}$ also exceeded the standard numerous times. The pollution degree of groundwater varies with the mining method (Figure 7). Acid leaching has a significant influence on groundwater. The highest content elements in the groundwater of the acid-leaching mining area are usually $SO_4{}^{2-}$, ∑Fe, Ca^{2+}, Mg^{2+}, and U. Other heavy metals will also be leached into the groundwater. Although the concentration is relatively low, it still far exceeds the groundwater quality standard, posing a potential threat to the groundwater environment. The low pH of groundwater is another problem. Compared with acid leaching, CO_2 leaching is more friendly to the groundwater environment because the weakly alkaline environment leads to the precipitation of pollutants.

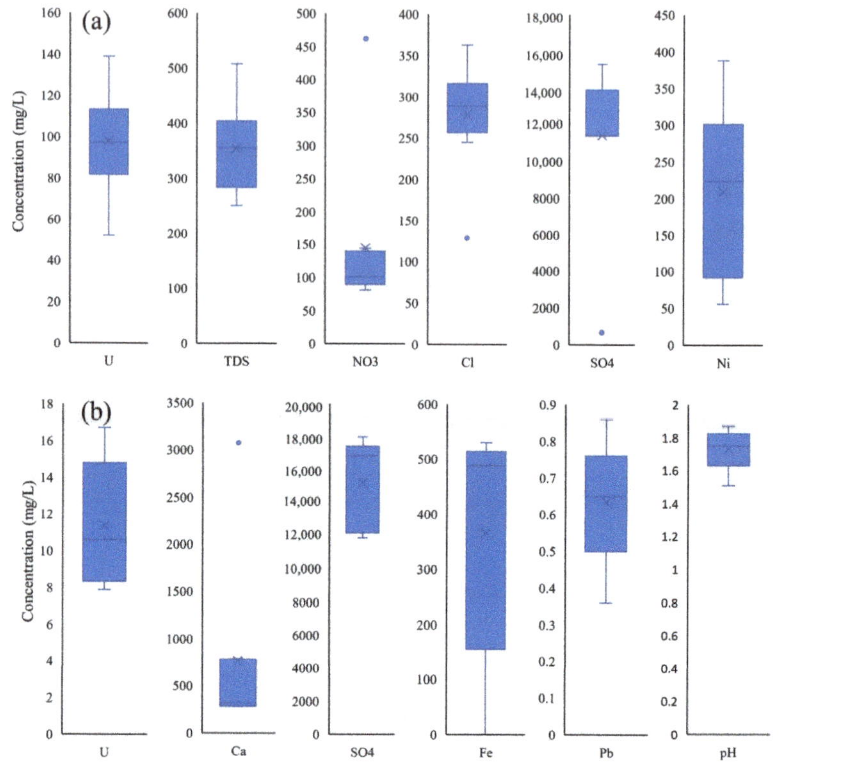

Figure 7. The statistical data of the main pollution ions in the acid-leaching (**a**) and CO_2-leaching (**b**) mining areas [89–91,100].

For example, in an acid-leaching area of a mixed mine in Ili (Table 3) [101], the pH of groundwater was 1.51–1.87, with an average of 1.73. The $\sum$Fe pollution was the most serious, exceeding the background value by more than 2400 times; the content of SO_4^{2-} ranged from 11,900 to 18,000 mg/L, with an average of 15,200 mg/L, which was 120 times more than the background value. The content of U (11231 μg/L) was approximately 114 times that before mining (98 μg/L). Other elements, such as Ca, Mn, Ni, Zn, Pb, As, and Cd, all exceeded the background value by dozens or even hundreds of times. Similar to the $CO_2 + O_2$ leaching area in the same mine [101], the $\sum$Fe in groundwater also decreased slightly after mining due to precipitation. The content of SO_4^{2-} was approximately 1/13 of the acid-leaching area, and Ca and Mg were approximately 1/2 and 1/3 of the acid-leaching area, respectively. The contents of other heavy metals were also significantly reduced. The U content increased, which was five times that of the acid-leaching area. However, CO_2 leaching can also lead to other chemical anomalies that are not available in the acid-leaching area. For example, the content of Cl^- (279 mg/L) was 3.2 times greater than that in the acid-leaching area (87 mg/L) and 14.5 times greater than the background value (19.43 mg/L). The HCO_3^- content was 148 mg/L, which was not detected in the acid-leaching area or before mining. In addition, ISL will use resin adsorption and nitrate leaching, resulting in nitrate exceeding the standard. Similar to a deposit in northwestern China, NO_3^- exceeded the background value by 140 times during the production period [80].

Table 3. Hydrochemical anomalies in some representative mining areas.

Location	Mining Method	pH	Concentration of Main Abnormal Elements (Background Value, mg/L)					Other Abnormal Indices		References
			Ca^{2+}	Mg^{2+}	U	SO_4^{2-}	Fe			
Ili Basin Mine No. 737, 738 and 739	Acid (Decommissioned)	/	360 (17)	252 (1)	3.9 (<0.1)	3920 (127)	275 (<1)	K, Na, Re, Th, Mn, Cu		[99]
Inner Mongolia	$CO_2 + O_2$	7.62 (8.76)	16.10 (5.24)	8.17 (3.35)	43.05 (0.14)	272 (64)	0.09 (0.09)	TDS, HCO_3^-, Cl^-		[90,91]
Ili Basin (A mixed leaching mine)	Acid	1.73	766	342	11.2	15222	363	NO_3^-, K, NaAl, Cu, Zn, Cd, Pb, Th, As		[101]
	$CO_2 + O_2$	7.32 (8.32)	342 (78)	99 (13)	60.0 (<0.1)	1139 (127)	0.04 (0.15)			
Ili Basin	Pollution halo	pH	Ca^{2+}	Mg^{2+}	U	SO_4^{2-}	Fe	NO_3^-	Cl^-	[102]
	Vertically adjacent aquifers	7.55 (8.04)	104.3 (78)	23.8 (0.03)	0.03 (/)	216.8 (83.5)	3.13 (60)	0.73 (0.84)	75.1 (24.8)	
	Upstream	7.43	91.5	28.6	0.026	202.4	2.38	2.81	42.35	
	Mining area	3.42	266.5	232.62	6.28	4029.8	209.83	111.04	74.67	
	200 m downstream	7.03	183.2	52.16	0.315	606.5	2.13	44.49	185.35	
	1000 m downstream	7.5	81.6	27.36	0.054	109.4	0.57	0.06	52.48	
	350 m from the side	7.37	121.1	33.01	0.022	414.5	1.13	0.06	77.25	

Horizontally, the pollution halo is distributed along the groundwater flow direction, mainly concentrated downstream of the mine with a spread to both sides. The impact on the upstream groundwater is minor [81,91,102]. Vertically, it can barely affect the aquifer beyond the mining layer due to the dense aquifer [102]. Monitoring wells in some Ili mines [81,91,102] showed that the content of SO_4^{2-} downstream (606.5 mg/L) was 3 times that upstream (202.4 mg/L), U (315.34 μg/L) was 16 times that upstream (25.9 μg/L), and NO_3^- (144.49 mg/L) was 51 times that upstream (2.81 mg/L) (Table 3).

3.4. Chemical Clogging and Its Effects

Most of the sandstone-type uranium ores in China have high carbonate content, low permeability and low grade, which readily cause chemical clogging during ISL, resulting in the reduction of rock porosity and permeability [28,34,36,61,66]. The dissolution of rocks leads to a large increase in Ca^{2+}, Mg^{2+} and other heavy metal ions in groundwater. These ions react with the solvent to form $CaCO_3$, $CaSO_4$ or metal precipitates [28,103]. For iron or a kind of metal X, some of the main chemical reactions are as follows.

$$XS_2 + 7/2O_2 + H_2O \rightarrow X^{2+} + 2SO_4^{2-} + 2H^+ \tag{1}$$

$$Fe^{2+} + 1/2O_2 + 2H^+ \rightarrow Fe^{3+} + H_2O \tag{2}$$

$$CaCO_3 + CO_2 + H_2O \rightarrow Ca^{2+} + 2HCO_3^- \tag{3}$$

$$X^{2+} + CO_3^{2-} \rightarrow XCO_3\downarrow \tag{4}$$

$$X^{2+} + 2OH^- \rightarrow X(OH)_2\downarrow \tag{5}$$

$$2Fe^{3+} + 3CO_3^{2-} \rightarrow Fe_2(CO_3)_3\downarrow \tag{6}$$

$$Fe^{3+} + 3OH^- \rightarrow Fe(OH)_3\downarrow \tag{7}$$

Spatially, chemical clogging mainly occurs at the periphery of the injection well. The injection well is usually not clogged due to the continuous injection of the solvent. At locations far from the injection well, as the concentration of the solvent decreases or the pH increases, a precipitate will form in the solution (Figure 8a) [28,76]. The chemical precipitation moves along the groundwater and accumulates downstream, resulting in increased blockage along the groundwater flow direction (Figure 8b). This dissolution-precipitation-redissolution process leads to significant stratification of porosity around the injection well [68,76].

Temporally, chemical clogging has some hysteresis (Figure 8b). In the initial stage, due to the dissolution of the solvent, the porosity of the rock expands. Rock dissolution may increase porosity by 1–3% [68,104]. As the solvent content decreases, the pH decreases, and the metal cations react with CO_3^-, SO_4^{2-} or OH^- to form chemical clogging (Figure 8c) [103]. Clogging usually occurs after tens to hundreds of days if there is no human intervention [28].

Limited by chemical clogging, the spread of pollution halos is usually in a small area (Table 3, Figure 9). For high concentrations of pollutants (SO_4^{2-}, Fe, U, etc.) Along the groundwater flow direction, the migration distance can exceed 1000 m (Figure 9a). Other minor pollutants (Cl^-, NO_3^- and other heavy metals) are mainly concentrated within 200 m (Figure 9b). The maximum diffusion range to both sides was approximately 350 m [102]. In mining areas with a poor permeability coefficient, the effect of chemical clogging is more pronounced. The diffusion range of the pollution halo will be smaller. Similar to a mine in Qianjiadian [89], the deposit has low permeability ($k < 0.2$ m/d). The average diffusion rate of pollutants during mining was only 0.0843 m/d. After six years of mining, the diffusion range of the pollution halo was within 130 m, which basically controlled the area around the mine.

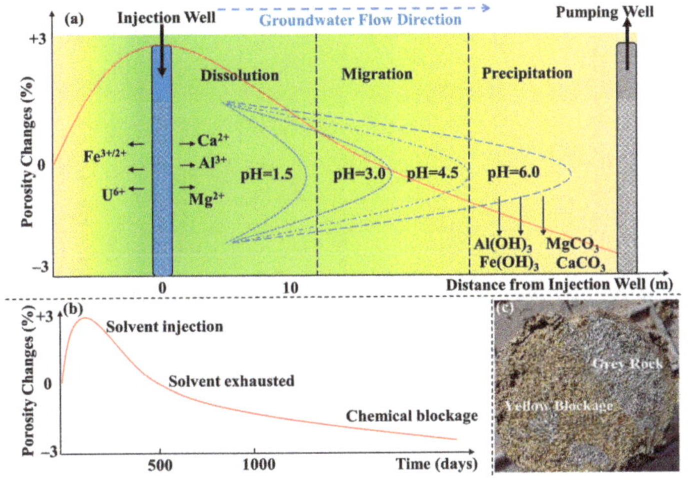

Figure 8. Chemical clogging during ISL. (**a**) Shows that chemical clogging occurs in space, while (**b**) shows that chemical clogging occurs over time; the red line shows the porosity changes; a positive value on the vertical axis indicates an increase, while a negative value indicates a decrease; (**c**) shows the yellow blockage in the gray rock, and the sample was collected in a decommissioned uranium mine in Northwest China.

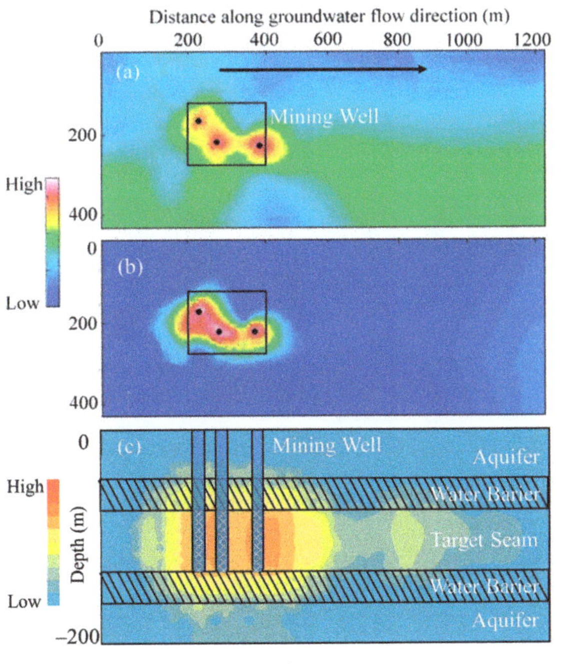

Figure 9. Schematic view of the groundwater pollution halo. The black square delineates the range of the mine, and the dots indicate the mining wells. (**a**) shows the pollution halo of primary high-concentration pollutants, such as SO_4^{2-}, Fe, and U; (**b**) shows the horizontal pollution halo of low-concentration pollutants, such as Cl^-, NO_3^- and other heavy metals; (**c**) shows the vertical pollution halo; the data were interpolated from the measurement of a decommissioned uranium mine in Northwest China.

3.5. Radiation Environment Impacts

Radon release and radiation exposure in uranium mines can result in radioactive abnormalities. The dissolution of the rock and soil by the solvent will result in the release of ^{226}Rn [105]. There are two main ways radon is released: evaporation of wastewater or waste liquid pools and precipitation of ^{226}Rn-content dust on the solid surface after uranium mine decommissioning [106,107]. The released content is mainly related to the concentration of ^{226}Rn in groundwater and leachate and the scale of the mine [106,107]. As in the No. 737 mine, mining produced approximately 1.22×10^{11} Bq of radon per 1 t of uranium produced. During the production of the Qian IV uranium mine in the Qianjiadian Basin, the ^{226}Rn released from the liquid collecting tank and the evaporation pond were 1.06×10^{-12} Bq/a and 1.48×10^{-7} Bq/a, respectively [100]. Due to the ventilation system and human management, radioactivity is typically controllable and will not cause a significant impact on the environment during production. However, for decommissioned uranium mines, the radioactivity of ^{226}Rn from heap leaching of tailings waste may far exceed the environmental limit. Similar to tailing in Liaoning Province, the ^{226}Rn radiation dose on the surface was approximately 3.14 Bq/($m^2 \cdot s$), which was far beyond the management limits [108]. Another tailing's radiation dose exceeded the background value by 8.21 times but was only concentrated within 10 cm of the surface [109].

3.6. Plant Accumulation of Nuclides

Sandstone-type uranium deposits in northern China are mostly distributed in areas with poor soils. In these areas, most of the plants are herbs, such as Poaceae, Compositae or Leguminosae [110,111]. Since these plants are usually not used as food or economic crops and have good heavy metal tolerance, they have little impact on human or ecological health. Plants are usually used for geo-environmental restoration due to their good enrichment [82,112]. The U distribution in plants is usually root > stem > leaf [83,113]. Similar to tailing [109], the U content in plant roots was 0.18–42.7 mg/kg. For another tailing, the highest activity of uranium in plants in the mining area was 784.58 Bq/kg, 291 times higher than the background value. The content of heavy metals in plants was consistent with the distribution of the soil pollution halo. Additionally, affected by wind direction, the U content in plants behaves as follows: inside the mining area > southeast > northwest.

4. Mitigation of Environmental Impacts

4.1. Environmental Mitigation during Mining

With the increasing emphasis on environmental protection in recent years, various measures have been adopted in the ISL process to protect the geo-environment of mines. First, in terms of mining methods, an increasing number of mines use weak-alkaline leaching. $CO_2 + O_2$ solvent with mild chemical properties has begun to replace acid leaching [28]. In Ili, Ordos, Qianjiadian, etc., most of the mines use the weak-alkali method when the geological conditions allow [28,102]. Second, closed-loop mining methods were used in the mining processes. In Qianjiadian, a closed-loop treatment system of "leachate extraction-surface treatment-tail liquid treatment and reinjection" was adopted [89]. At the same time, an anti-seepage device was introduced into the production to treat wastewater, which basically achieved no wastewater discharge and no waste residue stacking [114]. Finally, the government usually formulated environmental norms in the mining process and made demands on land area, waste storage and treatment, quality and safety, etc. An environmental monitoring system must be established around the mine (Figure 10).

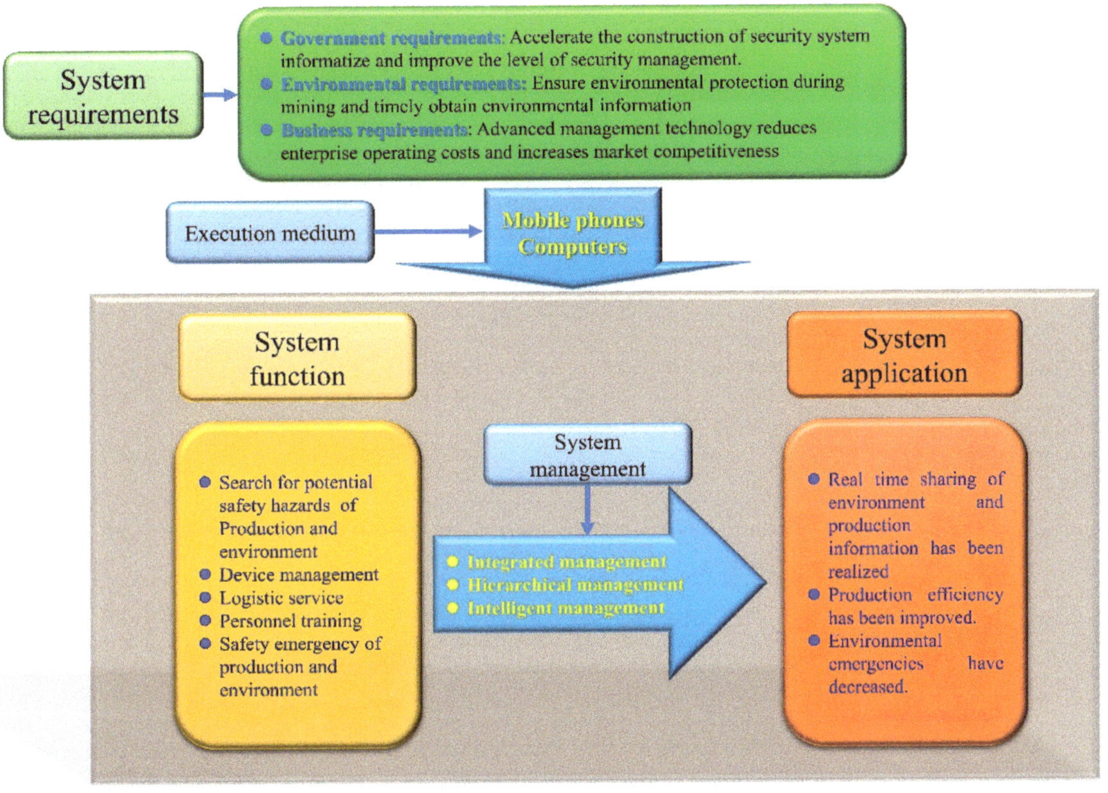

Figure 10. A demonstration security management system of ISL ore mine in Ordos.

4.2. Environmental Mitigation after Mining

At present, the main objects of remediation and treatment of decommissioned uranium mines in China are tailings storage, abandoned wells, wastewater and vegetation restoration. The assessment indicators of remediation and treatment usually focus on radiation reduction [79]. For tailings storage and abandoned wells, removal, landfill and plugging are the main methods for treatment, while for wastewater, artificial excretion is the main method [78]. The impact of the repaired mine on the surrounding environment can be reduced to a negligible level. Similar to the No. 440 uranium mine in Northeast China [115], there were 11 tailings heaps, 30 abandoned wells and tunnels, and more than 100,000 tons of waste residue after decommissioning. The average radiation dose in the area before treatment was approximately 557 nGy/h, 5 times more than the background value. By removing the waste and covering the site with compacted soil, the average environmental radiation dose after treatment was 60.54 nGy/h, which was even better than the level before mining. In addition, at the No. 781 mine in Henan Province, after decommissioning, U in groundwater exceeds the background value by 4.6 times and that in soil by 82 times. After the polluted water and soil were removed, 1.3 m thick planting soil was used for covering. Due to these remediation measures, the content of radioactive pollutants in water decreased by 82%, and the content of U in soil decreased by 98%, reaching the level before mining.

5. Environmental Management for Prospective Mining in the Future

5.1. Environmental Management during Uranium Mining

As described above, the pollution of uranium ISL is mainly concentrated in groundwater. In view of the difficult treatment of groundwater after pollution, it is necessary to strengthen groundwater protection during the mining process [77]. Measures can be taken from the following three phases. Before mining, the hydrogeological characteristics and chemical indicators of groundwater in the mining area should be investigated in detail, and the mine construction plan and decommissioning restoration goals should be formulated based on this. The drilling design of the mine should consider the groundwater flow direction and layout of wells from upstream to downstream, which can reduce the spread of the pollution halo to both sides. During operation, the volume of leaching solution extraction should be considered larger than the solvent injection. This can lead to negative pressure in the mining area and prevent pollutants from spreading outward. After the mine is exhausted, a monitoring system should be set up in the surrounding area, especially downstream. A periodic monitoring plan should be formulated to determine the range of pollution halo diffusion.

5.2. Environmental Management for Uranium Mine Closing

At present, the mainstream remediation methods of decommissioned uranium mines are to ensure the nonproliferation of pollution. Physical methods are used to isolate the contaminated area to prevent further damage to the nearby environment. However, U itself has not been repaired or removed, which should be considered in the future, especially under complex conditions. Conventional physical removal and chemical leaching are usually expensive and can easily cause secondary damage to the environment; phytoremediation is economical and environmentally friendly but is inefficient and limited by the capacity of enrichment [109,116,117]. These factors make it difficult to commercialize such methods on a large scale. Three methods can be considered for future application. The first is electric repair. Through electrodialysis, electromigration and other electrodynamic effects, uranium compounds migrate along the direction of the electric field and accumulate at the electrode, and then they can be removed by physicochemical methods (adsorption, ion exchange, precipitation, etc.) [75,84]. Especially in an acidic environment with a high iron content, the removal efficiency can reach 58–88% [75,84]. Second, microbial remediation can be considered. As a major participant in the biogeochemical cycle, microorganisms can reduce U(VI) to insoluble U(IV) precipitation through biological reduction, which can retard uranium migration in the environment [116,117]. Similar microbial reduction processes also exist in many uranium mineralization processes [118]. For example, with ethanol as the electron donor, the uranium concentration in groundwater can be reduced by more than 99.9% by microbial reduction [119]. Finally, for heavily polluted sites, PRB can be used for remediation. By placing the PRB wall downstream of the pollution halo, U can be removed by reacting with the medium in the wall. For the use of PRB, material selection is a challenge, such as zero-valent iron, activated carbon, slaked lime, quartz sand, etc. [120], which are all very promising materials but still need to be further verified in field trials.

6. Conclusions

In the context of China's carbon peak and carbon neutrality, the development of nuclear energy will reach its peak in the next few decades. As uranium is the main raw material for nuclear energy, environmental problems caused by mining should be taken seriously by government agencies. Sandstone-type uranium in northern China is mainly distributed in the Songliao Basin, Erlian Basin, Ordos Basin, Turpan–Hami Basin and Ili Basin from east to west. Most of the deposits are located in arid or semiarid regions with poor aquifer permeability, slow groundwater flow velocity and low hydraulic gradient. The type of uranium ore is usually dominated by pitchblende and is accompanied by pyrite, hematite, coal and organic matter. The main form of U is UO_2 in ore and uranyl carbonate in groundwater. In situ leaching is the main method of sandstone-type uranium mining. The dissolution of the rock by the solvent will increase the porosity in a short period of time. However, as leaching continues, the rock crevice becomes clogged by metal precipitation, resulting in a decrease in porosity. Moreover, dissolution also leads to the release of other enriched metals, which cause groundwater pollution. The groundwater pollution in the mining area is mainly SO_4^{2-}, Fe and U. Other elements, such as Ca, Mg, NO_3^- and TDS, also have a significant increase. The pollution halo mainly diffuses along the groundwater flow direction. Due to chemical clogging, the diffusion range usually does not exceed 1000 m. Compared with groundwater, soil pollution is often much lighter. Soil chemical anomalies usually come from heap leaching and wind erosion of tailings. The heavy metals in the tailings can spread for several kilometers by wind, but they are mainly concentrated within 20 cm of the surface. Finally, uranium mining and heap leaching also cause radiation anomalies and radioactive enrichment in plants. Through treatment, decommissioned mines can have a negligible impact on the surrounding environment. Treatment methods are mainly physical partitioning, and pollutants have not been removed. At present, the attention and treatment of uranium tailings are still in their infancy. In the future, scholars should focus on the impact of decommissioned mines on the environment, including collecting more data for treatment assessment and providing statistical data support for environmental recovery. Economic and feasible methods should be widely reviewed to assess their practicality, such as electric remediation, microbial remediation or PRB, which are worth considering.

Author Contributions: All authors contributed to the study's conception and design. Material preparation, data collection and analysis were performed by Y.T. and Y.Z.; The draft of the manuscript was written by F.Z.; Figures and tables were made by J.H. and R.Z.; Proofreading and perfecting of the manuscript were performed by J.D. All authors commented on previous versions of the manuscript. All authors have read and agreed to the published version of the manuscript.

Funding: This work was supported by The National Natural Science Foundation of China (41831283 and 42077170), and The National Key Research and Development Program of China (2020YFC1806604).

Institutional Review Board Statement: Not applicable.

Informed Consent Statement: Not applicable.

Data Availability Statement: Not applicable.

Conflicts of Interest: The authors have no relevant financial or non-financial interest to disclose.

References

1. Mallapaty, S. How China could be carbon neutral by mid-century. *Nature* **2020**, *586*, 482–483. [CrossRef] [PubMed]
2. Meng, M.; Yu, J. Chinese nuclear energy politics: Viewpoint on energy. *Energ. Source Part B* **2018**, *13*, 72–75. [CrossRef]
3. Dittmar, M. The end of cheap uranium. *Sci. Total Environ.* **2013**, *461*, 792–798. [CrossRef]
4. Xu, D.; Chi, G.; Nie, F.; Fayek, M.; Hu, R. Diversity of uranium deposits in China—An introduction to the Special Issue. *Ore. Geol. Rev.* **2021**, *129*, 103944. [CrossRef]
5. Jin, R.; Liu, H.; Li, X. Theoretical System of Sandstone-Type Uranium Deposits in Northern China. *J. Earth Sci.-China* **2022**, *33*, 257–277. [CrossRef]
6. Chen, Z.B.; Zhao, F.M.; Xiang, W.D.; Chen, Y.H. Uranium provinces in China. *Acta Geol. Sin.-Engl.* **2000**, *74*, 587–594.
7. Ilankoon, I.M.S.K.; Tang, Y.; Ghorbani, Y.; Northey, S.; Yellishetty, M.; Deng, X.; McBride, D. The current state and future directions of percolation leaching in the Chinese mining industry: Challenges and opportunities. *Min. Eng.* **2018**, *125*, 206–222. [CrossRef]
8. Hague, N.; Norgate, T. The greenhouse gas footprint of in-situ leaching of uranium, gold and copper in Australia. *J. Clean Prod.* **2014**, *84*, 382–390. [CrossRef]
9. Coral, T.; Placko, A.; Beaufort, D.; Tertre, E.; Bernier-Latmani, R.; Descostes, M.; De Boissezon, H.; Guillon, S.; Rossi, P. Biostimulation as a sustainable solution for acid neutralization and uranium immobilization post acidic in-situ recovery. *Sci. Total Environ.* **2022**, *822*, 153597. [CrossRef] [PubMed]
10. Pereira, R.; Antunes, S.C.; Marques, S.M.; Goncalves, F. Contribution for tier 1 of the ecological risk assessment of Cunha Baixa uranium mine (Central Portugal): I soil chemical characterization. *Sci. Total Environ.* **2008**, *390*, 377–386. [CrossRef]

11. Coral, T.; Descostes, M.; De Boissezon, H.; Bernier-Latmani, R.; de Alencastro, L.F.; Rossi, P. Microbial communities associated with uranium in-situ recovery mining process are related to acid mine drainage assemblages. *Sci. Total Environ.* **2018**, *628–629*, 26–35. [CrossRef] [PubMed]
12. Wanty, R.B.; Berger, B.R.; Plumlee, G.S.; King, T. Geoenvironmental models—An introduction. In *Deposit and Geoenvironmental Models for Resource Exploitation and Environmental Security*; Fabbri, A.G., Gaal, G., McCammon, R.B., Eds.; Springer: Matrahaza, Hungary, 2002; Volume 80, pp. 3–42.
13. Wang, M.; Ren, P.; Yan, G. The Updates on Geoenvironmental Study of Mineral Deposits in The Unite States and Propoed Strategy for China. *Adv. Earth Sci.* **2004**, *19*, 636–641.
14. Bussiere, B. Colloquium 2004: Hydrogeotechnical properties of hard rock tailings from metal mines and emerging geoenvironmental disposal approaches. *Can. Geotech. J.* **2007**, *44*, 1019–1052. [CrossRef]
15. Huang, J.; Huang, S. Regional metallogenic characteristics of China's uranium resources. *Uranium Geol.* **2005**, *21*, 129–138.
16. Nie, J.; Horton, B.K.; Saylor, J.E.; Mora, A.; Mange, M.; Garzione, C.N.; Basu, A.; Moreno, C.J.; Caballero, V.; Parra, M. Integrated provenance analysis of a convergent retroarc foreland system: U-Pb ages, heavy minerals, Nd isotopes, and sandstone compositions of the Middle Magdalena Valley basin, northern Andes, Colombia. *Earth-Sci. Rev.* **2012**, *110*, 111–126. [CrossRef]
17. Akhtar, S.; Yang, X.; Pirajno, F. Sandstone type uranium deposits in the Ordos Basin, Northwest China: A case study and an overview. *J. Asian Earth Sci.* **2017**, *146*, 367–382. [CrossRef]
18. Gomez, P.; Garralon, A.; Buil, B.; Turrero, M.J.; Sanchez, L.; De la Cruz, B. Modeling of geochemical processes related to uranium mobilization in the groundwater of a uranium mine (vol 300, pg 295, 2000). *Sci. Total Environ.* **2008**, *390*, 579. [CrossRef]
19. Cuvier, A.; Pourcelot, L.; Probst, A.; Prunier, J.; Le Roux, G. Trace elements and Pb isotopes in soils and sediments impacted by uranium mining. *Sci. Total Environ.* **2016**, *566*, 238–249. [CrossRef]
20. Bradshaw, P.; Lett, R. Geochemical-exploration for uranium using soils. *J. Geochem. Explor.* **1980**, *13*, 305–319. [CrossRef]
21. Cai, Y.; Zhang, J.; Li, Z.; Guo, Q.; Song, J.; Fan, H.; Liu, W.; Qi, F.; Zhang, M. Outline of Uranium Resources Characteristics and Metallogenetic Regularity in China. *Acta Geol. Sin.-Engl.* **2015**, *89*, 918–937. [CrossRef]
22. Li, Z.; Qin, M.; Fan, H.; Cai, Y.; Cheng, J.; Guo, D.; Ye, F.; Fan, G.; Liu, X. Main Progresses of Uranium Geology and Exploration Techniques for the Past Decade in China. *Bull. Mineral. Petrol. Geochem.* **2021**, *40*, 845–857.
23. Zhang, J. Progress and Outlook of Uranium Exploration in Meso-Cenozoic Basins in North China. *Uranium. Geol.* **2012**, *28*, 193–198.
24. Min, M.Z.; Chen, J.; Wang, J.P.; Wei, G.H.; Fayek, M. Mineral paragenesis and textures associated with sandstone-hosted roll-front uranium deposits, NW China. *Ore. Geol. Rev.* **2005**, *26*, 51–69. [CrossRef]
25. Min, M.Z.; Xu, H.F.; Chen, J.; Fayek, M. Evidence of uranium biomineralization in sandstone-hosted roll-front uranium deposits, northwestern China. *Ore. Geol. Rev.* **2005**, *26*, 198–206. [CrossRef]
26. Rong, H.; Jiao, Y.; Wu, L.; Wan, D.; Cui, Z.; Guo, X.; Jia, J. Origin of the carbonaceous debris and its implication for mineralization within the Qianjiadian uranium deposit, southern Songliao Basin. *Ore. Geol. Rev.* **2019**, *107*, 336–352. [CrossRef]
27. Shi, Z.; Chen, B.; Wang, Y.; Hou, M.; Jin, X.; Song, H.; Wang, X. A linkage between uranium mineralization and high diagenetic temperature caused by coal self-ignition in the southern Yili Basin, northwestern China. *Ore. Geol. Rev.* **2020**, *121*, 103443. [CrossRef]
28. Su, X.; Liu, Z.; Yao, Y.; Du, Z. Petrology, mineralogy, and ore leaching of sandstone-hosted uranium deposits in the Ordos Basin, North China. *Ore. Geol. Rev.* **2020**, *127*, 103768. [CrossRef]
29. Zhang, F.; Jiao, Y.; Wu, L.; Rong, H. Roles of dispersed organic matters in sandstone-type uranium mineralization: A review of geological and geochemical processes. *Ore. Geol. Rev.* **2021**, *139*, 104485. [CrossRef]
30. Fu, Y.; Wei, S.; Jin, R.; Li, J.; Ao, C. Current Status and Existing Problems of China' s Sandstone-type Uranium Deposits. *Acta Geol. Sin.* **2016**, *90*, 3519–3544.
31. Chen, X.; Romaniello, S.J.; Herrmann, A.D.; Hardisty, D.; Gill, B.C.; Anbar, A.D. Diagenetic effects on uranium isotope fractionation in carbonate sediments from the Bahamas. *Geochim. Cosmochim. Acta* **2018**, *237*, 294–311. [CrossRef]
32. Meunier, J.D.; TROUILLER, A.; Brulhet, J.; Pagel, M. Uranium and organic-matter in a paleodeltaic environment—The coutras deposit (gironde, france). *Chem. Geol.* **1988**, *70*, 189. [CrossRef]
33. Och, L.M.; Mueller, B.; Marz, C.; Wichser, A.; Vologina, E.G.; Sturm, M. Elevated uranium concentrations in Lake Baikal sediments: Burial and early diagenesis. *Chem. Geol.* **2016**, *441*, 92–105. [CrossRef]
34. Bonnetti, C.; Cuney, M.; Malartre, F.; Michels, R.; Liu, X.; Peng, Y. The Nuheting deposit, Erlian Basin, NE China: Synsedimentary to diagenetic uranium mineralization. *Ore. Geol. Rev.* **2015**, *69*, 118–139. [CrossRef]
35. Bonnetti, C.; Liu, X.; Mercadier, J.; Cuney, M.; Deloule, E.; Villeneuve, J.; Liu, W. The genesis of granite-related hydrothermal uranium deposits in the Xiazhuang and Zhuguang ore fields, North Guangdong Province, SE China: Insights from mineralogical, trace elements and U-Pb isotopes signatures of the U mineralisation. *Ore. Geol. Rev.* **2018**, *92*, 588–612. [CrossRef]
36. Qiu, J.; Mu, H.; Yu, X.; Rui, X.; Yang, Y.; Qiu, L. Identifying the principal factors controlling uranium enrichment: Semi-quantitative mineralogy and geochemistry of the sandstone-type Qianjiadian uranium deposit, northeast China. *Ore. Geol. Rev.* **2022**, *144*, 104807. [CrossRef]

37. Chen, J.; Chen, P.; Yao, D.; Huang, W.; Tang, S.; Wang, K.; Liu, W.; Hu, Y.; Zhang, B.; Sha, J. Abundance, Distribution, and Modes of Occurrence of Uranium in Chinese Coals. *Miner.-Basel.* **2017**, *7*, 239. [CrossRef]
38. Chen, J.; Bai, Q.; Tian, Y.; Lu, Y.; Niu, J.; Shi, L.; Yang, Y. Discussion on Prospecting Directions of Uranium Deposit in Coal Measure Strata in the North of Yining Area, Xinjiang. *Uranium. Geol.* **2021**, *37*, 160–170.
39. Huang, W.; Wan, H.; Finkelman, R.B.; Tang, X.; Zhao, Z. Distribution of uranium in the main coalfields of China. *Energ. Explor. Exploit.* **2012**, *30*, 819–835. [CrossRef]
40. Liu, X.; Yu, R.; Cao, H.; Zhu, Q.; Wang, S.; Yang, J. Metallogenic conditions of sandstone uranium deposits in the coal-bearing area on the southeastern margin of Ordos basin. *Coal. Geol. Explor.* **2018**, *46*, 26–32.
41. Tang, C.; Zhu, Q.; Feng, X.; Chen, Y.; Li, G.; Chen, L.; Liu, X.; Liu, X.; Hu, P. REE geochemical study on Dongsheng sandstone uranium deposit, Inner Mongolia, China. *Acta Mineral. Sin.* **2017**, *37*, 121–131.
42. Wang, Z.; Cao, S.; Pan, J.; Guan, T.; Zhang, G. Trace element geochemistry of No. 511 uranium ore deposit in Xinjiang. *Miner. Depos.* **2005**, *24*, 409–415.
43. Xue, W.; Xue, C.; Chi, G.; Peng, Y.; Wang, K. Trace Element and REE Geochemical Characteristics of Sandstone-type Uranium Deposit in the Dongsheng Area of the Ordos Basin, China. *Geoscience* **2010**, *24*, 776–784.
44. Nie, F.; Yan, Z.; Feng, Z.; Li, M.; Xia, F.; Zhang, C.; Wang, Y.; Yang, J.; Kang, S.; Shen, K. Genetic models and exploration implication of the paleochannel sandstone-type uranium deposits in the Erlian Basin, North China—A review and comparative study. *Ore. Geol. Rev.* **2020**, *127*, 103821. [CrossRef]
45. Wang, S.Y.; Cheng, Y.H.; Xu, D.H.; Miao, P.S.; Jin, R.S.; Zhang, T.F.; Xu, Z.L.; Cheng, X.Y.; Zhao, L.; Li, C.H.; et al. Late Cretaceous-Cenozoic tectonic-sedimentary evolution and U-enrichment in the southern Songliao Basin. *Ore. Geol. Rev.* **2020**, *126*, 103786. [CrossRef]
46. Liu, Z.; Yang, B.; Zhao, X.; Wang, T. Mineral Ingredient and Geochemical Characteristics of Uranium Mineralization at Shihongtan Uranium Deposit, Turpan-Hami Basin. *Xinjiang Geol.* **2018**, *36*, 257–261.
47. Chen, G.; Sun, Z.; Nie, F.; Li, C.; Zhen, Y.; Zhou, Z. Hydrogeochemical characteristics of the sandstone-hosted uranium mineralization in northern Ordos Basin, China. *Ore. Geol. Rev.* **2020**, *126*, 103769. [CrossRef]
48. Ren, J.; Xia, Y.; Zhang, Z.; Lin, Y. Study on the Hydrological Conditions of Qianjiadian Sandstone Type Uranium for In-situ Leaching Mining in Songliao Basin. *Uranium. Min. Metall.* **2022**, *41*, 21–29. [CrossRef]
49. Cao, L.; Qiao, H.; Zhou, W.; Wu, Z. Analysis on affecting factors of metallogenic difference in the southern ore belt of Shihongtan uranium deposit. *World Nucl. Geosci.* **2018**, *35*, 77–83.
50. Yue, S.; Wang, G. Relationship between the hydrogeochemical environment and sandstone-type uranium mineralization in the Ili basin, China. *Appl. Geochem.* **2011**, *26*, 133–139. [CrossRef]
51. Qiao, H.; Song, Z.; Zhang, J. The effect of gas from the groundwater in the forming of uranium deposit at Shihongtan. *Uranium. Geol.* **2009**, *25*, 222–227.
52. Qiao, H.; Zhang, T.; Zhang, F.; Shang, G.; Song, Z. Hydrogeochemical characteristics of Shihongtan uranium deposit in Turpan-Hami basin. *Uranium. Geol.* **2005**, *21*, 345–352.
53. Li, Y.; Li, S.; Ren, M. Study on the Relationship Between Hydrogeological Models with Uranium Mineralization, Take Uranium of Northern Ili Basin as an Example. *Xinjiang Geol.* **2019**, *37*, 107–111.
54. Lin, X.; Li, X.; Liu, W.; Xu, Q. Hydrogeological Conditions of Uranium Mineralization in Yihegaole Area, Erlian Basin. *Uranium. Geol.* **2020**, *36*, 464–476.
55. Lin, X.; Liu, H.; Wang, Z.; Hao, W. Hydrogeochemical Modeling Study on Mengqiguer Uranium Deposit in Yili Basin. *Uranium. Geol.* **2017**, *33*, 346–355.
56. Chen, G.; Sun, Z.; Wang, G.; Liu, J. The isotopic and chemical characteristics of groundwater from Shihongtan uranium deposit and its surrounding area. *Geochim. Cosmochim. Acta* **2009**, *73*, A212.
57. Chen, Z.; Man, A.; Chen, X.; Xiong, Y.; Shao, J.; Wang, L.; Sun, P. Relationship Between Hydrogeology and Sandstone Uranium Mineralization in Sedimentary Basin: A Case Study of Sandstone Uranium Survey in the Kailu Depression of the Songliao Basin. *Geotecton. Metallog.* **2021**, *45*, 1174–1184.
58. Iskander, F.Y. Assessment of trace elements in honey produced on uranium mining reclaimed land. *Sci. Total Environ.* **1996**, *192*, 119–122. [CrossRef]
59. Yao, Y.; Liu, D.; Yan, T. Geological and hydrogeological controls on the accumulation of coalbed methane in the Weibei field, southeastern Ordos Basin. *Int. J. Coal. Geol.* **2014**, *121*, 148–159. [CrossRef]
60. Gang, W.; Fa, Z.; Dapeng, H.; Ji'An, L. Analysis of In-situ Leaching Feasibility in Northern Belt of Shihongtan Uranium Deposits in Xinjiang Turpan. *J. East China Inst. Technol. (Nat. Sci. Ed.)* **2015**, *38*, 58–63.
61. Qiu, W.; Liu, Z.; Yang, Y.; Weng, H.; Jia, Y.; Wu, J.; Li, H.; Liu, S.; Wu, J. Reactive transport numerical modeling of CO_2 + O_2 in-situ leaching in sandstone-type uranium ore. *Sci. Sin. (Technol.)* **2022**, *52*, 627–638.
62. Hao, J. Application of Geochemical Model in the Prediction of Sandstone-Type Uranium Deposits in Eastern Erlian Basin. *Uranium. Geol.* **2013**, *29*, 24–30, 46.
63. Lin, N.; Shutao, H.; Guisheng, Y. The characteristics of uranium mineralization and cenesis of Nuheting uranium deposit in Erlian basin inner mongolian. *China Nucl. Sci. Technol. Rep.* **1995**, *00*, 1005–1017.

64. Zhang, P.; Nie, F.; Zhang, C.; Zhang, H.; Dong, Y.; Zhang, X. Characteristics and Its Implications for the Mineralization of the Sandstone from Uranium-Bearing Target Layer in Langka, Yili Basin. *Sci. Technol. Eng.* **2018**, *18*, 29–38.
65. Jiao, Y.; Chen, A.; Wang, M.; Wu, L.; Yuan, H.; Yang, Q.; Zhang, C.; Xu, Z. Genetic Analysis of the Bottom Sandstone of Zhiluo Formation, Northeastern Ordos Basin:Predictive base of spatial orientation of sandstone-type uranium deposit. *Acta Sedimentol. Sin.* **2005**, *23*, 371–379.
66. Ma, X.; Wang, K.; Jin, R.; Li, J.; Zhou, H.; Yang, H. Characteristics of fluid inclusions in the sandstone-hosted Qianjiadian uranium deposit, southwest Songliao Basin, northeastern China: Implications for the nature and evolution of ore-forming fluids. *Resour. Geol.* **2022**, *72*, e12281. [CrossRef]
67. Min, M.; Peng, X.; Zhou, X.; Qiao, H.; Wang, J.; Zhang, L. Hydrochemistry and isotope compositions of groundwater from the Shihongtan sandstone-hosted uranium deposit, Xinjiang, NW China. *J. Geochem. Explor.* **2007**, *93*, 91–108. [CrossRef]
68. Zhao, L.; Deng, J.; Xu, Y.; Zhang, C. Mineral alteration and pore-plugging caused by acid in situ leaching: A case study of the Wuyier uranium deposit, Xinjiang, NW China. *Arab. J. GeoSci.* **2018**, *11*, 707. [CrossRef]
69. Huang, G.; Yu, F.; Pan, J.; Cheng, Z.; Wu, D.; Xue, W.; Song, T.; Li, S. Genesis of pyrite from the Mengqiguer uranium deposit in Ili Basin of Xinjiang and its implication for mineralization. *Geol. China* **2021**, *48*, 507–519.
70. Zhang, X.; Nie, F.; Xia, F.; Zhang, C.; Feng, Z.; Ullah, R.; Zhang, P. Provenance constraints on the Xishanyao Formation, southern Yili Basin, northwest China: Evidence from petrology, geochemistry, and detrital zircon U-Pb geochronology. *Can. J. Earth Sci.* **2018**, *55*, 1020–1035. [CrossRef]
71. Gao, B.; Sun, Z.X.; Wang, G.C. Geochemical characteristics of the Shihongtan uranium deposit in Turpan-Hami basin, West Clehina. *Geochim. Cosmochim. Acta* **2008**, *72*, A292.
72. Liu, H.; Ding, B.; Liu, Z.; Zhang, X.; Pan, C. Genesis of strong kaolinization in ore-bearing sandstone from Mengqiguer uranium deposit, Yili Basin, China. *Acta Mineral. Sin.* **2017**, *37*, 40–48.
73. Zhu, X.; Wang, Y.; Wang, Z.; Zhang, C.; Liu, J. Trace element geochemistry of sandstone-type uranium deposits in dongsheng area. *Geol. -Geochem.* **2003**, *31*, 39–45.
74. Li, R.; Liu, C.; Li, Z.; Liang, L. Geochemical Characteristics of Trace Elements of Sandstone-type Uranium Deposit in Huhe Area, Bayintala Depression. *Uranium. Geol.* **2015**, *31*, 438–444, 466.
75. Kong, L.; Su, M.; Mai, Z.; Li, H.; Diao, Z.; Xiong, Y.; Chen, D. Removal of uranium from aqueous solution by two-dimensional electrosorption reactor. *Environ. Technol. Innov.* **2017**, *8*, 57–63. [CrossRef]
76. Zhao, L.; Cai, C.; Jin, R.; Li, J.; Li, H.; Wei, J.; Guo, H.; Zhang, B. Mineralogical and geochemical evidence for biogenic and petroleum-related uranium mineralization in the Qianjiadian deposit, NE China. *Ore. Geol. Rev.* **2018**, *101*, 273–292. [CrossRef]
77. Zhu, Q.; Yu, R.; Feng, X.; Li, J.; Sima, X.; Tang, C.; Xu, Z.; Liu, X.; Si, Q.; Li, G.; et al. Mineralogy, geochemistry, and fluid action process of uranium deposits in the Zhiluo Formation, Ordos Basin, China. *Ore. Geol. Rev.* **2019**, *111*, 102984. [CrossRef]
78. Qiu, G. Discussion on radioactive environmental impact assessment for the decommission of uranium mining and milling facilities. *Uranium. Geol.* **2008**, *24*, 188–192.
79. Mao, Y.; Yong, J.; Liu, Q.; Wu, B.; Chen, H.; Hu, Y.; Feng, G. Heavy Metals/Metalloids in Soil of a Uranium Tailings Pond in Northwest China: Distribution and Relationship with Soil Physicochemical Properties and Radionuclides. *Sustainability* **2022**, *14*, 5315. [CrossRef]
80. Wang, Q.; Hu, N.; Ding, D.; Zhang, G.; Hu, E.; Zhang, H.; Yang, Y.; Jiang, X. Removal of NO_3~–N From Polluted Groundwater in Decommissioned Mining Area in an In-situ Leach Uranium Mine by Denitrifying Bacteria Bioreactor. *At. Energy Sci. Technol.* **2013**, *47*, 1300–1306.
81. Dong, Y.; Xie, Y.; Li, G.; Zhang, J. Efficient natural attenuation of acidic contaminants in a confined aquifer. *Environ. Earth Sci.* **2016**, *75*, 595. [CrossRef]
82. Wang, Z.; Qin, H.; Wang, J. Accumulation of uranium and heavy metals in the soil-plant system in Xiazhuang uranium ore field, Guangdong Province, China. *Environ. Geochem. Health* **2019**, *41*, 2413–2423. [CrossRef] [PubMed]
83. Yu, X.; Song, Y.; Wei, G.; Yu, G.; Gao, Y.; Xiao, F.; Ma, M. Pollution Characteristics of U and Th in Soil and Plant Screening in Light Rare Earth Tailings. *Environ. Sci. Technol.* **2020**, *43*, 26–30.
84. Li, M.; Gao, F.; Zhang, X.; Lv, S.; Huang, J.; Wu, X.; Fang, Q. Recovery of uranium from low-grade tailings by electro-assisted leaching. *J. Clean Prod.* **2020**, *271*, 122639. [CrossRef]
85. Brown, P.L.; Guerin, M.; Hankin, S.I.; Lowson, R.T. Uranium and other contaminant migration in groundwater at a tropical Australian Uranium Mine. *J. Contam. Hydrol.* **1998**, *35*, 295–303. [CrossRef]
86. Wu, Y.; Li, J.; Wang, Y.; Xie, X. Variations of uranium concentrations in a multi-aquifer system under the impact of surface water-groundwater interaction. *J. Contam. Hydrol.* **2018**, *211*, 65–76. [CrossRef]
87. Atkins, M.L.; Santos, I.R.; Perkins, A.; Maher, D.T. Dissolved radon and uranium in groundwater in a potential coal seam gas development region (Richmond River Catchment, Australia). *J. Environ. Radioact.* **2016**, *154*, 83–92. [CrossRef]
88. Shi, S.; Tang, X.; Yang, Y.; Liu, Z. Biological effects of uranium in water, soil and rice in uranium deposits in southern China. *J. Radioanal. Nucl. Ch.* **2021**, *328*, 507–517. [CrossRef]
89. Zhang, Y.; Ma, L.; Zhang, B.; Li, X. Discussion the influence of the in-suit leaching process on groundwater environment in a low permeability uranium deposit. *Uranium. Min. Metall.* **2017**, *36*, 75–86. [CrossRef]

90. Dong, H.; Li, X.; Zhang, B. Experimental study on CO_2 + O_2 in-suit leaching of uranium in a uranium mine in Inner Mongolia. *Uranium. Min. Metall.* **2019**, *38*, 8–13. [CrossRef]
91. Long, H.; Shen, H.; He, X.; Chen, L.; Zuo, W. Practice of groundwater control in acid in-suit leaching of uranium. *Uranium. Min. Metall.* **2017**, *36*, 58–65. [CrossRef]
92. Zhang, W.; Liu, H.; Fan, X.; Zhuo, Z.; Guo, Y. Removal of uranium from aqueous solution by a permeable reactive barrier loaded with hydroxyapatite-coated quartz sand: Implication for groundwater remediation. *Geochemistry* **2020**, *80*, 125545. [CrossRef]
93. Kirsch, K.; Navarre-Sitchler, A.K.; Wunsch, A.; McCray, J.E. Metal Release from Sandstones under Experimentally and Numerically Simulated CO2 Leakage Conditions. *Environ. Sci. Technol.* **2014**, *48*, 1436–1442. [CrossRef] [PubMed]
94. Youlton, B.J.; Kinnaird, J.A. Gangue-reagent interactions during acid leaching of uranium. *Min. Eng.* **2013**, *52*, 62–73. [CrossRef]
95. Tan, K.; Li, C.; Liu, J.; Qu, H.; Xia, L.; Hu, Y.; Li, Y. A novel method using a complex surfactant for in-situ leaching of low permeable sandstone uranium deposits. *Hydrometallurgy* **2014**, *150*, 99–106. [CrossRef]
96. Ji, H.; Zhou, Y.; Sun, Z.; Yang, Y. Analysis of and Discussion on CO_2 + O_2 In-situ Leaching of Uranium Process at Mengqiguer Deposit. *Nonferrous Met. (Extr. Metall.)* **2018**, *10*, 55–59.
97. Guo, Y.; Wei, W.; Cheng, L.; Li, J.; Han, X. U, Th and ^{226}Ra Vertical Migration in Sub-Clay From an Uranium Tailings Impoundment. *Radiat. Prot.* **2005**, *25*, 24–30.
98. Yao, F.; Xiaolei, J.; Liangshu, X. Contamination Assessment to a Uranium Mine in China Based on the Fault Tree Analysis. *J. Univ. Souht China (Sci. Technol.)* **2015**, *29*, 22–26. [CrossRef]
99. Tian, X.; Diao, C. Analysis of environmental impact about in-situ leaching uranium mines of Xinjiang and measures of pollution prevention. *Uranium. Min. Metall.* **2010**, *29*, 29–32. [CrossRef]
100. Zhong, J.; Zhang, Z.; Xia, M. Research on in-suit leaching technological mineralogy in Qian IV block of Qianjiadian uranium deposit in southern Songliao Basin. *Miner. Explor.* **2022**, *13*, 192–199. [CrossRef]
101. He, C.; Tan, K.; Li, Y.; Liu, Z.; Zhang, Q.; Li, D.; Liu, Z. Pollution characteristics and mechanism of groundwater of acid and CO2 In-situ leaching mining area of uranium deposit in Xinjiang, China. *Nonferrous Met. (Extr. Metall.)* **2021**, *6*, 53–59.
102. Zuo, W.; Tan, K. The characteristics of groundwater contamination of a decommissioned wellfield in some in suit leaching uranium min, Xinjiang, China. *J. Univ. South China (Sci. Technol.)* **2014**, *28*, 28–34. [CrossRef]
103. Zeng, S.; Shen, Y.; Sun, B.; Zhang, N.; Zhang, S.; Feng, S. Pore structure evolution characteristics of sandstone uranium ore during acid leaching. *Nucl. Eng. Technol.* **2021**, *53*, 4033–4041. [CrossRef]
104. Cheng, Y.; Hu, X.; Kong, Y.; Dong, Y.; Guo, J.; Wang, L.; Wang, J. Imaging acidic contaminants in a confined aquifer using electromagnetic geophysical method constrained by hydrochemical data. *J. Hydrol.* **2022**, *609*, 127704. [CrossRef]
105. Darby, S.C.; Whitley, E.; Howe, G.R.; Hutchings, S.J.; Kusiak, R.A.; Lubin, J.H.; Morrison, H.I.; Tirmarche, M.; Tomasek, L.; Radford, E.P.; et al. Radon and cancers other than lung-cancer in underground miners—A collaborative analysis of 11 studies. *Jnci.-J. Natl. Cancer I* **1995**, *87*, 378–384. [CrossRef]
106. Bernhard, S.; Legac, J.A.; Zettwoog, P.R.; Sequin, H. Occupational exposure to rn in non-uranium mines in the european-community. *Health Phys.* **1985**, *49*, 999–1002. [PubMed]
107. Lespukh, E.; Stegnar, P.; Yunusov, M.; Tilloboev, H.; Zyazev, G.; Kayukov, P.; Hosseini, A.; Stromman, G.; Salbu, B. Assessment of the radiological impact of gamma and radon dose rates at former U mining sites in Tajikistan. *J. Environ. Radioact.* **2013**, *126*, 147–155. [CrossRef]
108. Hu, L.; Guo, D.; Li, Z.; Li, Y.; Zhang, N. Design of cover system for decommissioning of uranium mill tailings pond. *Uranium. Min. Metall.* **2021**, *40*, 70–75. [CrossRef]
109. Jiang, X.; Yan, D.; He, Y.; Zhao, X.; Ding, K. Survey of radioactivity level of uranium in plants within uranium tailings waste backfilling area. *Radiat. Prot.* **2018**, *38*, 132–136.
110. Luo, Y.; Peng, Q.; Li, K.; Gong, Y.; Liu, Y.; Han, W. Patterns of nitrogen and phosphorus stoichiometry among leaf, stem and root of desert plants and responses to climate and soil factors in Xinjiang, China. *Catena* **2021**, *199*, 105100. [CrossRef]
111. Lu, L.; Mao, L.; Yang, T.; Ye, J.; Liu, B.; Li, H.; Sun, M.; Miller, J.T.; Mathews, S.; Hu, H.; et al. Evolutionary history of the angiosperm flora of China. *Nature* **2018**, *554*, 234. [CrossRef]
112. Feng, Y.; Hu, X.; Lu, L.; Zhou, W.; Zhang, Y. The Migration of U and Th in the Tailings-plants System. *Earth Environ.* **2014**, *42*, 201–206.
113. Yan, D.; Ding, K.; He, Y.; Fan, L.; Zuo, R.; Jiang, X. Uptake of uranium by plants growing around an old uranium mill tailing in northeast China. *Fresen. Environ. Bull* **2018**, *27*, 3102–3106.
114. Tang, Q.; Yang, J.; Tian, C.; Jin, R.; Li, J.; Ao, C.; Zhang, Z. In-situ leaching for sandstone type uranium deposit in Songliao basin and ecological conservation. *North China Geol.* **2016**, *39*, 136–139.
115. Long, B.; Lu, X.; Li, Y.; Tao, J. Decommissioning governance and significance of geological exploration engineering facilities of a military uranium mine in Heilongjiang Province. *Jilin Geol.* **2021**, *40*, 77–79.
116. Gavrilescu, M.; Pavel, L.V.; Cretescu, I. Characterization and remediation of soils contaminated with uranium. *J. Hazard Mater.* **2009**, *163*, 475–510. [CrossRef]
117. Lovley, D.R.; Phillips, E.; Gorby, Y.A.; Landa, E.R. Microbial Reduction of Uranium. *Nature* **1991**, *350*, 413–416. [CrossRef]
118. Borch, T.; Kretzschmar, R.; Kappler, A.; Van Cappellen, P.; Ginder-Vogel, M.; Voegelin, A.; Campbell, K. Biogeochemical Redox Processes and their Impact on Contaminant Dynamics. *Environ. Sci. Technol.* **2010**, *44*, 15–23. [CrossRef]

119. Wu, W.; Carley, J.; Gentry, T.; Ginder-Vogel, M.A.; Fienen, M.; Mehlhorn, T.; Yan, H.; Caroll, S.; Pace, M.N.; Nyman, J.; et al. Pilot-scale in situ bioremedation of uranium in a highly contaminated aquifer. 2. Reduction of U(VI) and geochemical control of U(VI) bioavailability. *Environ. Sci. Technol.* **2006**, *40*, 3986–3995. [CrossRef]
120. Xiao, J.; Pang, Z.; Zhou, S.; Chu, L.; Rong, L.; Liu, Y.; Li, J.; Tian, L. The mechanism of acid-washed zero-valent iron/activated carbon as permeable reactive barrier enhanced electrokinetic remediation of uranium-contaminated soil. *Sep. Purif. Technol.* **2020**, *244*, 116667. [CrossRef]

water

MDPI

Article

Investigating the Change Pattern in Adsorption Properties of Soil Media for Non-Polar Organic Contaminants under the Impact of Freezing and Thawing

Jingjing Huang [1,2,3], Rong Zhong [1,2,3] and Hang Lyu [1,2,3,*]

1 Key Laboratory of Groundwater Resources and Environment, Ministry of Education, Jilin University, Changchun 130026, China; huangjj9920@mails.jlu.edu.cn (J.H.); zhongrong21@mails.jlu.edu.cn (R.Z.)
2 Jilin Provincial Key Laboratory of Water Resources and Environment, Jilin University, Changchun 130026, China
3 Institute of Water Resources and Environment, Jilin University, Changchun 130021, China
* Correspondence: lvhangmail@163.com; Tel.: +137-5602-9934

Abstract: The adsorption of petroleum hydrocarbons by soils in the unsaturated zone determines the amount that goes into the groundwater. However, the intricate behavior of petroleum hydrocarbon adsorption in soil media under the influence of freeze–thaw conditions in globally prevalent seasonally frozen regions remains unclear. Alkanes as a non-polar compound are an important part of petroleum hydrocarbons. We conducted field-scale seasonal freeze-thaw experiments using n-dodecane to quantify the dynamic patterns and influencing factors of the physicochemical properties of soil media and their adsorption capacity for petroleum hydrocarbons during different freeze–thaw cycles. Our findings demonstrated that, as the number of natural freeze–thaw cycles increased, the proportion of soil micro-agglomerates rose rapidly, thereby expanding the available adsorption sites and enhancing the adsorption capacity for non-polar organic pollutants. The rise in sorption capacity for the outdoor freeze–thaw experimental group surpassed that of the indoor room-temperature control group by an impressive 75.57%, showing the enhancement of the adsorption capacity for non-polar organic pollutants. Conversely, the decline in soil organic matter content during the later stages of the freeze–thaw process hampered its adsorption performance for non-polar organic pollutants. The decrease in sorption capacity for the outdoor freeze–thaw experimental group surpassed that of the indoor room temperature control group by 77.97%. By shedding light on the adsorption mechanisms of non-polar organic pollutants in soils subjected to freeze–thaw conditions, our research facilitated a comprehensive understanding and predictive modeling of this process. Furthermore, our study provided a scientific foundation for exploring the convergence and migration transformation patterns of other organic compounds in petroleum-contaminated areas within seasonally frozen regions.

Keywords: freezing and thawing; non-polar organic contaminants; adsorption; soil physicochemical properties

Citation: Huang, J.; Zhong, R.; Lyu, H. Investigating the Change Pattern in Adsorption Properties of Soil Media for Non-Polar Organic Contaminants under the Impact of Freezing and Thawing. *Water* **2023**, *15*, 2515. https://doi.org/10.3390/w15142515

Academic Editors: Maurizio Barbieri, Yuanzheng Zhai and Jin Wu

Received: 2 June 2023
Revised: 28 June 2023
Accepted: 5 July 2023
Published: 9 July 2023

1. Introduction

Petroleum, as a versatile raw material for numerous industries, currently stands as one of the foremost energy sources [1]. However, the irrational extraction and utilization of petroleum can lead to inadvertent leaks and releases, permeating the underground soil and water environment via urban pipelines, underground storage tanks, and other pathways [2]. Extensive research has demonstrated that adsorption plays a pivotal role in the entrapment of petroleum hydrocarbons within the soil, significantly influencing their transport dynamics [3,4]. This phenomenon greatly determines the sequestration process of contaminants in the unsaturated zone, as well as the threat posed to groundwater resources. For instance, the Golmud region in China exhibits elevated concentrations of petroleum hydrocarbons in its groundwater. Through the exclusion of human activities as a factor, it has been observed that the aquifer lithology in this region is characterized by loose, coarse-grained, and highly permeable characteristics, which can impede the adsorption of petroleum hydrocarbons and render them more prone to enter the groundwater system [5]. Petroleum hydrocarbons are categorized as alkanes, cycloalkanes, and aromatics. Alkanes, as non-polar compounds, are an important part of petroleum hydrocarbons [6]. The results of our testing and analysis of the composition and content of petroleum hydrocarbon contamination at the site found that alkanes accounted for the highest percentage of total petroleum hydrocarbons in groundwater, with an average of 65.6% of the total oil percentage, specifically including dodecane and pentadecane. Therefore, studying the adsorption processes of non-polar organic contaminants within the subsurface unsaturated zone can be of paramount importance in safeguarding groundwater environments and effectively managing petroleum pollution.

The adsorption mechanisms of organic pollutants in the soil encompass a range of phenomena, such as electrostatic interactions, hydrogen bonding, hydrophobic bonding, van der Waals forces, donor-acceptor mechanisms, and ligand exchange, etc. These mechanisms synergistically operate and compete, ultimately determining the fate of organic pollutant adsorption in soil [7]. Non-polar organic contaminants, classified as hydrophobic organic chemicals, are subject to the influence of both organic matter content and environmental factors present within the soil [8–14]. Soil organic matter, characterized by a heterogeneous mixture of diverse organic compounds consisting of hydrophilic and hydrophobic constituents, exerts a significant impact on the adsorption behavior of non-polar organic contaminants. The hydrophobic effect induces the formation of complexes between organic matter and hydrogen bonding, thus increasing the availability of adsorption sites within the soil. Consequently, as the hydrophobicity of organic matter intensifies, the soil exhibits a heightened capacity for adsorbing non-polar organic contaminants [15]. The pH levels also exert an influence on the charge of soil colloids and the distribution patterns of petroleum contaminants. Studies have revealed that, in acidic solutions, contaminants can establish hydrogen bonds with carboxyl and phenolic hydroxyl groups present in organic matter, thereby facilitating the adsorption process [7,16]. Furthermore, the stability of soil aggregates plays a crucial role in retaining soil organic matter. The enhanced stability of soil aggregates correlates with higher levels of organic matter content within the soil [17]. Overall, variations in soil pH and aggregate stability can impact the organic matter content of the soil, which can subsequently influence its affinity for organic pollutants, thereby modulating the adsorption properties of the soil medium.

Seasonal permafrost is extensively distributed worldwide, exhibiting an average of 50.5% of the total land area in the Northern Hemisphere, as well as reaching 81% during extremely cold years [18,19]. Freeze–thaw cycles, as abiotic stressors impacting the soil, possess the capacity to basically alter soil states to influence soil structure and water distribution characteristics, which can subsequently affect the sorption processes of contaminants. Although current research primarily focuses on the adsorption processes under freeze–thaw conditions for heavy metals, studies involving organic pollutants are relatively limited. Dang et al. [20] discovered that freeze–thaw cycles can significantly impact dissolved organic matter, agglomerates, and pore ratios in soils, leading to reduced sorption capacity for Cd, thereby promoting vertical migration and morphological transformations. Wang, X et al. [21] observed through laboratory batch experiments that pH, organic matter content, and free iron oxide content exhibited negative correlations with freeze-thaw frequency, resulting in diminished sorption capacity for soil Pb and Cd compared to unfrozen soil. This can consequently raise the risk of heavy metal contamination in groundwater. Yang et al. [22] revealed that the number and duration of freeze–thaw cycles influenced the strength properties and chemical stability of solidified/stabilized lead-contaminated soil. The freezing and thawing processes can induce alterations in the physicochemical properties of soil media. Existing research on the transport of organic pollutants under freezing and thawing conditions primarily concentrates on indoor experiments [23–25]. However, there remains a relative dearth of research examining the effects of freeze–thaw action on soil adsorption of organic pollutants under natural conditions. Neglecting the influence of freezing and thawing on the organic pollutant adsorption process may significantly cause the predicted result range of soil groundwater organic pollution to deviate from the actual scenario, substantially impacting the risk assessment and remediation strategies for organic pollutants contamination in groundwater.

Building upon this foundation, this study delved into the changes in the adsorption properties of soil media regarding non-polar organic pollutants under the influence of the natural freeze–thaw processes. Through the execution of field-based freeze–thaw experiments within a seasonally frozen region, we conducted adsorption tests on soil samples across various freeze–thaw stages while also assessing the physical and chemical property indices of the soil. The research objectives were to comprehensively characterize the alterations in non-polar organic pollutant adsorption properties within soils at different freeze–thaw stages under natural conditions and to discern the key factors influencing the adsorption capacity of soils for non-polar organic pollutants in the context of freeze–thaw action.

2. Experimental Materials and Methods

2.1. Materials

To better simulate the change pattern of non-polar organic pollutants adsorption under real soil conditions, the unpolluted areas at the designated field test site, characterized by a relatively flat and stable terrain with flourishing vegetation, were selected for conducting the random sampling process. Considering that the adsorption of non-polar organic contaminants by soil occurs mostly on the soil surface, three sampling points were randomly identified with the selection of surface soil samples from a depth of 0 to 20 cm. Subsequently, the soil samples obtained from different sampling points were thoroughly mixed. These samples were then transported to a shaded indoor laboratory for the air-drying process. During this process, visible debris, such as fallen leaves and roots, were meticulously removed from the soil. The dried soil was sieved through a 2-mm mesh and stored in sealed plastic bags for subsequent analysis. Because the highest percentage of alkanes in the total petroleum hydrocarbons in the groundwater of the sampling site, accounting for 65.6% of the total oil percentage on average, specifically including dodecane and pentadecane, the study selected dodecane as the typical alkane pollutant of petroleum hydrocarbon at the site. For the experiments, methanol (Tianjin Fuyu Fine Chemical Co., Ltd., Tinjin, China), n-dodecane (Tianjin Kemiou Chemical Reagent Co., Ltd., Tianjin, China), and dichloromethane (Tianjin Fuyu Fine Chemical Co., Ltd., Tinjin, Chian) of analytical purity were employed as reagents, whereby n-dodecane was dissolved in methanol to prepare the stock solution, and di-chloromethane was used to extract n-dodecane from the aqueous solution, while the water used was all ultrapure water.

2.2. Experimental Design

The experimental setup consisted of two distinct components, including an outdoor experimental group and an indoor control group, each comprising five soil samples. The outdoor group aimed to observe the dynamic changes in soil properties and its adsorption capacity for pollutants over the course of the winter season under natural conditions. In contrast, the indoor control group was established to isolate the effects of non-freezing and thawing conditions. Uniform filling methods were employed for both the outdoor and indoor soil columns, with each column containing 500 g of dry soil (including the amount used for parallel indicator samples). The soil columns were filled in accordance with the volume corresponding to the desired moisture content of 15%. Layered additions of soil and water were created during the filling process, followed by a 24-h equilibration period at room temperature to ensure uniform moisture distribution. In total, 11 soil columns were filled, with one column designated for the determination of initial values; as well, the remaining columns were divided into two groups. For the latter two-group soil columns, five of them for the outdoor group were deployed at the field test site on 19 November 2022 (at the beginning of November, the temperature at the test site gradually dropped to below zero, and the samples were made and placed on the site on 19 November), while the remaining five columns were placed in the indoor laboratory as control samples. Probes were utilized to monitor both site temperature and soil temperature changes, and the sampling interval was adjusted accordingly. Probes were utilized to monitor both site temperature and soil temperature changes, and the sampling interval was adjusted accordingly. The first sample was collected on 29 November 2022, followed by subsequent samples on 9 December 2022, 9 January 2023, 9 February 2023, and 12 March 2023. To facilitate the comparison of the results of different groups, the outdoor samples were labeled as K1–K5, whereas the indoor samples were labeled as D1–D5. Following retrieval, the samples were dried in shaded conditions in preparation for the subsequent processing. The samples after retrieval at different time points needed to be dried indoors for subsequent analytical processing.

2.3. *Measurement of Sample Indicators*

2.3.1. Adsorption Kinetics Parameters

The equilibrium adsorption quantities and corresponding adsorption rates of non-polar organic contaminants adsorbed by soil were determined through adsorption kinetic experiments. These experiments aimed to analyze the process of non-polar organic contaminant adsorption by soil under both freeze–thaw conditions and indoor ambient conditions. To ensure the stability of n-dodecane in the solvent, methanol was selected as the solvent to prepare a stock solution of n-dodecane with a concentration of 20 g/L. Additionally, a 0.01 mol/L $CaCl_2$ solution was prepared to establish ionic equilibrium in the soil prior to commencing the adsorption experiments. Based on preliminary experiments, a soil-to-water ratio of 1:25 was determined as the appropriate ratio. For each experimental run, 2 g of treated soil were weighed into a series of 50 mL centrifuge tubes, followed by the addition of 50 mL of the prepared $CaCl_2$ solution to minimize headspace. The tubes were then placed on an oscillator and shaken overnight to allow the soil to achieve ionic equilibrium prior to the initiation of adsorption. Subsequently, the n-dodecane stock solution was added to achieve an initial concentration of 0.2 g/L in each centrifuge tube; after that, they were put in the oscillator again. In sampling after a specific interval (5, 15, 35, 65, 125, 275, 515, 1440, and 2880 min), the supernatant of the centrifuged sample was extracted and dichloromethane was added to extract the n-dodecane from the supernatant, during which the sample was fully extracted using ultrasound. The resulting samples were transferred to 1.5 mL agilent vials, and the concentration of n-dodecane was determined using gas chromatography (Agilent 7890B/5977B). Parallel experiments were conducted under identical conditions for comparison and validation.

2.3.2. Physical and Chemical Properties of Media

To further analyze the pattern of variations in the adsorption performance, several parameters, including pH, agglomerate composition, and the soil organic carbon (SOC) index, were measured in the soil samples from both the outdoor and control groups. A 10 g soil sample was weighed in a beaker, and 25 mL of ultrapure water were added. The mixture was vigorously stirred for 2 min and left undisturbed for 30 min. Within 1 h, the pH of the solution was measured using a pH meter (Thunder PHSJ 6L). To assess the changes in soil agglomerates during the winter season, the soil was divided into three particle-size ranges: 2–0.3 mm, 0.3–0.25 mm, and <0.25 mm. The agglomerates of each particle-size range were individually weighed (Sieve Shaker, POWTEQ SS2000). The SOC content was determined using the Elementar vario MACRO cube (German), which provided an accurate measurement of soil organic carbon.

2.4. Data Analysis

To examine the variations in the sorption process of non-polar organic contaminants by soil across various sampling nodes and under varying conditions, the sorption quantity was employed to represent the sorption at a specific time node. In addition, the sorption kinetic equation was utilized to analyze the sorption characteristics of organic pollutants. The intraparticle diffusion adsorption kinetics model is one of the most widely used adsorption kinetics models (also cited as the "Weber–Morris model" and the "IPD model" in some published papers) [17,26,27]. Furthermore, the mean weight diameter (MWD) of soil agglomerates pertaining to various particle sizes served as an indicator of soil agglomerate stability. The WDM was calculated by referring to the method of Yonker et al. [28]. The calculation of the adsorption kinetics parameters and MWD were conducted using the following formula:

Adsorption amount:

$$q_c = \frac{(C_0 - C_t)V}{M} \tag{1}$$

pseudo-first-order kinetics:

$$lg(q_e - q_c) = lgq_e - \frac{k_1}{2.303}t \tag{2}$$

pseudo-second-order kinetics:

$$\frac{t}{q_c} = \frac{1}{k_2 q_e^2} + \frac{1}{q_e}t \tag{3}$$

intraparticle diffusion model:

$$q_c = k_{id}t^{1/2} + C \tag{4}$$

where C_0 represents the initial concentration of n-dodecane (g/L); C_t represents the concentration of the n-dodecane at moment t (g/L); V denotes the volume of the solution (L); M represents the mass of the soil (kg); t represents the time (min); q_c represents the amount of n-dodecane adsorbed at moment t (g/kg); q_e represents the amount of n-dodecane adsorbed on the soil surface at equilibrium, i.e., the equilibrium adsorption capacity (g/kg); k_1 represents the rate constant of pseudo-first-order sorption (min^{-1}); k_2 represents the rate constant of pseudo-second-order sorption (kg·g^{-1}·min^{-1}); k_{id} represents the intraparticle diffusion rate constant g/(kg·min$^{-1/2}$); and C represents the intercept.

MWD:

$$\text{MWD} = \sum\nolimits_1^n (x_i \times W_i) \tag{5}$$

where x_i, W_i, and n denote the average diameter of soil agglomerates (mm), the proportion of each particle-size agglomerate to the total sample, and the number of agglomerate components, respectively.

3. Results and Discussion

3.1. Patterns of Air and Soil Temperature Changes during the Freeze–Thaw Period

The air temperature and surface soil temperature exhibited typical seasonal dynamics at the field experiment site spanning from 1 November 2022 to 15 March 2023 (Figure 1). Notably, temperatures started to decrease significantly on November 11, reaching sub-zero levels on 25 November, as well as with the lowest temperature recorded on 24 January 2023. Temperature fluctuations persisted until March 2023, when temperatures began to rise above zero again. The temperature fluctuations observed in different soil depths were generally consistent, with the surface layer (0.2 cm) displaying significantly smaller temperature variations than the overall temperature changes. The soil temperature decreased to 0 °C for the first time on 5 December, with a 10-day delay compared to the air temperature. The soil temperature at the depth of 10 cm was 3–4 °C higher than that at the depth of 0.2 cm, indicating a gradual decline in heat transfer between the atmosphere and the soil. Although the soil temperature was influenced by the air temperature, there existed a certain lag and diminishing response of soil temperature to changes in air temperature. Freezing was considered to occur from the surface downwards, and the adsorption of surface non-polar organic contaminants in the soil primarily occurred in the soil surface layer. Based on the 0.2 cm soil temperature, the winter–spring phases of freeze–thaw cycles in 2022–2023 were categorized as follows: the first freeze-thaw cycle period from 5 December to 14 December 2022; the freezing period from 15 December 2022 to 27 February 2023; and the second freeze–thaw cycle period from 28 February to 5 March.

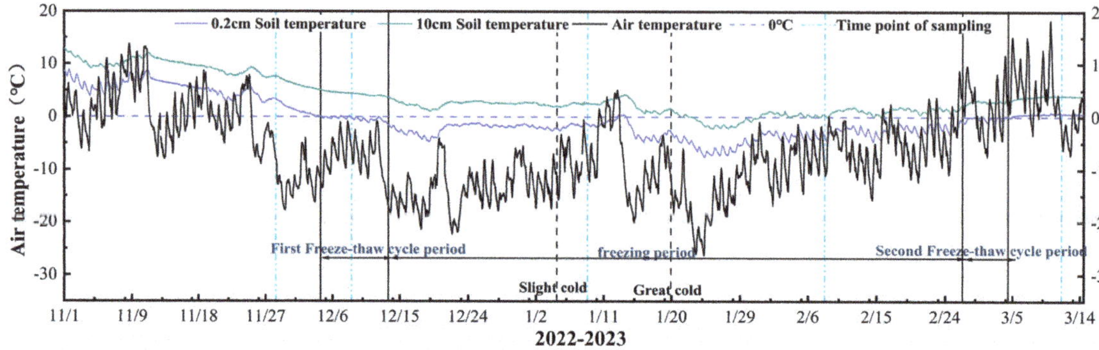

Figure 1. Schematic diagram of air temperature and soil temperature (2022–2023).

During the first freeze–thaw cycle, the soil surface temperature exhibited constant fluctuations around 0 °C. This freeze–thaw period spanned 9 days, with a temperature range between −0.6 °C and 0.2 °C. Overall, four freeze–thaw cycle processes occurred. Upon entering the subsequent freezing period, the soil temperature remained below 0 °C. However, there was an exceptional weather event during the period between Lesser Cold (5 January) and Greater Cold (20 January), when the temperature suddenly rose above 0 °C. Consequently, the 0.2 cm soil temperature also increased above zero. However, this occurrence lasted only 2 days, without subsequent fluctuations above or below 0 °C. Therefore, the impact of temperature changes on the soil during this period was disregarded. Following a soil freezing period of 54 days, the temperature gradually rose from sub-zero levels to above zero, signifying the entering of the second freeze–thaw cycle period. At this stage, a total of five freeze–thaw cycles were observed, spanning six days, with a freeze–thaw temperature range of −0.9 °C–0.5 °C.

3.2. *Changes in Sorption Capacity of Non-Polar Organic Contaminants in Soils during Freeze–Thaw Conditions*

Both the outdoor experimental samples and the indoor control samples reached adsorption equilibrium within 1500 min at different times during the freeze–thaw period. The measured equilibrium sorption amounts for each month, from 29 November 2022 to 12 March 2023, were 973.1, 996.1, 992.5, 996.0, and 920.0 g/kg for the outdoor experimental group, as well as 976.9, 990.0, 995.3, 990.0, and 931.0 g/kg for the indoor ambient control group during the same period, respectively. In general, the equilibrium sorption capacities of both the outdoor experimental group and the indoor control group exhibited a decreasing trend over time, with the final equilibrium sorption capacity of the outdoor experimental group being lower than that of the indoor control group (Figure 2). This indicates that freeze–thaw action can affect the non-polar organic contaminants' sorption performance of the soil media. To explore the adsorption mechanism of non-polar organic contaminants adsorbed by soils under freeze–thaw conditions, we employed the pseudo-first order kinetics model and pseudo-second order kinetics model. The correlation between $lg(q_e q_t)$ and t for both the outdoor freeze–thaw test group and the indoor room temperature control group did not conform to a linear relationship. Therefore, it was concluded that the process of soil adsorption of non-polar organic contaminants did not follow a pseudo-first-order kinetics model, but rather a pseudo-second-order kinetics model (Table 1). The pseudo-first-order kinetics model involved a physical process as the rate-limiting step, while the pseudo-second-order kinetics model relied on chemical processes [29]. Organic pollutants underwent transfer, exchange, or electron sharing with molecules or atoms on the soil surface, leading to the formation of adsorption chemical bonds. To further understand the key processes controlling the adsorption of non-polar organic contaminants from soils, we investigated the relationship between q_c and $t^{1/2}$ based on the theoretical intraparticle diffusion model proposed by Weber and Morris [17]. If the graph of $t^{1/2}$-q_c followed a curve passing through the origin, it indicated that soil adsorption of non-polar organic contaminants was primarily accomplished by intra-particle diffusion, where the rate of inter-particle diffusion affected the rate of soil adsorption of non-polar organic contaminants [30,31]. The change patterns of the inter-particle diffusion model plots for both the outdoor experimental group and the indoor control group were consistent (Figure 3). These plots were not linear over time while they exhibited two linear straight lines, which suggested that the adsorption process of non-polar organic contaminants in soil during the freeze–thaw period can be divided into two steps: surface adsorption and intra-particle diffusion. Initially, non-polar organic contaminants were adsorbed on the outer surface of the soil, and when the outer surface became saturated, the hydrocarbons penetrated the inner soil particles and were adsorbed by the inner surface [32]. Hence, the adsorption rate of soil non-polar organic contaminants was influenced by both the adsorption rate on the soil surface and the intra-particle adsorption rate. Soil adsorption of non-polar organic contaminants was a chemisorption process involving adsorption on the soil surface and diffusion adsorption within the particles, where molecules or atoms on the soil surface and within the particles formed chemisorption bonds with non-polar organic contaminants.

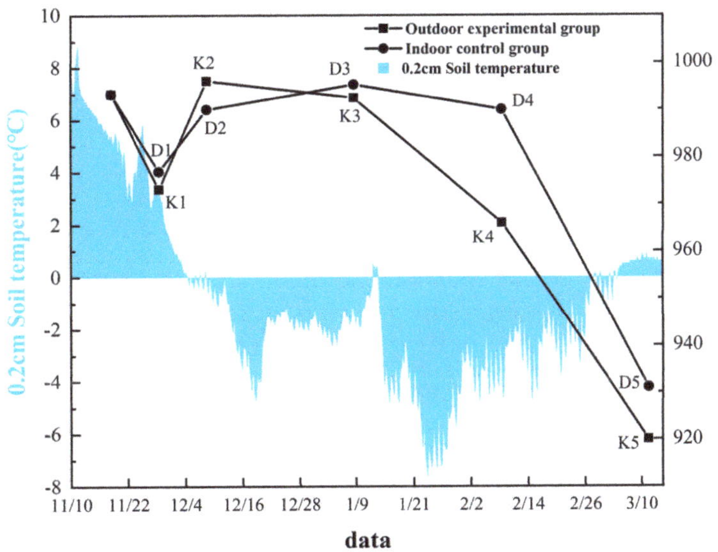

Figure 2. Time-dependent equilibrium adsorption amount.

Table 1. Kinetic constants and correlation coefficients for kinetic models.

Pseudo-First Order										Pseudo-Second Order									
Outdoor Experimental Group					Indoor Control Group					Outdoor Experimental Group					Indoor Control Group				
Time	$q_{e,exp}$	$q_{e,cal}$	k_2	R^2	Time	$q_{e,exp}$	$q_{e,cal}$	k_2	R^2	Time	$q_{e,exp}$	$q_{e,cal}$	k_2	R^2	Time	$q_{e,exp}$	$q_{e,cal}$	k_2	R^2
initial	993.30	63.35	2.50×10^{-3}	0.39982	initial	993.30	63.35	2.50×10^{-3}	0.39982	initial	993.30	1000.00	1.85×10^{-4}	0.99997	initial	993.30	1000.00	1.85×10^{-4}	0.99990
K1	973.10	144.66	1.02×10^{-5}	0.23509	D1	976.90	146.08	9.42×10^{-6}	0.13102	K1	973.10	961.54	5.85×10^{-5}	0.99883	D1	976.90	990.10	1.63×10^{-4}	0.99990
K2	996.10	178.62	3.20×10^{-3}	0.82961	D2	990.00	139.88	2.70×10^{-3}	0.76586	K2	996.10	1000.00	1.25×10^{-4}	0.99998	D2	990.00	961.54	7.69×10^{-5}	0.99980
K3	992.50	221.89	3.70×10^{-3}	0.85827	D3	995.30	143.41	3.30×10^{-3}	0.78954	K3	992.50	1000.00	7.58×10^{-5}	0.99989	D3	995.30	1000.00	1.40×10^{-4}	0.99990
K4	966.00	269.26	4.40×10^{-3}	0.85287	D4	990.00	140.20	2.50×10^{-3}	0.54376	K4	966.00	970.87	9.71×10^{-5}	0.99997	D4	990.00	990.10	1.07×10^{-4}	0.99990
K5	920.00	267.33	2.90×10^{-3}	0.81663	D5	931.00	459.07	2.50×10^{-3}	0.84052	K5	920.00	925.93	9.36×10^{-5}	0.99965	D5	931.00	943.40	2.29×10^{-4}	0.99640

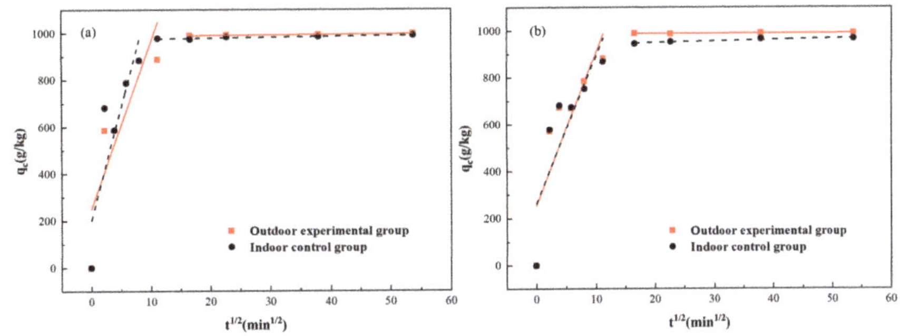

Figure 3. Intraparticle diffusion model (**a**) 9 December 2022; (**b**) 9 February 2023.

Prior to entering the freeze–thaw period, the equilibrium sorption capacity of both the outdoor experimental group and the indoor ambient group exhibited a decline with the continuous decrease in soil temperature. However, it is noteworthy that the equilibrium adsorption capacity of D1 (29th November) under indoor room temperature conditions only decreased by 16.4 g/kg compared to the initial value. In contrast, K1 under outdoor freeze–thaw conditions experienced a decrease of 20.2 g/kg compared to the initial value during the same period, representing a 23.17% larger decrease than that observed for D1. As the outdoor temperature gradually decreased prior to the freeze–thaw period, the soil underwent a chemisorption process to adsorb non-polar organic contaminants. The lowered temperature resulted in a decrease in energy and a slowdown in the movement of non-polar organic contaminant molecules. Consequently, the adsorption between non-polar organic contaminants and soil was weakened, as evidenced by the 23.17% reduction in the equilibrium adsorption capacity of the outdoor experimental group compared to the indoor room temperature group.

During the initial freeze–thaw cycle, K2 (9th December) exposed to outdoor freeze–thaw conditions experienced three consecutive freeze–thaw cycles, exhibiting an equilibrium sorption capacity of 996.1 g/kg. Conversely, D2, subjected to indoor ambient conditions during the same period, demonstrated a sorption capacity of 990.0 g/kg. Remarkably, the rise in sorption capacity for K2 surpassed that of the D2 group by an impressive 75.57%. Throughout the first freeze–thaw cycle, soil temperature fluctuated intermittently above and below the freezing point (0 °C). This dynamic temperature variation induced continuous transitions between solid and liquid states of soil moisture, consequently disrupting the structure of soil agglomerates. The resulting fragmentation led to the formation of smaller agglomerates, which in turn increased the availability of adsorption sites on the soil surface. Consequently, this facilitated the enhanced adsorption of non-polar organic contaminants by the soil.

Following the freezing period and two subsequent freeze–thaw cycle phases, K5 (12th March) exposed to outdoor freeze–thaw conditions underwent a total of 9 freeze–thaw cycles, ultimately attaining a final equilibrium adsorption capacity of 920.0 g/kg. In contrast, D5, placed under indoor room temperature conditions during the same period, achieved a final equilibrium adsorption capacity of 931.0 g/kg. Notably, the prolonged freezing and freeze–thawing of K5 resulted in a significantly lower equilibrium sorption capacity than D5. The extended exposure to a low-temperature environment with multiple freeze–thaw cycles induced alterations in various physicochemical properties of the soil, resulting in a diminished adsorption capability for non-polar organic contaminants. When the soil was frozen, its hydraulic conductivity experienced a substantial reduction, leading to the formation of an impermeable ice layer. This impeded the diffusion of organic pollutants within the soil and restricted their interaction with the soil matrix, thereby diminishing the sorption of non-polar organic contaminants in the outdoor experimental group during the freezing period. Furthermore, the occurrence of multiple freeze–thaw cycles can disrupt the soil structure, resulting in reduced soil stability and a decline in organic matter content. These factors collectively contributed to a decrease in adsorption capacity. The formation of an ice barrier during the freezing period, coupled with the decrease in soil organic matter content due to multiple freeze–thaw cycles, synergistically reduced the sorption properties of the soil matrix towards non-polar organic contaminants under final outdoor freeze–thaw conditions.

3.3. Analysis of Factors Influencing Sorption Capacity of Soils for Non-Polar Organic Contaminants during the Freeze–Thaw Period

In the conducted experiment, the pH fluctuation range observed under outdoor freeze–thaw conditions was narrower compared to indoor room temperature conditions. Notably, the final pH of K5 (12th March) exposed to outdoor freeze–thaw conditions was lower than that of D5 subjected to indoor room temperature conditions (Figure 4b). The agglomerates present in the soil were classified into three categories based on particle size: large agglomerates (2–0.3 mm), small agglomerates (0.3–0.25 mm), and micro-agglomerates (less than 0.25 mm). Throughout the experiment, there was a consistent overall increase in the percentage of micro-agglomerates under outdoor freeze–thaw conditions, while the percentage of large agglomerates exhibited a slight decrease (Figure 4a). Specifically, after the first freeze–thaw cycle period, the percentage of micro-agglomerates reached its peak value (65.66%) at K2 (9th December) under outdoor freeze–thaw conditions, whereas the percentage of large agglomerates dropped to its minimum value (14.41%). Moreover, the WMD of the agglomerates also reached its lowest value (30.26 mm) and subsequently fluctuated upwards. The overall SOC content under outdoor freeze–thaw conditions was lower compared to that of indoor room temperature conditions, primarily due to the destabilization of soil agglomerates. Specifically, the final SOC content of K5 (12th March) exposed to outdoor freeze–thaw conditions was 48.47 mg/kg, which was lower than the SOC content of D5 (51.80 mg/kg) subjected to indoor room temperature conditions during the same period.

During the initial freeze–thaw cycle, K2 subjected to outdoor freeze–thaw conditions underwent three consecutive freeze–thaw cycles, resulting in an increase in the proportion of small and micro-agglomerates within the soil. This led to a peak adsorption capacity of non-polar organic contaminants at 996.1 g/kg. Concurrently, the pH of the soil rose from 7.83 to 7.92, while the SOC content decreased from 47.73 mg/kg to 43.23 mg/kg. The process of freeze–thawing promoted the preferential adsorption of H^+ ions by organic matter, thereby contributing to an elevation in soil pH. Moreover, the freeze–thaw cycle induced the continuous transition of soil water between liquid and solid states, ultimately resulting in the disruption of soil aggregate structure. This led to the fragmentation of large aggregates into micro-aggregates and the subsequent loss of cementation between agglomerate particles, thereby reducing their stability [33]. Since SOC can be predominantly stored within agglomerates, the decrease in agglomerate stability led to a decline in SOC content [34]. On December 9th, the percentage of micro-agglomerates in K2 reached 65.66%, while large agglomerates accounted for only 14.41% under outdoor freeze–thaw conditions. Similarly, under indoor room temperature conditions, the percentage of micro-agglomerates in D3 was 57.44%, with large agglomerates comprising 20.13%. It can be evident that, following the first freeze–thaw cycle period, the increase in pH and decrease in SOC content under outdoor freeze–thaw conditions mitigated the organic matter's ability to adsorb non-polar organic contaminants. However, the rise in the proportion of micro-agglomerates within the soil provided additional adsorption sites for non-polar organic contaminants, collectively enhancing the soil's capacity to adsorb these hydrocarbons.

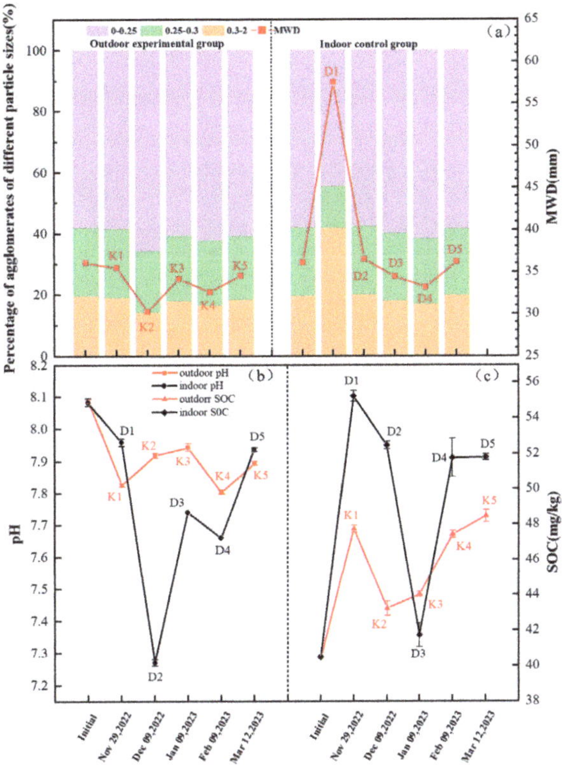

Figure 4. Changes in physical and chemical properties of soil indoors and outdoors. (**a**) Indoor and outdoor changes in the ratio of different particle-size agglomerates to MWD. (**b**) Indoor and outdoor soil pH changes. (**c**) Indoor and outdoor SOM content changes.

Following the completion of the second freeze–thaw cycle period, K5 (12 March) exposed to outdoor freeze–thaw conditions experienced a total of 9 freeze–thaw cycles. It was observed that the soil agglomerate stability remained lower compared to D5 subjected to indoor ambient conditions during the same period, indicating that multiple freeze–thaw cycles reduced agglomerate stability. Concurrently, the pH and SOC contents of the final K5 were lower than those of D5 under room temperature conditions in the corresponding period. However, the percentage of agglomerates did not exhibit a significant difference between the two groups. The freeze–thaw cycle disrupted soil aggregates, releasing dissolved organic matter and soil colloids. Additionally, it stimulated low-temperature stress responses in soil microorganisms, promoting nitrification reactions and the release of dissolved organic acids. This, in turn, increased H+ concentration in the soil solution [35]. The decrease in soil pH enhanced the adsorption of soil organic carbon to hydrophobic organic matter. This occurred due to the dissociation of carboxyl and hydroxyl groups of organic matter under alkaline conditions, resulting in an increase in net negative charge and enhancing the hydrophilicity of soil organic matter. Consequently, the affinity of soil organic matter towards hydrophobic organic matter was reduced [13,36,37]. The SOC, a crucial component of soil organic matter (SOM), declined because of freeze–thaw cycles. The SOM was recognized as a vital adsorption medium for organic compounds in soil [38,39]. The lower the organic matter content in the soil, the weaker the adsorption properties of the soil towards non-polar organic contaminants. Although the proportion of agglomerates of different particle sizes did not differ significantly, the decrease in pH caused by freeze–thaw action can enhance the adsorption affinity of soil organic matter towards non-polar organic contaminants. At the same time, the reduction in soil stability and organic carbon content led to a decrease in SOM, consequently resulting in a decline in the soil's adsorption capacity for non-polar organic contaminants. Hence, after the completion of two freeze–thaw cycle periods, the equilibrium adsorption capacity of the outdoor experimental group was significantly lower than that of the indoor room temperature group.

In summary, the freeze–thaw cycle exerted an impact on the distribution of agglomerates of various particle sizes within the soil. It induced the fragmentation of large agglomerates and consequently increased the proportion of micro-agglomerates. This, in turn, enhanced the number of adsorption sites for non-polar organic contaminants in the soil, thereby improving the overall adsorption performance of the soil media for these hydrocarbons. Additionally, the freeze–thaw process influenced the pH levels in the soil. The breakup of agglomerates during freeze–thaw released organic acids, leading to an elevation in H^+ concentration in the soil. As a result, the adsorption affinity of organic matter towards non-polar organic contaminants was enhanced. However, the freeze–thaw cycle destabilizes soil agglomerates, leading to a reduction in organic matter content and subsequently decreasing the amount of non-polar organic contaminants adsorbed by the organic matter. Following three freeze–thaw cycles, there was a notable increase in the proportion of soil micro-agglomerates, thereby augmenting the adsorption sites for non-polar organic contaminants in the soil and resulting in improved adsorption properties. However, after nine freeze–thaw cycles, the ratios of soil agglomerates stabilized, and the decline in SOC content led to a decrease in sorption performance. In the pre-freeze–thaw period, the increased proportion of micro-agglomerates increased the sorption of non-polar organic pollutants by the soil, but at the same time the increase in pH and the decrease in SOC content weakened the sorption capacity of the soil. The combined effect of these two factors resulted in a small difference between the indoor and outdoor soil sorption capacity during the pre-freeze–thaw period. While the ratio of micro-agglomerates in indoor and outdoor soils remained stable during the late freeze–thaw cycle, the reduction of SOC content in the outdoor group greatly weakened the sorption capacity of its soils. This makes the difference between the sorption capacity of indoor and outdoor soils for non-polar organic pollutants in the late stage. Overall, as the number of freeze–thaw cycles increased, alterations in soil physicochemical properties initially enhanced the non-polar organic contaminant adsorption properties of the soil, while eventually they contributed to a decline in these properties.

4. Conclusions

In this study, we conducted monitoring of physical and chemical properties of soils and their variations in a non-polar organic contaminant adsorption capacity during the freezing and thawing cycles in a seasonal frozen region (from November 2022 to March 2023). In the pre-freeze–thaw period, the proportion of ruptured microaggregates in the soil increased, leading to the release of organic acids and lowering the pH. The acidic environment enhanced the ability of soil organic matter to adsorb non-polar organic pollutants. Meanwhile, the increased proportion of micro-agglomerates provided additional adsorption sites for non-polar organic contaminants, thereby improving the overall adsorption performance of the soil medium. However, with continued freeze–thaw cycles, the stability of the agglomerates decreased, resulting in a reduction in organic matter content within the soil. This, in turn, mitigated the adsorption capacity of organic matter for non-polar organic contaminants, leading to a decrease in the soil's overall adsorption performance for these hydrocarbons. In summary, the freeze–thaw process demonstrated a significant impact on the distribution of aggregates of varying particle sizes and the organic matter content within the soil, consequently influencing the sorption of non-polar organic contaminants by the soil. In the pre-freeze–thaw period, the effect of the soil on non-polar organic pollutants is mainly influenced by the proportion of soil aggregates, while in the later period, it is mainly influenced by the organic matter content of the soil. This research shed light on the adsorption patterns of non-polar organic contaminants under freeze–thaw conditions, which can be vital for the prediction of groundwater petroleum hydrocarbon pollution in seasonal permafrost areas as well as in designing effective remediation strategies. Moreover, it can help reduce prediction errors resulting from the effects of freeze–thaw cycles. However, the coupling between the changes of the physicochemical properties of the packaged gas zone media and the adsorption of petroleum hydrocarbons under the freeze–thaw effect needs to be further investigated through indoor experiments.

Author Contributions: Conceptualization, H.L.; Data curation, R.Z.; Formal analysis, J.H.; Investigation, J.H. and R.Z.; Methodology, H.L., J.H. and R.Z.; Resources: J.H. and R.Z.; Writing—original draft: J.H.; Writing—review & editing: H.L. and R.Z.; Supervision: H.L. All authors have read and agreed to the published version of the manuscript.

Funding: This research is supported by the National Natural Science Foundation of China (U19A20107, 42172267, 42230204).

Data Availability Statement: Not applicable.

Conflicts of Interest: The authors declare no conflict of interest.

References

1. Varjani, S.J.; Upasani, V.N. Carbon spectrum utilization by an indigenous strain of *Pseudomonas aeruginosa* NCIM 5514: Production, characterization and surface active properties of biosurfactant. *Bioresour. Technol.* **2016**, *221*, 510–516. [CrossRef] [PubMed]
2. Aislabie, J.; Saul, D.J.; Foght, J.M. Bioremediation of hydrocarbon-contaminated polar soils. *Extremophiles* **2006**, *10*, 171–179. [CrossRef] [PubMed]
3. Pu, X.C.; Cutright, T.J. Sorption–desorption behavior of PCP on soil organic matter and clay minerals. *Chemosphere* **2006**, *64*, 972–983. [CrossRef] [PubMed]
4. An, S.; Woo, H.; Kim, S.H.; Yun, S.T.; Chung, J.; Lee, S. Complex behavior of petroleum hydrocarbons in vadose zone: A holistic analysis using unsaturated soil columns. *Chemosphere* **2023**, *326*, 138417. [CrossRef] [PubMed]
5. Qiao, X.C.; Zheng, B.H.; Li, X.; Zhao, X.R.; Dionysiou, D.D.; Liu, Y. Influencing factors and health risk assessment of polycyclic aromatic hydrocarbons in groundwater in China. *J. Hazard. Mater.* **2021**, *402*, 123419. [CrossRef]
6. Wang, M.; Ding, M.; Yuan, Y. Bioengineering for the Microbial Degradation of Petroleum Hydrocarbon Contaminants. *Bioengineering* **2023**, *10*, 347. [CrossRef]
7. Ertli, T.; Marton, A.; Földényi, R. Effect of pH and the role of organic matter in the adsorption of isoproturon on soils. *Chemosphere* **2004**, *57*, 771–779. [CrossRef]
8. Cornelissen, G.; Rigterink, H.; Ferdinandy, M.M.A.; Van Noort, P.C.M. Rapidly desorbing fractions of PAHs in contaminated sediments as a predictor of the extent of bioremediation. *Environ. Sci. Technol.* **1998**, *32*, 966–970. [CrossRef]
9. Li, Q.; Zhao, D.; Yin, J.; Zhou, X.; Li, Y.; Chi, P.; Han, Y.; Ansari, U.; Cheng, Y. Sediment instability caused by gas production from hydrate-bearing sediment in Northern South China Sea by horizontal wellbore: Evolution and mechanism. *Nat. Resour. Res.* **2023**, *32*, 1595–1620. [CrossRef]
10. Li, Q.; Zhang, C.; Yang, Y.; Ansari, U.; Han, Y.; Li, X.; Cheng, Y. Preliminary experimental investigation on long-term fracture conductivity for evaluating the feasibility and efficiency of fracturing operation in offshore hydrate-bearing sediments. *Ocean Eng.* **2023**, *281*, 114949. [CrossRef]
11. Wang, F.; Liu, X.; Jiang, B.; Zhuo, H.; Chen, W.; Chen, Y.; Li, X. Low-loading Pt nanoparticles combined with the atomically dispersed FeN4 sites supported by FeSA-NC for improved activity and stability towards oxygen reduction reaction/hydrogen evolution reaction in acid and alkaline media. *J. Colloid Interface Sci.* **2023**, *635*, 514–523. [CrossRef]
12. Luthy, R.G.; Aiken, G.R.; Brusseau, M.L.; Cunningham, S.D.; Gschwend, P.M.; Pignatello, J.J.; Reinhard, M.; Traina, S.J.; Weber, W.J. Sequestration of hydrophobic organic contaminants by geosorbents. *Environ. Sci. Technol.* **1997**, *31*, 3341–3347. [CrossRef]
13. Pil-Gon, K.I.M.; Tarafdar, A.; Jung-Hwan, K. Effect of soil pH on the sorption capacity of soil organic matter for polycyclic aromatic hydrocarbons in unsaturated soils. *Pedosphere* **2023**, *33*, 365–371.
14. Lamoureux, E.M.; Brownawell, B.J. Chemical and biological availability of sediment-sorbed hydrophobic organic contaminants. *Environ. Toxicol. Chem.* **1999**, *18*, 1733–1741. [CrossRef]
15. Ehlers, G.A.C.; Forrester, S.T.; Scherr, K.E.; Loibner, A.P.; Janik, L.J. Influence of the nature of soil organic matter on the sorption behaviour of pentadecane as determined by PLS analysis of mid-infrared DRIFT and solid-state 13C NMR spectra. *Environ. Pollut.* **2010**, *158*, 285–291. [CrossRef]
16. Chiou, C.T.; Porter, P.E.; Schmedding, D.W. Partition equilibriums of nonionic organic compounds between soil organic matter and water. *Environ. Sci. Technol.* **1983**, *17*, 227–231. [CrossRef]
17. Weber Jr, W.J.; Morris, J.C. Kinetics of adsorption on carbon from solution. *J. Sanit. Eng. Div.* **1963**, *89*, 31–59. [CrossRef]
18. Park, J.; Kwon, O.S.; Di Sarno, L. Influence of seasonal soil temperature variation and global warming on the seismic response of frozen soils in permafrost regions. *Earthq. Eng. Struct. Dyn.* **2021**, *50*, 3855–3871. [CrossRef]

19. Tokuda, D.; Kim, H.; Yamazaki, D.; Oki, T. Development of a global river water temperature model considering fluvial dynamics and seasonal freeze-thaw cycle. *Water Resour. Res.* **2019**, *55*, 1366–1383. [CrossRef]
20. Dang, X.L.; Zhang, Y.L.; Yu, N.; Zhang, Y.L. Cadmium adsorption-desorption of brown soil with freeze–thaw cycles in northeast China. *Int. J. Environ. Pollut.* **2012**, *49*, 89–99. [CrossRef]
21. Wang, X.; Li, Y.M.; Mao, N.; Zhou, Y.Q.; Guo, P. The adsorption behavior of Pb^{2+} and Cd^{2+} in the treated black soils with different freeze-thaw frequencies. *Water Air Soil Pollut.* **2017**, *228*, 193. [CrossRef]
22. Yang, Z.P.; Zhang, K.S.; Li, X.Y.; Ren, S.P.; Li, P. The effects of long-term freezing–thawing on the strength properties and the chemical stability of compound solidified/stabilized lead-contaminated soil. *Environ. Sci. Pollut. Res.* **2023**, *30*, 38185–38201. [CrossRef]
23. Zhao, Q.; Xing, B.S.; Tai, P.D.; Yang, K.; Li, H.; Zhang, L.Z.; Lin, G.; Li, P.J. Effect of freeze–thawing cycles on aging behavior of phenanthrene, pyrene and their mixture in soil. *Sci. Total Environ.* **2013**, *452*, 246–252. [CrossRef] [PubMed]
24. Zhao, Q.; Li, P.J.; Stagnitti, F.; Ye, J.; Dong, D.B.; Zhang, Y.Q.; Li, P. Effects of aging and freeze-thawing on extractability of pyrene in soil. *Chemosphere* **2009**, *76*, 447–452. [CrossRef]
25. Zhao, Q.; Weise, L.; Li, P.J.; Yang, K.; Zhang, Y.Q.; Dong, D.B.; Li, P.; Li, X.J. Ageing behavior of phenanthrene and pyrene in soils: A study using sodium dodecylbenzenesulfonate extraction. *J. Hazard. Mater.* **2010**, *183*, 881–887. [CrossRef] [PubMed]
26. Erdoğan, B.C.; Ülkü, S. Cr (VI) sorption by using clinoptilolite and bacteria loaded clinoptilolite rich mineral. *Microporous Mesoporous Mater.* **2012**, *152*, 253–261. [CrossRef]
27. Zhuang, S.; Zhu, K.; Xu, L.; Hu, J.; Wang, J. Adsorption of Co^{2+} and Sr^{2+} in aqueous solution by a novel fibrous chitosan biosorbent. *Sci. Total Environ.* **2022**, *825*, 153998. [CrossRef]
28. Yonker, R.E.; McGuinness, J.L. A Short Method of Obtaining Mean Weight Diameter Values of Aggregate Analysis of Soil. *Soil Sci.* **1957**, *83*, 291–294.
29. Mohan, D.; Singh, K.P.; Singh, V.K. Trivalent chromium removal from wastewater using low cost activated carbon derived from agricultural waste material and activated carbon fabric cloth. *J. Hazard. Mater.* **2006**, *135*, 280–295. [CrossRef]
30. Wang, J.L.; Guo, X. Rethinking of the intraparticle diffusion adsorption kinetics model: Interpretation, solving methods and applications. *Chemosphere* **2022**, *309*, 136732. [CrossRef]
31. Saleh, T.A.; Siddiqui, M.N.; Al-Arfaj, A.A. Kinetic and intraparticle diffusion studies of carbon nanotubes-titania for desulfurization of fuels. *Pet. Sci. Technol.* **2016**, *34*, 1468–1474. [CrossRef]
32. Söenmezay, A.; Öncel, M.S.; Bektas, N. Adsorption of lead and cadmium ions from aqueous solutions using manganoxide minerals. *Trans. Nonferr. Metal. Soc. China* **2012**, *22*, 3131–3139. [CrossRef]
33. Edwards, L.M. The effect of alternate freezing and thawing on aggregate stability and aggregate size distribution of some Prince Edward Island soils. *J. Soil Sci.* **1991**, *42*, 193–204. [CrossRef]
34. Guan, S.; Dou, S.; Chen, G.; Wang, G.; Zhuang, J. Isotopic characterization of sequestration and transformation of plant residue carbon in relation to soil aggregation dynamics. *Appl. Soil Ecol.* **2015**, *96*, 18–24. [CrossRef]
35. van Bochove, E.; Prévost, D.; Pelletier, F. Effects of freeze–thaw and soil structure on nitrous oxide produced in a clay soil. *Soil Sci. Soc. Am. J.* **2000**, *64*, 1638–1643. [CrossRef]
36. Yang, F.; Wang, M.; Wang, Z.Y. Sorption behavior of 17 phthalic acid esters on three soils: Effects of pH and dissolved organic matter, sorption coefficient measurement and QSPR study. *Chemosphere* **2013**, *93*, 82–89. [CrossRef]
37. Li, B.H.; Qian, Y.; Bi, E.P.; Chen, H.H.; Schmidt, T.C. Sorption behavior of phthalic acid esters on reference soils evaluated by soil column chromatography. *Clean* **2010**, *38*, 425–429. [CrossRef]
38. Wang, X.L.; Guo, X.Y.; Yang, Y.; Tao, S.; Xing, B.S. Sorption mechanisms of phenanthrene, lindane, and atrazine with various humic acid fractions from a single soil sample. *Environ. Sci. Technol.* **2011**, *45*, 2124–2130. [CrossRef]
39. Kołtowski, M.; Hilber, I.; Bucheli, T.D.; Oleszczuk, P. Effect of activated carbon and biochars on the bioavailability of polycyclic aromatic hydrocarbons in different industrially contaminated soils. *Environ. Sci. Pollut. Res.* **2016**, *23*, 11058–11068. [CrossRef]

water

MDPI

Article

Migration Law of LNAPLs in the Groundwater Level Fluctuation Zone Affected by Freezing and Thawing

Jing Zhou [1,2,3], Minghao Pan [1,2,3], Chuping Chang [1,2,3], Ao Wang [4], Yongqi Wang [1,2,3] and Hang Lyu [1,2,3,*]

1 Key Laboratory of Groundwater Resources and Environment, Ministry of Education, Jilin University, Changchun 130026, China; zj982702836@163.com (J.Z.); 202021470021@mail.bnu.edu.cn (M.P.); bcyz010438@163.com (C.C.); yqwang21@mails.jlu.edu.cn (Y.W.)
2 Jilin Provincial Key Laboratory of Water Resources and Environment, Jilin University, Changchun 130026, China
3 Institute of Water Resources and Environment, Jilin University, Changchun 130021, China
4 Jilin Provincial Bureau of Hydrology and Water Resources Changchun Branch, Changchun 130028, China; xiaojiuwo2022@163.com
* Correspondence: lvhangmail@163.com

Abstract: Freezing and thawing can cause dynamic fluctuations of the groundwater level, resulting in the migration and retention of LNAPLs. However, this process is difficult to observe visually, and a suitable simulation method for its quantitative calculation is lacking. In this study, a numerical simulation is established by coupling the HYDRUS-1D software and the TOUGH program to realize dynamic simulation of the entire process of soil temperature changes, water migration, water level fluctuation, and redistribution of LNAPLs during the freeze–thaw process. The results of the study show that the process of soil freezing and thawing causes water migration, which in turn causes groundwater level fluctuation, leading to the migration and redistribution of LNAPLs within the water level fluctuation zone. In this process, the soil particle size and porosity control the response degree and speed of the water level under freezing and thawing and the spatiotemporal distribution of LNAPLs by affecting the soil temperature, capillary force, and water migration. The migration ability of free LNAPLs is determined by their own density and viscosity; the retention of residual LNAPLs is affected by soil porosity and permeability as well as LNAPL viscosity. The results of this study can not only be used to develop a simulation method for the migration and retention mechanism of LNAPLs in cold regions but also serve as a scientific and theoretical basis for LNAPL pollution control in seasonal frozen soil regions.

Keywords: freeze–thaw process; LNAPLs; fluctuation zone of groundwater level; numerical simulation

Citation: Zhou, J.; Pan, M.; Chang, C.; Wang, A.; Wang, Y.; Lyu, H. Migration Law of LNAPLs in the Groundwater Level Fluctuation Zone Affected by Freezing and Thawing. *Water* **2022**, *14*, 1289. https://doi.org/10.3390/w14081289

Academic Editor: Andrzej Witkowski

Received: 4 March 2022
Accepted: 12 April 2022
Published: 15 April 2022

1. Introduction

The exploration, development, transportation, and utilization of petroleum are often accompanied by serious groundwater and soil pollution problems. The main pollutants are petroleum hydrocarbons in the form of light nonaqueous phase liquids (LNAPLs), which are carcinogenic and teratogenic to the human body and pose a continuous threat to the ecological environment [1]. After the pollution of LNAPLs leaks on the surface, under the action of gravity and capillary force it will gradually diffuse and migrate to the deep layer of the vadose zone and its surroundings, eventually making it difficult to use groundwater resources and soil [2,3].

Seasonal frozen soil is a type of soil in which the surface layer is frozen in winter and thawed in summer. Approximately 30% of the total land area of the northern hemisphere consists of this type of soil, and there are a large amount of oil and oil fields located in these seasonal frozen soil areas which are contributing to the pollution of LNAPLs in the freeze–thaw environment [4,5]. Compared with non-frozen soil areas, seasonal frozen soil areas experience more than 100 days of surface freezing and thawing processes each year. The soil temperature during this process will transmit downward, thereby causing a large amount of moisture migration in the range above the groundwater level, which is mainly carried out in the vertical direction [6–10]. The drastic changes in water content and heat will become involved in the process of temperature-dependent redistribution of LNAPLs in the groundwater level fluctuation zone, and will result in more complicated migration and retention of LNAPLs [11–15]. Ignoring the important effects of freezing and thawing on the dynamics and distribution of LNAPLs will result in significant deviations in the assessment results of the damage, diffusion range, and remediation efficiency of the LNAPL pollution in the groundwater and soil [15,16]. This will severely restrict the rationality of LNAPL pollution prevention and control planning, control objectives, and remediation plans in seasonal frozen soil areas. Therefore, it is necessary to focus on the freeze–thaw environment and study the migration law of LNAPLs in the groundwater level fluctuation zone.

Monitoring LNAPLs in the underground environment in cold regions is difficult; for this reason, numerical simulation has become a relatively effective method [17–19]. Currently, scholars have established advanced research on the simulation of water and heat transfer in the freeze–thaw process, but the existing research does not consider both the freeze–thaw process and the migration and retention law of LNAPLs in the water level fluctuation zone [17–34]. There is no single piece of software that can simulate the entire process of soil temperature changes, water migration, water level fluctuation, and redistribution of LNAPLs during the freeze–thaw process [20–22]. And in order to solve more complicated simulation problems than the water and heat transfer under freezing and thawing, it is essential to establish a coupling model [23–26]. To study the migration of LNAPL with the change of water levels under freezing and thawing, the established model needs to couple the water and heat transfer under freezing and thawing and the dynamics of LNAPL with the water level.

In existing research, the water and heat transfer in the freezing and thawing process of the coupled model is mainly simulated by HYDRUS [23–26]. Compared with other finite element numerical codes for simulating freezing and thawing processes such as FEFlow and OpenGeoSys [27–35], HYDRUS is more suitable for freeze-thaw simulation in coupled models. This is because the HYDRUS model can be used for both direct problems and inverse problems when certain parameters are calibrated from observation data, which is very important for the fitting of coupled models [30]. And for soil hydrodynamics and heat conduction, HYDURS-1D is also the most satisfactory [31,32]. Ce et al. [23] developed a fully coupled numerical module and incorporated it into the Hydrus-1D software to simulate the simultaneous movement of water, steam and heat during freezing and thawing, and the simulation result restores the moisture and temperature data of the frozen site well. However, for LNAPLs in the fluctuation zone of water levels in seasonal frozen soil regions, relying solely on the HYDRUS model can only reflect the changes in the water level and temperature with freezing and thawing and does not elucidate on the migration and retention of LNAPLs during water level fluctuations.

Meanwhile, TOUGH is the most suitable numerical simulation method for studying the migration process of LNAPLs [36–38]. Tao et al. [39] used the T2VOC module in the TOUGH2 program and constructed a numerical model to simulate the migration and redistribution of LNAPLs after they leak from the surface. The results showed that water level fluctuation will affect the continuous migration of LNAPLs; further, the pollution level will expand in the vertical and horizontal directions until it reaches the entire water level fluctuation zone, and the structural complexity of the polluted area will increase. Yang [40] used the TMVOC module in the TOUGH program to discover that water level fluctuations lead to the mutual displacement of the water, gas, and NAPL phases in the porous medium, and the saturation of the three phases shows a complementary relationship between growth and decline. Dafny [41] used the TMVOC module to quantify the distribution, partitioning, and fate of NAPL in the vadose zone. However, because the TOUGH model has limitations in the simulation of multiphase flow in sub-zero temperature environments, this type of research can only focus on the migration of LNAPLs in the water level fluctuation zone under normal temperature environments and is not suitable for seasonal frozen soil regions [42–45].

Based on the foundation and deficiencies of related research, in this study, HYDRUS-1D, which can simulate the water and heat transfer in the freeze–thaw process [46], is coupled with the TOUGH program, which can simulate the distribution of LNAPLs with water level fluctuation. Through this coupled simulation method, under the conditions of different media and different types of LNAPLs, the migration and retention law of LNAPLs with the water level fluctuation during freeze–thaw periods is comprehensively analyzed, and the factors affecting it are evaluated. This research is expected to provide the theoretical methods and promotional basis for the numerical simulation of LNAPL migration in the water level fluctuation zone of seasonal frozen soil regions as well as serve as a reference for pollution evaluation, management, and the restoration of LNAPLs.

2. Materials and Methods

2.1. Soil Column Experiment of Water and Heat Transport

The equipment used in the experiment consists of a freeze–thaw cycle box, a refrigeration system, moisture content and temperature probes, a pressure measuring tube, and a water supply system with a constant water head. The experimental setup is shown in Figure 1.

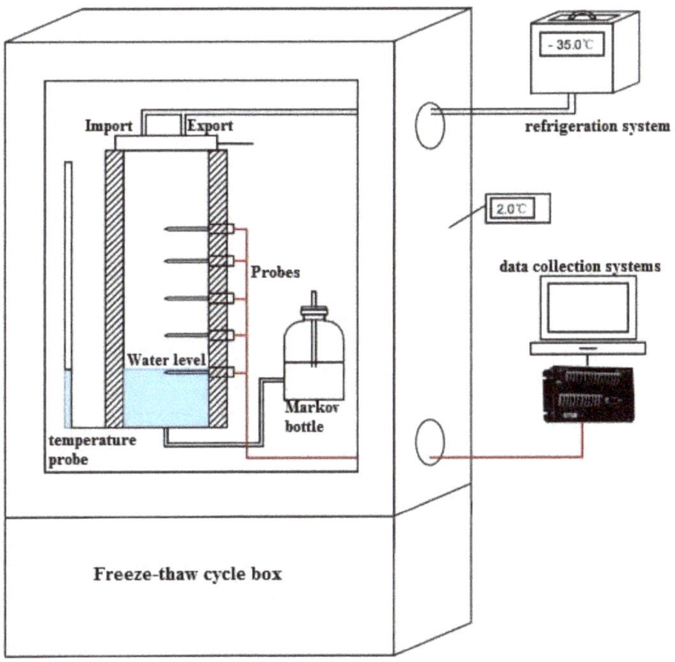

Figure 1. Schematic diagram of the experimental setup.

The test soil column is a plexiglass container with a height of 30 cm, and medium sand (average particle size 0.5–0.2 mm) was selected as the representative test medium. A water supply pipe and a Markov bottle were used to supply water at a fixed head to the soil column, and the initial water level was set to 16 cm from the top of the soil.

A total of five monitoring holes were set on the wall of the column at an interval of 4 cm to monitor the soil temperature and moisture content in the soil column in real time. Each monitoring hole is equipped with a CR300 soil temperature and humidity sensor. The piezometer reading was periodically taken to obtain the dynamic data of the water level. After the start of the experiment, the piezometer is read every 10 min, and the temperature and water content are automatically monitored and outputted every 2 min through the sensor.

The experimental soil column was placed in the freeze–thaw cycle box, which adjusts the temperature around the soil column to be constant at 2 °C. The temperature control cover on the top of the soil is used to control the freezing and thawing of the soil from the top to the bottom in stages. The minimum temperature in the freezing period is −35 °C, and the temperature in the thawing period is 5 °C. The temperature control process is shown in Figure 2.

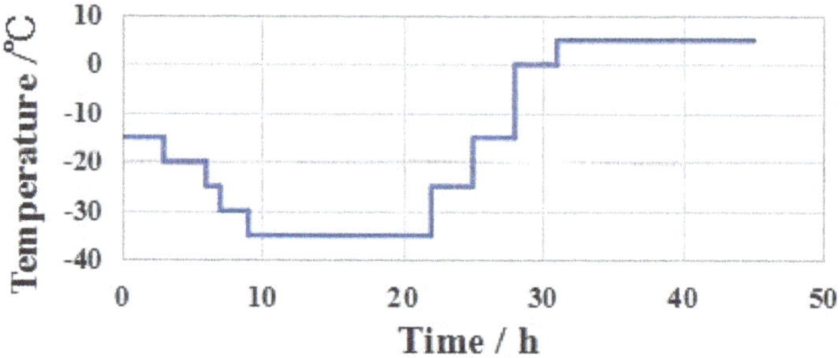

Figure 2. Freeze–thaw temperature control process.

The resulting data from the soil column experiment will be used to adjust and fit the parameters in the subsequent HYDRUS-1D model. Taking the empirical value of the hydrothermal parameters in HYDRUS-1D as the initial value [46], the hydraulic parameters related to the soil hydraulic conductivity in the medium sand obtained by adjusting and fitting are shown in Table 1, and the parameters related to the soil heat conductivity are shown in Table 2. And for the data of the soil column experiment, please refer to Table S1 in the Supplementary Materials.

Table 1. Soil hydraulic parameters.

Parameter	θ_r (cm^3/cm^3)	θ_s (cm^3/cm^3)	α (cm^{-1})	n	K (cm/s)	l
medium sand	0.040	0.300	0.094	2.05	9.72×10^{-3}	0.5

Table 2. Soil thermodynamic parameters.

Parameter	Solid	Org.	D_L (cm^2·s^{-1})	b1	b2	b3
medium sand	0.59	0.00	5.00	1.06376×10^{12}	-1.12254×10^{10}	2.48832×10^{11}

2.2. Model Method

HYDRUS-1D and TOUGH are coupled to establish a model to simulate the effect of the freeze–thaw cycle on temperature, moisture migration, water level fluctuation, and LNAPL migration and retention. The coupling process mainly considers the consistency of temperature and water level between the two codes. The simulation range of the HYDRUS-1D model is from the top of the soil column (0 cm) to 20 cm below, and the initial groundwater level is located 16 cm below the column surface. The upper and lower boundaries of the water flow are zero flux boundaries, and the upper and lower boundaries of heat transfer are temperature boundaries that change with time. Since TOUGH is not suitable for simulation at a temperature below 0 °C, the TOUGH simulation range is the area within −10 cm to −30 cm in the soil column experiment, and the soil temperature change within this depth range is consistently above 0 °C. And order to comply with the aforementioned soil column experiment and the HYDRUS-1D numerical model, the top is set as the Dirichlet boundary, with constant atmospheric pressure, no water infiltration, and zero water flux. The bottom is set as a waterproof boundary condition. The initial temperature of the topper and lower are set to 6.55 °C and 7 °C, respectively, according to the experimental results, and the temperature changes with time during the simulation. And since the heat and water migration in the freezing and thawing process are mainly reflected in the vertical direction, the boundary conditions in the coupling model are assumed to be horizontally closed [6–10].The left boundary (x = 0–0.05 m) and the right boundary (x = 0.85–0.90 cm) of the simulation area are both physical boundaries, which are set as zero flux boundary conditions [39].

Coupled simulation is achieved through the transfer of heat and water volume time by time between HYDRUS-1D and TOUGH, and the temperature and water content at each depth are verified by measured data. The interface for the water flux interaction of the coupled model is the layer above the water level fluctuation in the TOUGH model. In order to realize the mutual coupling of water flow, 16 cells in this layer are used as the water injection and pumping cells to control the water level changes, and the changed flux is entered into the source and sink of the cell. Figure 3 shows the water level control obtained by TOUGH. On the other hand, the interaction of heat in the coupled model is achieved jointly by each layer of cells in TOUGH. The heat source–sink items are added to each cell after the division in TOUGH, and the corresponding conversion relationship between the rate of heat addition and the temperature change at each representative layer is calculated. Thus, the coupling of water and heat between the two models is realized.

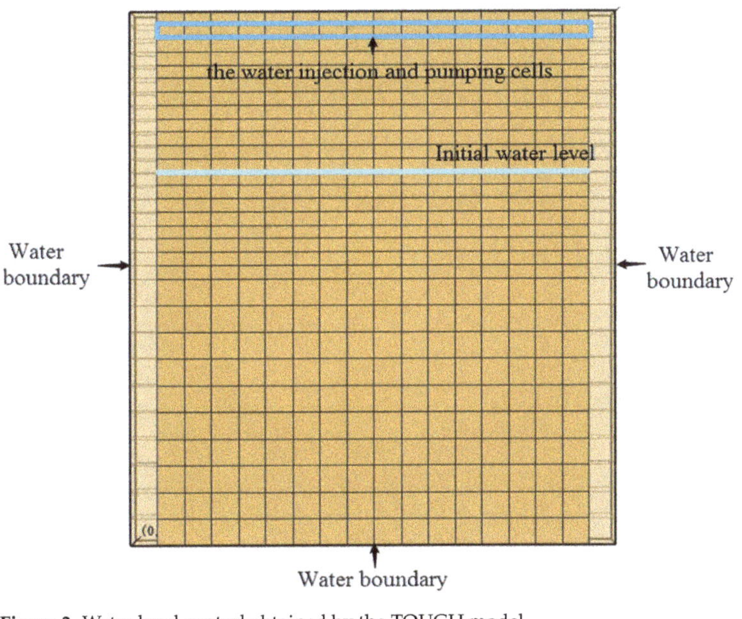

Figure 3. Water level control obtained by the TOUGH model.

In the water and heat transfer model, The modified Richards' equation [43] is used as the basic equation of water migration:

$$\frac{\partial \theta u(h)}{\partial t} + \frac{\rho_i}{\rho_w}\frac{\partial \theta_i(T)}{\partial t} = \frac{\partial}{\partial z}\left[K_{Lh}(h)\frac{\partial h}{\partial z} + K_{Lh}(h) + K_{LT}(h)\frac{\partial T}{\partial z} + K_{Vh}(\theta)\frac{\partial h}{\partial z} + K_{vT}(\theta)\frac{\partial T}{\partial z}\right] - S \tag{1}$$

where θ_u is the volumetric unfrozen water content (L^3 L^{-3}) (= θ + θ_u), θ is the volumetric liquid water content (L^3 L^{-3}), θ_v is the volumetric vapor content expressed as an equivalent water content (L^3 L^{-3}), θ_i is the volumetric ice content (L^3 L^{-3}), t is time (T), z is the spatial coordinate positive upward (L), ρ_i is the density of ice (M L^{-3}) (931 kg m^{-3}), ρ_w is the density of liquid water (M L^{-3}) (approximately 1000 kg m^{-3}), h is the pressure head (L), T is the temperature (K), and S is a sink term (T^{-1}) usually accounting for root water uptake. Water flow in Equation (1) is assumed to be caused by five different processes with corresponding hydraulic conductivities. The first three terms on the right-hand side of Equation (1) represent liquid flows due to a pressure head gradient (K_{Lh}, [LT^{-1}]), gravity, and a temperature gradient (K_{LT}, [L^2 T^{-1} K^{-1}]), respectively. The next two terms represent vapor flows due to pressure head (K_{vh}, [LT^{-1}]) and temperature (K_{vT}, [L^2 T^{-1} K^{-1}]) gradients, respectively.

The soil hydraulic conduction module uses the Van Genuchten–Mualem equation, which is generally recognized in soil water calculations. The basic equation of soil heat transfer [28] is given as follows.

$$\frac{\partial C_p T}{\partial t} - L_f \rho_i \frac{\partial \theta_i}{\partial t} + L_0(T)\frac{\partial \theta_v(T)}{\partial t} = \frac{\partial}{\partial z}\left[\lambda(\theta)\frac{\partial T}{\partial z}\right] - C_\omega \frac{\partial q_l T}{\partial z} - C_v \frac{\partial q_v T}{\partial z} - L_0(T)\frac{\partial q_v}{\partial z} - C_\omega S T \quad (2)$$

where the first term on the left-hand side represents changes in the energy content, and the second and third terms represent changes in the latent heat of the frozen and vapor phases, respectively. The terms on the right-hand side represent soil heat flow by conduction, convection of sensible heat with flowing water, transfer of sensible heat by diffusion of water vapor, transfer of latent heat by diffusion of water vapor, and uptake of energy associated with root water uptake, respectively. The phase change between water and ice is controlled by the generalized Clapeyron equation, which defines a relationship between the liquid pressure head and temperature when ice is present in the porous material [32]. Hence, the unfrozen water content can be derived from the liquid pressure head as a function of temperature alone when ice and pure water coexist in the soil [43]. Further, ice volume fraction is determined by the difference between initial water content and liquid water content. The volumetric heat capacity of the soil, C_p (J m^{-3} K^{-1}, M L^{-1} T^{-2} K^{-1}), in Equation (2) is defined as the sum of the volumetric heat capacities of the solid (C_n), liquid (C_w), vapor (C_v), and ice (C_i) phases multiplied by their respective volumetric fractions:

$$C_p = C_n\theta_n + C_w\theta + C_v\theta_v + C_i\theta_i \quad (3)$$

Furthermore, in Equation (2), L_o is the volumetric latent heat of vaporization of water (J m^{-3}, M L^{-1} T^{-2}), L_o = L_W ρ_W, L_w is the latent heat of vaporization of water (J kg^{-1}) (=2.501 × 10^6 − 2369.2 T [°C]), and L_f is the latent heat of freezing (J kg^{-1}, L^2 T^{-2}) (approximately 3.34 × 10^5 J kg^{-1}).

In the simulation of LNAPL migration and retention with water level fluctuation, the mass and energy conservation equations [37] are used as the basic governing equations of the components in the multiphase flow (i.e., water, gas, and NAPL phases). Parker's equation [39] is used as the relative permeability coefficient equation in the multiphase flow problem. Van Genuchten's equation [39] is used as the capillary pressure governing equation of the NAPL–water phase. Finally, the migration and flow equation of free LNAPLs in the soil medium [38] is shown in Equation (4).

$$\frac{\partial}{\partial t}(\varphi S_N \rho_N) = -\frac{\partial}{\partial x}(\varphi S_N \rho_N V) - \frac{\partial}{\partial z}(\varphi S_N \rho_N V) + I \quad (4)$$

where ϕ is the porosity, S_N is the NAPL phase saturation, ρ_N is the NAPL phase density, and V is the Darcy flow velocity of the immiscible phase. The last term I on the right side of the equation is the source (sink) term (=$\rho_N Q_N$ + E), Q_N is the volume of NAPL existing or leaked in the unit time, and E represents the mass exchange between phases. Such source and sink terms are not considered in this model, so I = 0.

3. Results

3.1. Law of Soil Water and Heat Transfer during Freezing and Thawing

During the freeze–thaw experiment, the water content within 4 cm of the top of the soil column changes drastically, becoming weaker with increasing depth (Figure 4(c1)). This is mainly related to the mutual transformation of the water and ice phases. During the freezing process, as shown in Figure 4(b1,c1), the upper part of the soil body freezes first, decreasing the water content. The negative temperature transfers from top to bottom and the water migrates upward, causing the water level to drop, as shown in Figure 4(d1). In the process of melting, the ice phase on the upper part of the soil melts, the water content increases, and the water moves downward, which increases the groundwater level. The HYDRUS-1D model also shows the changes of temperature, water content and water level in the medium-sand medium during the freezing and thawing process (Figure 4(a′2–d′2)), and its water and heat transfer law is consistent with the results of the soil column experiment: Moisture migration due to freezing and thawing of the upper soil is responsible for changes in the water table.

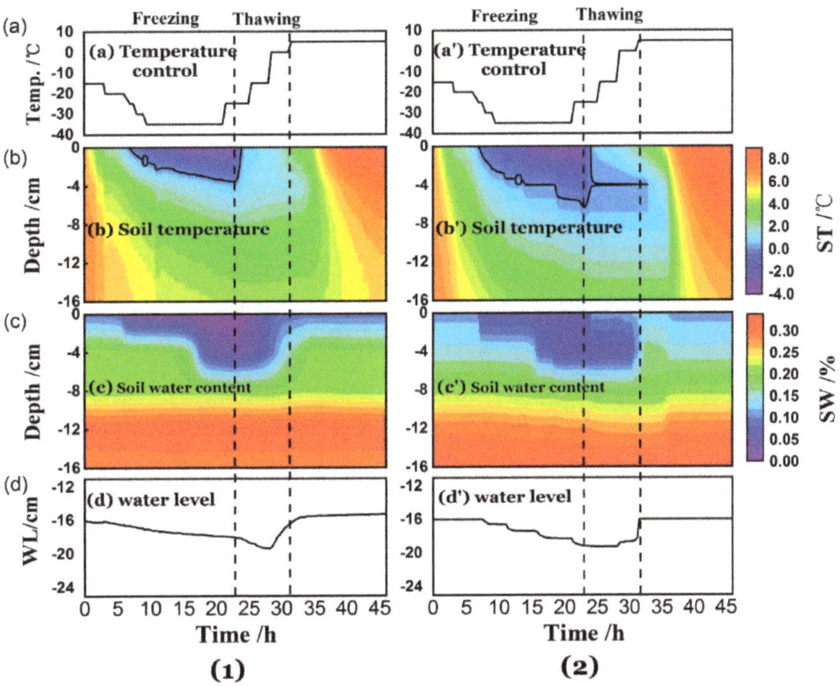

Figure 4. Water and heat transporting results of (**1**) medium-sand medium soil column and (**2**) Hydrus-1D model. (**a**,**a′**) temperature control process; (**b**,**b′**) change of soil temperature; (**c**,**c′**) change of soil moisture content; (**d**,**d′**) change of groundwater level.

As shown in Figure 4(1,2), In the HYDRUS-1D model, the changes of soil temperature and water content at each depth are similar to the soil column experiment results. Although the water level changes stepwise, the overall trend and range of change are consistent with the soil column experiment results. In order to evaluate the accuracy of the Hydrus-1D model, two indicators, the percentage of deviation (PBIAS) and the root mean square error (RMSE), were used to quantitatively evaluate the simulation data of HYDRUS-1D for medium-sand media. The PBIAS absolute value of the water level and temperature data in the HYDRUS-1D model is less than 10%, and RMSE is within 0.6, indicating that the water level and soil temperature simulation results are good, and they are overall consistent with the experimental results and have high accuracy. Figure 5a–f show the changes in the water level and temperature obtained by the HYDRUS-1D model with different media. The response rate of soil temperature conduction and water level to the freeze–thaw process is in the order of silty clay > fine sand > medium sand > coarse sand. Compared with the other three media, the soil temperature at the middle depth level of silty clay responds more quickly to the temperature changes in the upper part of the soil. This is because the soil solid particles are arranged closer together, allowing the upper temperature to diffuse downward during the freeze–thaw process [16] and thus improving the overall heat transfer properties of the soil. Moreover, because the temperature of silty clay soil changes rapidly, the capillary force of the upper part of the soil also changes faster, and the time of drop and rise of the water level is earlier than that in the case of coarse sand, medium sand, and fine sand.

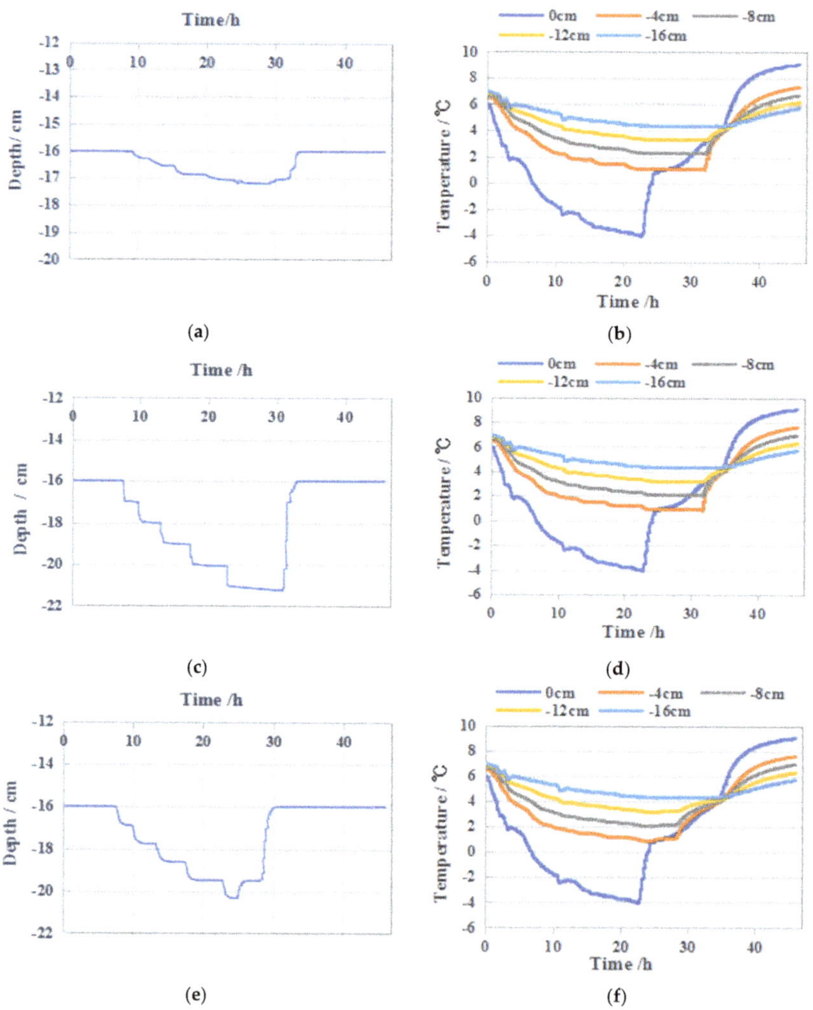

Figure 5. Comparison of water level and temperature changes obtained under various media. (**a**) Water level change and (**b**) temperature change obtained by the coarse sand model. (**c**) Water level change and (**d**) temperature change obtained by the fine sand model. (**e**) Water level change and (**f**) temperature change obtained by the silty clay model.

The magnitude of the water level change in different media is in the order of fine sand > silty clay > medium sand > coarse sand. For the same temperature gradient, the smaller the particle size of the soil, the better is its water holding capacity and the greater is the capillary force produced during freezing. Therefore, in the freezing process, the amount of water migration on the soil profile in the fine sand medium is greater than that in the medium sand and coarse sand, which increases the water level change.

Compared with fine sand, silty clay has smaller particles; however, owing to its lower residual moisture content and greater soil porosity, silty clay has lower capillary force generated during the freeze–thaw process. Consequently, the water migration and the water level fluctuation range on the soil profile are smaller than those in the case of fine sand. In addition, the water level change range in silty clay is larger than that in coarse sand and medium sand. The reason is similar to that for fine sand: the smaller the soil particle size, the better is the water holding capacity and the greater is the capillary force generated during freezing.

Taken together, in the same temperature gradient, the capillary force generated by the medium during the freeze–thaw process and the thermal conductivity of the medium are important factors that affect the water migration, range, and speed of the water level change.

3.2. *Migration and Retention Process of LNAPLs Affected by Freezing and Thawing*

Figure 6 shows the migration and retention of LNAPLs with water level fluctuation (TOUGH model with toluene as the LNAPL and medium sand as the medium). The residual saturation of the LNAPL phase is 0.05; therefore, when the saturation is less than 0.05, the LNAPLs will remain under the control of capillary force and transform into a residual state. As the water level drops, the free-state LNAPLs migrate downward; simultaneously, their saturation continues to decrease and gradually expands to the entire level in the horizontal direction. When the LNAPLs drop to the lowest position with the water level, a part of the LNAPLs with saturation close to 0.05 are distributed in the descending range. This indicates that a considerable part of the free-state LNAPLs changes into the residual state and remains in the process of the water level drop. In addition, there are some LNAPLs below the water level during the water level drop. This proves that LNAPLs will preferentially displace water when the water level drops [40].

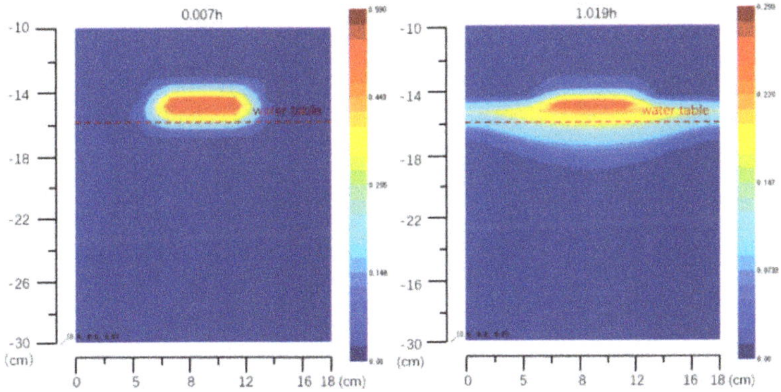

Figure 6. *Cont.*

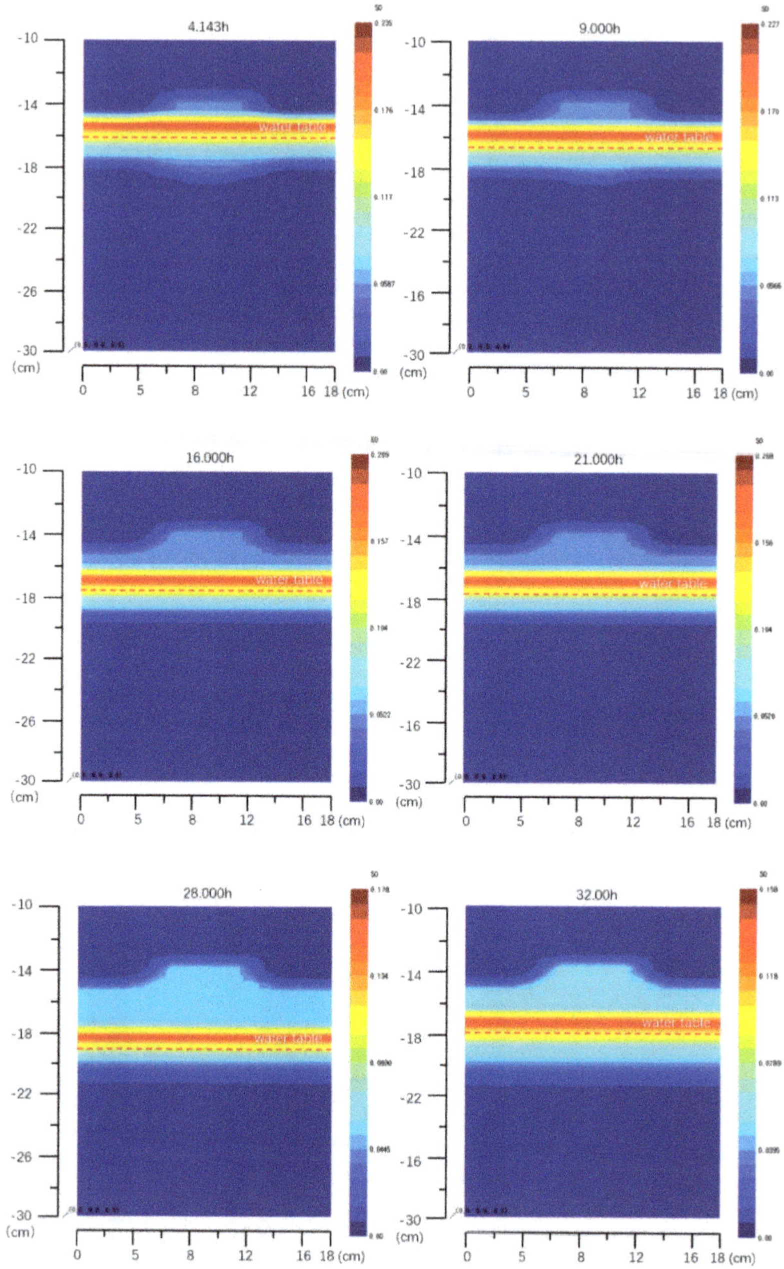

Figure 6. *Cont.*

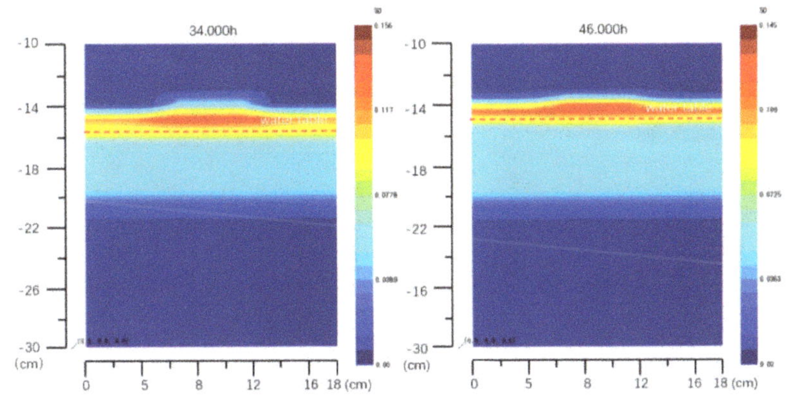

Figure 6. Redistribution of toluene in medium sand.

When the water level rises, the free LNAPLs also rise with the water level, and, at the same time, their saturation continues to decrease. In this process, residual LNAPLs continuously stay below the groundwater surface. Therefore, the drop and rise of the water level during the entire freeze–thaw process will cause the original LNAPLs to undergo a significant redistribution process. Furthermore, the pollution range of LNAPLs continues to expand both vertically and horizontally with decreasing water level, while the pollution range of the area above the initial water level increases as the water level increases. During the whole process, the vertical fluctuation range of the water level is approximately 3.3 cm, and the vertical pollution range caused by the migration and retention of LNAPLs is approximately 6 cm.

Taken together, the mechanism of the freeze–thaw process on the migration and retention of LNAPLs is as follows: soil temperature changes cause water phase changes, resulting in water migration, which in turn causes water level fluctuations and the migration and retention of LNAPLs within the water level fluctuation zone. The temperature at the top of the soil is the most fundamental controlling factor in this process. During the freeze–thaw cycle, when LNAPLs rose and fell with the water level, a considerable part of the LNAPLs was retained and transformed into a residual state. Therefore, the effect of retention cannot be ignored. Both migration and retention effects lead to the continuous expansion of the pollution range of LNAPLs in the water level fluctuation zone, resulting in the redistribution of LNAPLs with the water level. Therefore, compared with that in areas not affected by freeze–thaw, the pollution range of LNAPLs in the water level fluctuation zone will be larger in the seasonal frozen soil area.

3.3. Analysis of Factors Affecting the Migration of LNAPLs

After successfully simulating the migration and retention model of LNAPLs in the freeze–thaw water level fluctuation zone by combining HYDRUS-1D and TOUGH, the model is extended to different media conditions and different types of LNAPLs to study their effect on the migration and retention of LNAPLs. Table 3 shows the comparison of model results with different media conditions and different types of LNAPLs.

The response of water level and LNAPL position to temperature changes during freeze–thaw is affected by the medium condition; the response speed in different media is in the order of fine sand < silty clay < medium sand < coarse sand. This indicates that the smaller the soil particles, the smaller the porosity. That is, with closer arrangement of the soil particles, the heat conduction properties of the soil improves and the hysteresis of the soil temperature change at the middle depth of the soil decreases. As a result, water migration can quickly respond to changes in the top surface temperature of the soil during the freeze–thaw process, and the start of the fluctuation of free LNAPLs near the water level occurs earlier.

Table 3. Comparison of influencing factors.

Influencing Factors		Corresponding Time of Water Level Change/h	Water Level Fluctuation Range/cm	LNAPLs Pollution Range/cm	The Speed of Migration of LNAPLs	Maximum Saturation of LNAPLs after Freeze-Thaw
Media	coarse sand	9.300	1.210	4 cm	fast	0.227
	Medium sand	7.540	3.300	6 cm	slow	0.145
	Fine sand	7.380	5.250	8 cm	Extremely slow	0.115
	Silty clay	7.510	4.250	7 cm	slow	0.195
LNAPLs	Toluene	7.540	3.300	6 cm	slow	0.145
	Chloroethane	7.540	3.300	6 cm	fast	0.134
	Ethylbenzene	7.540	3.300	6 cm	slow	0.15

The water level fluctuation and the final pollution depth range of LNAPLs in different media are in the order of fine sand > silty clay > medium sand > coarse sand. This phenomenon should be related to the porosity of the soil. The smaller the porosity of the soil, the greater is the capillary force generated in the unsaturated zone during freezing, resulting in the increase of the amount of water migration on the soil profile as well as the ranges of water level fluctuation and LNAPL pollution.

The rate of LNAPL migration and expansion is related to the porosity, permeability, and properties of the medium. With higher porosity and permeability of the soil or lower viscosity of LNAPLs, the migration speed of LNAPLs as well as the distribution thickness of free-state LNAPLs that can move with the water level increase. Conversely, with lower soil porosity and permeability or greater viscosity of LNAPLs, the migration speed of LNAPLs in the vertical direction decreases and more LNAPLs will expand and be retained in the lateral direction. Thus, the maximum saturation of LNAPLs becomes lower at the end of the freeze–thaw process. Therefore, at the lowest water level, in the fine sand medium with the least porosity, the residual LNAPLs near the initial position have the largest retention range. Among different LNAPLs, there is less residual chloroethane in the initial position because of the lower viscosity of chloroethane.

In addition, some LNAPLs exist below the water level during the freezing period, and from the perspective of saturation, this part of the LNAPLs is not in a stagnant state. This is possibly because LNAPLs will preferentially displace water when the water level decreases [42]. In each medium, this part of LNAPLs does not have much difference in the range below the water level; however, comparing the three LNAPLs, the denser chloroethane has a wider range below the water level. Therefore, LNAPLs with high density can more easily displace water below the water level.

In the same TOUGH model, each medium is simulated at a constant temperature, only changing the non-isothermal mode to the isothermal mode. The soil temperature was constant at 22 °C, with no change in other conditions such as the water level. The results of isothermal simulation are compared with the results during the freeze–thaw process. The comparison is shown in Table 4.

Table 4. Comparison of LNAPL migration status during isothermal and the freeze–thaw process.

Medium	LNAPLs Contamination Range (Isothermal/Freeze-Thaw)	Status of LNAPLs Near the Water Level	Maximum LNAPLs Saturation (Isothermal/Freeze-Thaw)
coarse sand	4.3 cm/4 cm	similar	0.137/0.127
Medium sand	6.2 cm/6 cm	similar	0.161/0.145
Fine sand	8.0 cm/8 cm	similar	0.130/0.115
Silty clay	7.0 cm/7 cm	similar	0.214/0.195

The migration distribution and pollution range of LNAPLs in both isothermal and non-isothermal models do not show any considerable difference. The only difference is that the maximum saturation of LNAPLs in the isothermal condition is approximately 0.01–0.02

larger than that during the freeze–thaw process; this difference is relatively small. This confirms that temperature does not directly control the migration and retention of LANPLs, but indirectly leads to the redistribution of LNAPLs by affecting water level changes.

4. Conclusions

In this study, the HYDRUS-1D software and TOUGH program were combined to successfully establish a migration and retention model of LNAPLs with water level fluctuation under the effect of the freeze–thaw cycle. This model can thus be applied to study the interaction mechanism of freezing and thawing temperature-dependent LNAPL redistribution in the water level fluctuation zone. Based on this model, the migration models under different media, LNAPLs, and temperature conditions were established. The following conclusions can be drawn:

(1) The temperature changes during freezing and thawing is the most fundamental factor controlling the migration of LNAPLs in the water level fluctuation zone. The LNAPLs near the water level during the freeze–thaw process migrate or remain with the fluctuation of the water level. Thus, the pollution range and diffusion degree of LNAPLs in the seasonal frozen soil area are greater than those in the non-frozen soil area.

(2) The particle size and porosity of the medium as well as LNAPL composition type are the main factors that affect the migration of LNAPLs. In general, the finer and more closely packed the medium particles, the faster is the response of soil temperature and water level changes to freezing and thawing. The smaller the porosity, the greater is the capillary force generated above the water level during the freeze–thaw process and the greater are the amount of water migration, the amplitude of the water level, and the migration and diffusion range of LNAPLs. By contrast, the composition type of LNAPLs determines their retention process and phase change.

(3) The coupling model in this study can provide a simulation method for the analysis of the migration and retention of LNAPLs in the water level fluctuation zone of seasonal frozen soil regions. The research results on the migration and retention mechanism and the factors affecting LNAPLs can be used as a theoretical basis for LNAPL pollution evaluation and treatment in areas subjected to freeze–thaw cycles. Considering that T2VOC lacks simulation functions for chemical adsorption and microbial degradation processes, the effect of temperature on chemical and biological processes is not reflected and requires further research.

Supplementary Materials: The following are available online at https://www.mdpi.com/article/10.3390/w14081289/s1, Table S1: Result data of water and heat transfer in medium sand medium soil column experiment, Table S2: HYDRUS-1D simulation results.

Author Contributions: All authors made a substantial contribution to this paper. Conceptualization, H.L., A.W. and Y.W.; methodology, J.Z., C.C. and H.L.; software, J.Z. and M.P.; data curation, M.P., A.W. and Y.W.; writing—original draft preparation, J.Z.; writing—review and editing, J.Z. and H.L.; supervision, H.L. All authors have read and agreed to the published version of the manuscript.

Funding: This research received no external funding.

Data Availability Statement: Not applicable.

Acknowledgments: This research is supported by the National Natural Science Foundation of China (42172267, U19A20107).

Conflicts of Interest: The authors declare that they have no conflicts of interest.

References

1. Kim, J.; Corapcioglu, M.Y. Modeling dissolution and volatilization of LNAPL sources migrating on the groundwater table. *J. Contam. Hydrol.* **2003**, *65*, 137–158. [CrossRef]
2. Singh, K.; Niven, R.K. Non-aqueous Phase Liquid Spills in Freezing and Thawing Soils: Critical Analysis of Pore-Scale Processes. *Crit. Rev. Environ. Sci. Technol.* **2013**, *43*, 551–597. [CrossRef]
3. Lari, K.S.; Davis, G.B.; Rayner, J.L.; Bastow, T.P.; Puzon, G.J. Natural source zone depletion of LNAPL: A critical review supporting modelling approaches. *Water Res.* **2019**, *157*, 630–646. [CrossRef] [PubMed]
4. Zhang, T.; Barry, R.G.; Knowles, K.; Ling, F.; Armstrong, R.L. Distribution of seasonally and perennially frozen ground in the Northern Hemisphere. In Proceedings of the 8th International Conference on Permafrost, Zurich, Switzerland, 21–25 July 2003; A.A. Balkema Publishers: Amsterdam, The Netherland, 2003; Volume 2, pp. 1289–1294.
5. Zhai, Y.; Han, Y.; Xia, X.; Li, X.; Lu, H.; Teng, Y.; Wang, J. Anthropogenic Organic Pollutants in Groundwater Increase Releases of Fe and Mn from Aquifer Sediments: Impacts of Pollution Degree, Mineral Content, and pH. *Water* **2021**, *13*, 1920. [CrossRef]
6. Wei, C.; Yu, S.; Jichun, W.; Yaling, C.; Erxing, P.; Leonid, G. Soil hydrological process and migration mode influenced by the freeze-thaw process in the activity layer of permafrost regions in Qinghai-Tibet Plateau. *Cold Reg. Sci. Technol.* **2021**, *184*, 103236. [CrossRef]
7. Weaver, J. Assessment of sub-permafrost groundwater conditions at the Red Dog Mine, Alaska. In Proceedings of the Permafrost Vols 1 and 2, Zurich, Switzerland, 21–25 July 2003; pp. 1223–1228.
8. Mohammed, A.A.; Bense, V.F.; Kurylyk, B.L.; Jamieson, R.C.; Johnston, L.H.; Jackson, A.J. Modeling Reactive Solute Transport in Permafrost-Affected Groundwater Systems. *Water Resour. Res.* **2021**, *57*, e2020WR028771. [CrossRef]
9. Wang, T.H.; Zhang, H.; Lu, X.S. Experimental study on moisture migration in unsaturated loess under freezing effect. *Appl. Mech. Mater.* **2012**, *204–208*, 552–556. [CrossRef]
10. Satoi, A.; Suzuki, T.; Nishimoto, S.; Yamashita, S. Improvement of high-moisture soil by repeated freezing and thawing. *New Front. Chin. Jpn. Geotech.* **2007**, 471–478.
11. Daniel, J.A.; Staricka, J.A. Frozen Soil Impact on Ground Water–Surface Water Interaction. *Am. Water Resour. Assoc.* **2000**, *36*, 151–160. [CrossRef]
12. Lari, K.S.; Rayner, J.L.; Davis, G.B. A computational assessment of representative sampling of soil gas using existing groundwater monitoring wells screened across the water table. *J. Hazard. Mater.* **2017**, *335*, 197–207. [CrossRef]
13. Lari, K.S.; Rayner, J.L.; Davis, G.B. Towards characterizing LNAPL remediation endpoints. *J. Environ. Manag.* **2018**, *224*, 97–105. [CrossRef]
14. Sweeney, R.E.; Ririe, G.T. Small purge method to sample vapor from groundwater monitoring wells screened across the water table. *Ground Water Monit. Remediat.* **2017**, *37*, 51–59. [CrossRef]
15. McDonald, R.; Knox, O.G.G. Cold region bioremediation of hydrocarbon contaminated soil: Do we know enough? *Environ. Sci. Technol.* **2014**, *48*, 9980–9981. [CrossRef]
16. Camenzuli, D.; Wise, L.E.; Stokes, A.J.; Gore, D. Treatment of soil co-contaminated with inorganics and petroleum hydrocarbons using silica: Implications for remediation in cold regions. *Cold Reg. Sci. Technol.* **2017**, *135*, 8–15. [CrossRef]
17. Chen, J.; Zheng, X.; Zang, H.; Liu, P.; Sun, M. Numerical Simulation of Moisture and Heat Coupled Migration in Seasonal Freeze-thaw Soil Media. *J. Pure Appl. Microbiol.* **2013**, *7*, 151–158.
18. Dai, L.; Guo, X.; Du, Y.; Zhang, F.; Ke, X.; Cao, Y.; Li, Y.; Li, Q.; Lin, L.; Cao, G. The Response of Shallow Groundwater Tables to Soil Freeze-Thaw Process on the Qinghai-Tibet Plateau. *Groundwater* **2018**, *57*, 602–611. [CrossRef]
19. Iwakun, O.; Biggar, K.; Sego, D. Influence of cyclic freeze-thaw on the mobilization of LNAPLs and soluble oil in a porous media. *Cold Reg. Sci. Technol.* **2010**, *64*, 9–18. [CrossRef]
20. Lari, K.S.; Davis, G.B.; Rayner, J.L. Towards a digital twin for characterising natural source zone depletion: A feasibility study based on the Bemidji site. *Water Res.* **2021**, *208*, 117853. [CrossRef]
21. Ming, F.; Li, D.Q. One-dimensional water-heat coupling model and experiment of unsaturated frozen soil. *J. Cent. South Univ.* **2014**, *45*, 889–894.
22. Tian, Z.; Lu, Y.; Horton, R.; Ren, T. A simplified de Vries-based model to estimate thermal conductivity of unfrozen and frozen soil. *Eur. J. Soil Sci.* **2016**, *67*, 564–572. [CrossRef]
23. Zheng, C.; Šimůnek, J.; Zhao, Y.; Lu, Y.; Liu, X.; Shi, C.; Li, H.; Yu, L.; Zeng, Y.; Su, Z. Development of the Hydrus-1D freezing module and its application in simulating the coupled movement of water, vapor, and heat. *J. Hydrol.* **2021**, *598*, 126250. [CrossRef]
24. Twarakavi, N.K.C.; Simunek, J.; Seo, S. Evaluating interactions between groundwater and vadose zone using the HYDRUS-based flow package for MODFLOW. *Vadose Zone J.* **2019**, *7*, 757–768. [CrossRef]
25. Rahbeh, M.; Srinivasan, R.; Mohtar, R. Numerical and conceptual evaluation of preferential flow in Zarqa River Basin, Jordan. *Ecohydrol. Hydrobiol.* **2018**, *19*, 224–237. [CrossRef]
26. Shelia, V.; Simunek, J.; Hoogenbooom, G. Coupling DSSAT and HYDRUS-1D for simulations of soil water dynamics in the soil-plant-atmosphere system. *J. Hydrol. Hydromech.* **2017**, *2066*, 232–245. [CrossRef]
27. Ling, M.; Heglie, J.; Mickelson, B. Estimate Dilution Factor with SEAWAT Modeling to Calculate Groundwater Cleanup Goals. In Proceedings of the World Environmental and Water Resources Congress 2017: Groundwater, Sustainability, and Hydro-Climate/Climate Change, Sacramento, CA, USA, 21–25 May 2017; pp. 36–50. [CrossRef]

28. Steiner, C.; Heimlich, K.; Hilberg, S. Comparison of heat-plume predictions using two large-scale models of groundwater heatpump installations: FEFLOW vs. the A-WAV-model. *Grundwasser* **2016**, *21*, 173–185. [CrossRef]
29. Pfeiffer, W.T.; Graupner, B.; Bauer, S. The coupled non-isothermal, multiphase-multicomponent flow and reactive transport simulator OpenGeoSys-ECLIPSE for porous media gas storage. *Environ. Earth Sci.* **2016**, *75*, 1347. [CrossRef]
30. Imunek, J.; Van Genuchten, M.T.; Ejna, M. MTHYDRUS: Model Use, Calibration, and Validation. *Trans. ASABE* **2016**, *55*, 1261–1274. [CrossRef]
31. Kanda, E.K.; Mabhaudhi, T.; Senzanje, A. Coupling Hydrological and Crop Models for Improved Agricultural Water Management—A Reveiw. *Bulg. J. Agric. Sci.* **2020**, *24*, 380–390.
32. Zhao, Y.; Horn, R. Modeling of Coupled Water and Heat Transfer in Freezing and Thawing Soils, Inner Mongolia. *Water* **2016**, *8*, 424. [CrossRef]
33. Zheng, F.; Zhai, Y.; Xia, X.; Yin, Z.; Du, Q.; Zuo, R.; Wang, J.; Teng, Y.; Xu, M. Simulation of Trinitrogen Migration and Transformation in the Unsaturated Zone at a Desert Contaminant Site (NW China) Using HYDRUS-2D. *Water* **2018**, *10*, 1363. [CrossRef]
34. Xue, K.; Wen, Z.; Zhang, M.; Li, D.; Gao, Q. Relationship between matric potential, moisture migration and frost heave in freezing process of soil. *Trans. Chin. Soc. Agric. Eng. (Trans. CSAE)* **2017**, *33*, 176–183.
35. Ge, S.; Mckenzie, J.; Voss, C.; Wu, Q. Exchange of groundwater and surface water mediated by permafrost response to seasonal and long-term air temperature variation. *Geophys. Res. Lett.* **2011**, *38*, 3138–3142. [CrossRef]
36. Pruess, K.; Oldenburg, C.; Moridis, G. *TOUGH2 User's Guide, Version 2.0*; Earth Sciences Division, Lawrence Berkeley National Laboratory: Berkeley, CA, USA, 1999; Volume 94720, p. 198. [CrossRef]
37. Zhang, K.; Yamamoto, H.; Pruess, K. *TMVOC-MP: A Parallel Numerical Simulator for Three-PhaseNon-Isothermal Flows of Multicomponent Hydrocarbon Mixtures Inporous/Fractured Media*; U.S. Department of Energy Office of Scientific and Technical Information: Washington, DC, USA, 2008. [CrossRef]
38. Guarnaccia, J.; Pinder, G.; Fishman, M. *NAPL: Simulator Documentation*; EPA/600/R-97/102; National Risk Management Research Laboratory, USEPA: Ada, OK, USA, 1997; Volume 74820.
39. Tao, J.H.; Shi, X.Q.; Kang, X.Y.; Xu, H.X.; Wu, J.C. Numerical analysis of influencing factors on the structure of light non-aqueous liquid pollution source area. *Hydrogeol. Eng. Geol.* **2018**, *45*, 132–140. (In Chinese) [CrossRef]
40. Yang, M.X. Study on the Differentiation and Evolution Mechanism of Petroleum Organic Pollution Components in the Water Level Fluctuation Zone. Master's Thesis, Jilin University, Changchun, China, 2014. (In Chinese)
41. Dafny, E. TCE longevity in the vadose zone and loading to the groundwater-The case of episodic NAPL releases from near-surface source. *Environ. Technol. Innov.* **2017**, *7*, 128–140. [CrossRef]
42. Pruess, K.; Battistelli, A. *TMVOC, a Numerical Simulator for Three-Phase Non-Isothermal Flows of Multicomponent Hydrocarbon Mixtures in Saturated-Unsaturated Heterogeneous Media*; Report LBNL-49375; Lawrence Berkeley National Laboratory: Berkeley, CA, USA, 2002. [CrossRef]
43. Hansson, K.; Šimůnek, J.; Mizoguchi, M.; Lundin, L.; van Genuchten, M. Water Flow and Heat Transport in Frozen Soil: Numerical Solution and Freeze-Thaw Applications. *Vadose Zone J.* **2004**, *3*, 693–704. [CrossRef]
44. Lekmine, G.; Lari, K.S.; Johnston, C.D.; Bastow, T.P.; Rayner, J.L.; Davis, G.B. Evaluating the reliability of equilibrium dissolution assumption from residual gasoline in contact with water saturated sands. *J. Contam. Hydrol.* **2016**, *196*, 30–42. [CrossRef]
45. Larsen, K.S.; Jonasson, S.; Michelsen, A. Repeated freeze-thaw cycles and their effects on biological processes in two arctic ecosystem types. *Appl. Soil Ecol.* **2002**, *21*, 187–195. [CrossRef]
46. Simunek, J.; Van Genuchten, M.T.; Sejna, M. *The HYDRUS-1D Software Package for Simulating the One-Dimensional Movement of Water, Heat, and Multiple Solutes in Variably-Saturated Media*; University of California, Department of Environmental Sciences: Berkeley, CA, USA, 2013.

water

MDPI

Article

Heavy Metals in Sediments of Hulun Lake in Inner Mongolia: Spatial-Temporal Distributions, Contamination Assessment and Source Apportionment

Tong Liu [1,†], Dasheng Zhang [1,†], Weifeng Yue [2,*], Boxin Wang [1], Litao Huo [1], Kuo Liu [1] and Bo-Tao Zhang [2,*]

1 Hebei Institute of Water Science, Shijiazhuang 050051, China; tongliu2020@163.com (T.L.)
2 College of Water Sciences, Beijing Normal University, Beijing 100875, China
* Correspondence: yuewf@bnu.edu.cn (W.Y.); zhangbotao@126.com (B.-T.Z.)
† These authors contributed equally to this work.

Abstract: The spatial and temporal distributions, contamination evaluation, and source apportionment of Cu, Zn, As, Pb, Cd, and Cr in the sediments of Hulun Lake were explored in this work. The pollution characteristics of six heavy metals were assessed by single factor pollution index (PI) and geo-accumulation index (Igeo). The sources of heavy metals in the surface sediments were analyzed by the positive definite matrix factorization (PMF) and Pearson correlation analysis. The sedimentary records of heavy metals in core sediments were reproduced by radioisotopes. The average concentrations of 6 heavy metals except Cd were lower than the corresponding background values. The spatial distributions of Cu, Zn, Cr, Cd and As were generally similar and showed higher abundances in the southwestern part of the lake. With the use and import of heavy metals, the concentration of heavy metals in core sediments increased with the fluctuation of years. The peak of heavy metal concentration was related to the high growth rate of gross domestic product in 2003-2008. The single factor pollution index and geo accumulation index results showed that the surface sediment was mainly polluted by Cd, followed by Zn and As. Natural parent material, agricultural activities and industrial activities were the main sources of heavy metal pollution in the sediments, accounting for 17.03%, 26.34%, and 56.63% of the total heavy metal accumulation, respectively. Pb was derived mainly from natural parent material. Cd and As were closely associated with agricultural activities. Cu and Zn were mainly attributed to industrial mining activities. Source apportionment of the ecological risks of heavy metals illustrated that industrial sources were the primary ecosystem risk sources (66.1%), followed by agricultural sources (23.75%) and natural sources (10.15%). The results will also provide reference data for future studies of heavy metals pollution in sediments from Hulun Lake and other lakes.

Keywords: heavy metal; sediment; spatiotemporal distribution; source apportionment; risk assessment Hulun Lake

Citation: Liu, T.; Zhang, D.; Yue, W.; Wang, B.; Huo, L.; Liu, K.; Zhang, B.-T. Heavy Metals in Sediments of Hulun Lake in Inner Mongolia: Spatial-Temporal Distributions, Contamination Assessment and Source Apportionment. *Water* **2023**, *15*, 1329. https://doi.org/10.3390/w15071329

Academic Editors: José Ángel Fernández and Paul B. Tchounwou

Received: 20 January 2023
Revised: 16 February 2023
Accepted: 24 March 2023
Published: 28 March 2023

1. Introduction

With the rapid development of industrialization and urbanization in recent decades, the increase of metal pollution in aquatic ecosystems has become one of the critical environmental problems that attract extensive attention because of their toxicity, bioavailability and mobility [1,2]. The hazards of heavy metals in the environment are not only related to their total concentration, but are also attributed to chemical partitioning. The sources of heavy metals in the aqueous environment are broadly classified as natural or anthropogenic. Weathering of bedrock and volcanic eruptions are categorized as natural sources. Anthropogenic sources include metal ore and coal mining, production and domestic wastewater, fertilization, fossil fuel combustion, tourism, aquaculture and fisheries, etc. [3–5]. The primary sources of heavy metals vary on a continental scale. In Africa, heavy metals mainly come from bedrock weathering, metallurgy, industry, and domestic wastewater [6,7]. The density of heavy metals (such as Cu, Pb, Zn, Cr, As and Cd) usually exceeds 4.5 g/cm^3 [7]. Some heavy metals (e.g., Pb and Cd) are mainly classified as toxic substances with serious effects on the aquatic environment and human health. Although V, Cr, Co, Ni and Zn are essential for specific biological processes, they are defined as toxic at higher concentrations [8,9]. Enrichment of heavy metals in lakes causes fatal harm to human health and aquatic organisms. Therefore, it is necessary to place a high premium on the heavy metals pollution in lakes. Heavy metal pollution control in lakes should receive more and more attention [10].

The prevention and control of heavy metal pollution in lake sediments is one of the best strategies due to the difficult and time-consuming degradation of heavy metals from the sediments [11]. To present, many studies have been carried out on the potential risk assessment and sources of heavy metals in lake sediments [12]. To quantify the contamination and ecological risk of metals in sediments, several indices have been proposed, which can be summarized as the geoaccumulation index (Igeo), the pollution index (PI) and the potential ecological risk Index (PERI) methods. Analyzing the distribution characteristics, PERI and potential sources of heavy metals in lake sediments is a necessary prerequisite for taking effective pollution control measures. The source of heavy metals is varied, such as machinery manufacturing, pesticides, fertilizers, exhaust emissions, mining, and natural parent materials. Hence, more research should be devoted to quantifying the source of heavy metals and preventing them from entering lakes [12]. The positive matrix factorization (PMF) is a quantitative method for pollutant source matrix analysis based on a receptor model that is not limited by the composition of a single pollutant source. In addition, it has been extensively used in the identification and quantification of heavy metals pollution sources [13].

Hulun Lake is located in the western part of Hulunbeier Grassland, known as the lung of the grassland, which brings together the rivers of Kelulun, Wuerxun, and Xinkai [14]. In recent years, with the economic and social development around the lake and the impact of global warming, the lake surface has been shrinking, grassland degradation and ecological environment and the water quality is deteriorating, fishery resources are on the verge of exhaustion, and the wetland ecological environment is in urgent need of protection [15]. At present, research on heavy metals in Hulun Lake mainly focuses on content distribution and pollution assessment, while previous studies focused on source analysis and vertical distribution characteristics.

To better protect fishery resources and environment in Hulun Lake waters and maintain aquatic environment, the objectives of this research are (1) to illustrate heavy metals content in the surface sediments of Hulun Lake, (2) to analyze the spatial distribution of heavy metals in surface sediments, (3) to assess heavy metals pollution in surface sediments, (4) to identify and analyse the potential sources of heavy metals and the contributions of the sources to ecological risk by the PMF model, and (5) to disclose heavy metals temporal variation in the core sediments. The overall objective of this study was to assess the spatial and temporal distribution of heavy metals in sediments, contamination evaluation, and the preferential contribution to ecological risk and source allocation. The results should provide basic data for the health of Hulun Lake aquatic ecosystem.

2. Materials and Methods

2.1. Study Area

Hulun Lake basin (including Hailar River basin and Halaha River basin) is located in China and Mongolia, with a drainage area of 2.92 × 10^5 km^2. With an area of 2339 square kilometers and an average water depth of 5.6 m, Hulun Lake is the fourth largest freshwater lake in China [16]. The lake water is mainly supplied by precipitation, surface runoff, and groundwater. The runoff of the Wuerxun River and the Kelulun River is the primary water source of Hulun Lake, and the Xinkai River is the outlet river. The Hailar basin around the lake is active in magmatic activity during the geological period and contains rich metal mineral resources such as Cu, Zn, Pb, Mn, Ag, Fe, etc.

2.2. Sample Collection and Physicochemical Analysis

In this study, Figure 1 shows the sample site locations of Hulun Lake in Inner Mongolia. The sediment samples from Hulun Lake (n = 12) were collected in August 2020. The surface sediment samples were gathered at a depth of 2 cm by a stainless steel sampler (JDHC-200A, Jintan, Changzhou, China). The core sediment sample C1 was gathered from the central part of the lake, cut sequentially at 2 cm thickness. Each sample was then transferred to a clean polyethylene sample bag and labeled, transported as quickly as possible back to the lab, and sand, gravel, and plant roots were removed. Stored at 4 °C until further analysis.

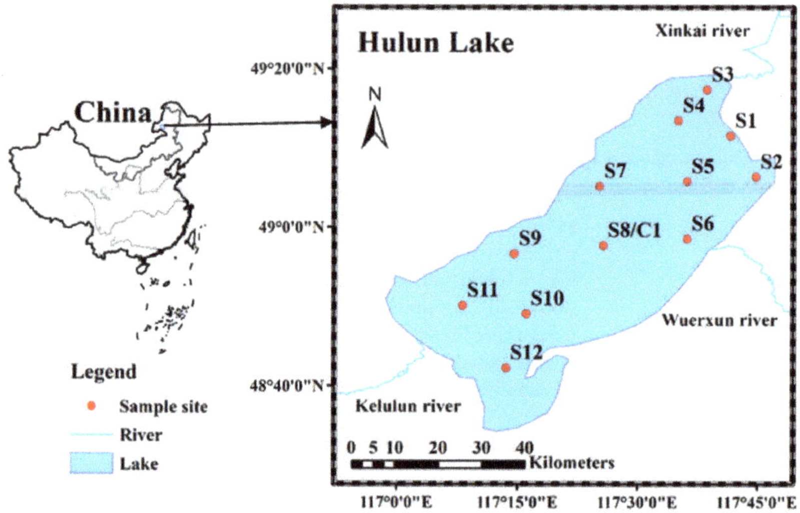

Figure 1. Locations of the sampling sites in Hulun Lake.

To detect heavy metals concentrations, each 8 g of sample was weighed with an electronic scale (XPR204S/AC) and dried in an oven (GN-53A) at 100 °C for 12 h, and then ground and sieved through a 100-mesh sieve (GILSON ISO 565). Eventually, Heavy metal content is measured using X-ray fluorescence spectrometry (Hitachi, Shanghai, China) and each sample was repeated three times. The core sediments' ^{210}Pb and ^{137}Cs specific activities were measured by a GWL high-purity germanium gamma spectrometer (AMETEK-AMT Ortec Co., Santiago, MN, USA). The China Institute of Atomic Energy supplied standard isotopic samples. Sediment age series calculated by constant recharge rate model.

2.3. Assessment Method

2.3.1. Single Factor Pollution Index (PI)

The single factor pollution index (PI) method refers to the method for assessing the pollution degree of a certain heavy metals in soil or sediment [17], as shown in Equation (1):

$$PI = CI/SI \tag{1}$$

where *CI* and *SI* mean actual content and the background value of heavy metals i, respectively.

2.3.2. Geo-Accumulation Index (Igeo)

The Igeo was originally proposed by Muller for application to bottom sediments [18]. This method has also been applied to identify anthropogenic sources of heavy metals in sediments [19]. The Igeo of heavy metals was calculated according to the follows:

$$Igeo = \log_2[C_n/(1.5B_n)] \tag{2}$$

where Cn is the concentration of heavy metals measured in sediment and Bn is their baseline shown in Table 1. 1.5 is the background matrix correction coefficient. The geo-accumulation index consists of 7 grades or classes as shown in Table S1 [20].

Table 1. Statistical results of heavy metals in Hulun Lake sediments.

Category	Heavy Metals (mg/kg, n = 12)					
	Cu	Pb	Cr	Zn	Cd	As
Maximum	30.21	28.93	34.11	116.10	0.85	14.8
Minimum	2.96	5.39	9.20	24.50	ND	1.50
Mean	19.32	17.63	21.90	71.36	0.33	7.22
S.D.	10.77	8.38	9.21	31.56	0.25	4.8
CV(%)	55	47	41	43	75	66
LEL	16	31	26	120	0.6	6
SEL	110	250	110	820	10	33
Background value [a]	35	35	90	100	0.2	12

S.D.: Standard deviation; CV: Coefficients of variation; ND: Not detected; LEL: Lowest effect level; SEL: Severe effect level [21]; [a] Soil background concentrations of heavy metals in Hulun Lake basin [15].

2.4. Receptor Model PMF 5.0

The PMF model was used as the source analysis method of heavy metals in sediments [22,23]. The calculation method is shown in Equation (3):

$$X_{ij} = \sum_{k=1}^{p} g_{ik} f_{kj} + e_{ij} \tag{3}$$

where X_{ij} means content of the j_{th} heavy metals at the i_{th} sampling site, g_{ik} represents the contribution of the k_{th} source to the i_{th} sample, f_{kj} means the content of the element j from the k_{th} source, and e_{ij} means the residual error matrix. The objective function Q is obtained in Equation (4):

$$Q = \sum_{i=1}^{n} \sum_{j=1}^{m} \left(\frac{e_{ij}}{u_{ij}} \right)^2 \tag{4}$$

where u_{ij} denotes the uncertainty of the j_{th} heavy metal in the i_{th} sample, calculated from the species-specific method detection limit (MDL), the concentration and the error component provided [24]. If the heavy metals concentration exceeds the MDL, the u_{ij} is obtained as described in Equation (5) [25]:

$$u_{ij} = \sqrt{(errorfraction \times concentrations)^2 + (MDL)^2} \tag{5}$$

2.5. Statistical Analysis

The spatial distribution pattern of PAEs in the sediments was mapped by kriging spatial interpolation analysis using ArcGIS 10.6 software. Data processing was completed by Origin Pro 9.0. Pearson correlation analysis with IBM SPSS Statistics 25 was used to investigate the relationship between metal content.

3. Results and Discussions

3.1. Heavy Metals Content in the Surface Sediments of Hulun Lake

Table 1 described the concentration of 6 heavy metals in the surface sediments of Hulun Lake. The ranges of Cu, Pb, Cr, Zn, Cd, and As content were 2.96–30.21 mg/kg, 5.39–28.93 mg/kg, 9.20–34.11 mg/kg, 24.50–116.10 mg/kg, ND-0.85 mg/kg and 1.50–14.8 mg/kg, respectively. Except for Cd, the average concentrations of these heavy metals were below the corresponding background values. The maximum concentrations of Zn and As were higher than the background concentrations. The coefficient of variation (CV) can visually indicate the changes in the regional distribution of heavy metals. The higher CV value, the stronger the disturbance of anthropogenic activities on the regional distribution of heavy metals. The CV of Cu, Cd, and As were all greater than 50%, indicating that the regional distribution of these elements was highly different in sediments.

The SQG of the Canadian Council of Ministers of the Environment (CCME) in Ontario classifies the impact of heavy metals on sediments as No Effect Level (NEL), Lowest Effect Level (LEL), and Severe Effect Level (SEL) [21]. This research has concluded that the LEL and SEL have the highest quantities of Cu, Cd, Cr, and As. In comparison to LEL, Pb and Zn have lower maximum concentrations.

To compare the pollution level of Hulun Lake, mean concentration of heavy metals in sediment of different lake systems are summarized in Table 2 following the order of sampling time. The concentrations of Cu, Pb, Cr, Zn, Cd, and As are lower than in lakes such as Dongjiang Lake, Hongze Lake, Taihu Lake, and Poyang Lake, which were probably related to population density and anthropogenic activities intensity. Taihu Lake has the highest concentration of heavy metals compared with other domestic lakes. The Taihu Lake basin is one of the regions with the most concentrated population and the fastest urbanization process in China. This region's rapid economic and social development has caused increasingly major water pollution problems. Although the catchment area of Hulun Lake is 256,000 km^2, the basin is remote and sparsely populated, with over 60% of the Hulun Lake basin located in Mongolia. The population density in China's watershed areas is close to ~$1.6/km^2$, which is much lower than other lake basins.

Table 2. Mean concentration of heavy metals in sediments of the lake systems following the order of sampling time.

Lake	Sampling Year	Heavy Metals (mg/kg)						References
		Cu	Pb	Cr	Zn	Cd	As	
Dongjiang Lake, China	2021	33.01	47.40	67.58	113.9	2.25	80.80	[26]
Kangryong River, North Korea	2019	32.70	30.90	53.90	132.50	0.28	14.30	[27]
Hongze Lake, China	2018	25.35	27.2	66.78	74.77	0.23	16.55	[28]
East Dongting Lake, China	2017	42.60	58.60	42.9	88.80	5.50	ND	[29]
Taihu Lake, China	2017	44.71	37.00	102.46	163.62	0.80	13.34	[30]
Persian Gulf, Iran	2017	14.87	7.42	59.22	32.51	0.15	5.92	[31]
Lake Emerald, India	2017	611.32	34.04	411.48	174.40	ND	ND	[32]
Inle Lake, Myanmar	2017	7.92	9.64	19.24	18.80	0.05	6.59	[2]
Los Molinos Lake, Argentina	2015	4.67	6.45	2.50	9.80	0.06	1.74	[33]
Poyang Lake, China	2014	35.17	32.63	81.39	104.17	0.66	11.34	[34]
The present study	2020	19.32	17.63	21.90	71.36	0.33	7.22	-

3.2. Spatial Distribution of Heavy Metals in Surface Sediments

The continuous spatial distribution profiles of heavy metals in the surface sediments were analyzed using ArcGIS to visualize the concentration level of heavy metals between different sampling sites, as shown in Figure 2. The distribution characteristics of heavy metal contents in the surface sediments of the study area are different. The concentration of Pb is highest in the Wuerxun estuary. Cu, Zn, Cr, Cd, and As exhibited roughly similar spatial distributions, showing higher abundance in the southwest of the lake district. There was a tourist resort and No. 5 fishing ground southwest of Hulun Lake, and the frequent anthropogenic activities might be the main cause. There were more than 70 oil wells around Wuerxun River, and the higher Pb concentration might originate from exploitation [35,36]. The results show that pollution sources and artificial activities impact the spatial distribution of heavy metals in the catchment area of specific lakes. As a result, most of the heavy metals were likely to be contaminants of terrestrial origin, and surface runoff and adsorption to suspended particulate matter might be the pathways that carry contaminants from the catchment to the lake.

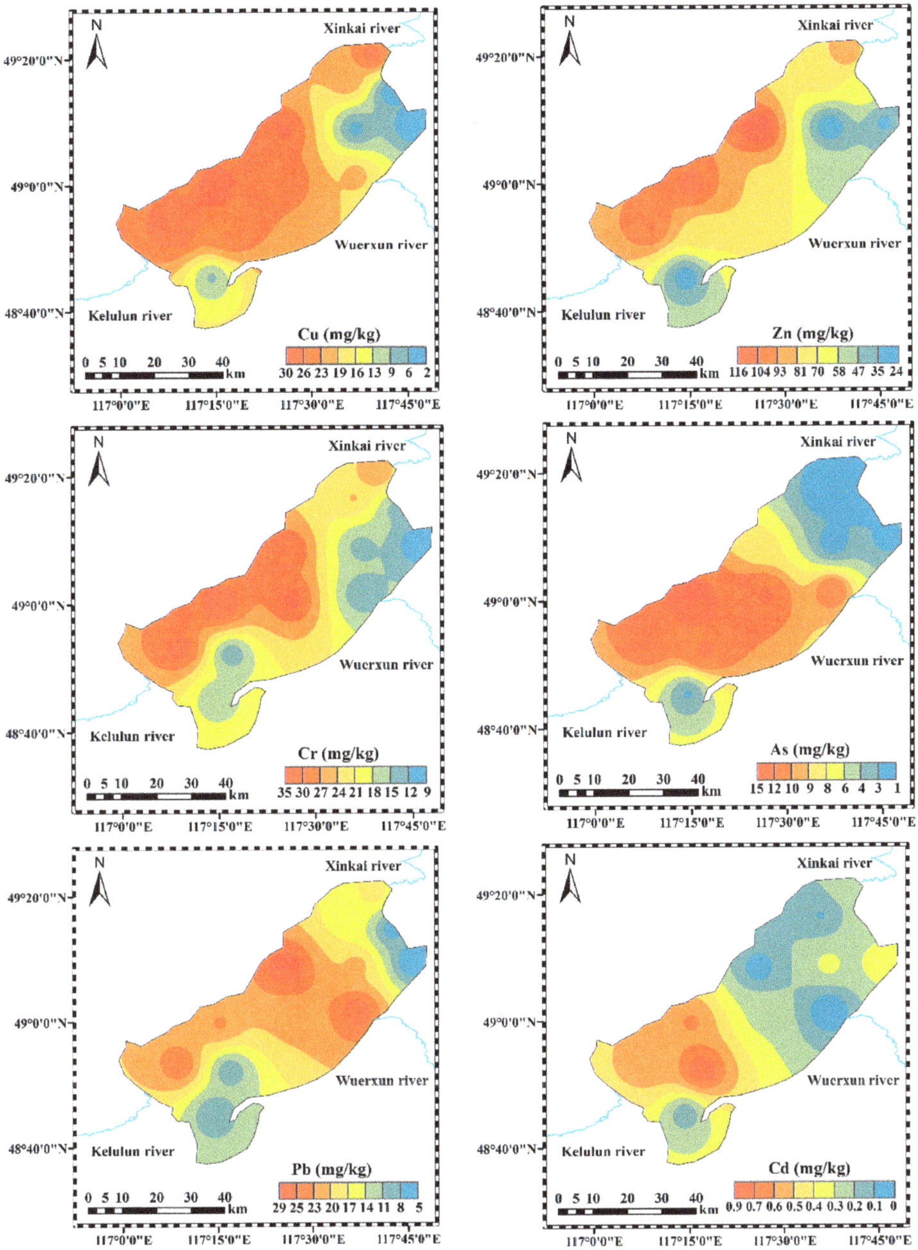

Figure 2. Spatial distribution patterns of heavy metals in the surface sediments of the Hulun Lake.

The CV of heavy metals pollution is a statistic reflecting the distribution uniformity and varying degree of heavy metals elements in the study area. The greater the CV value, the greater the disruptions of anthropogenic factors on its content distribution and the greater the difference in its spatial distribution [37]. Under the value of CV, the heavy metals variability in sediments could be divided into low variation (≤15%), medium variation (15–35%), and high variation (>35%) [38]. The CV values of all heavy metals in the surface sediments of the study area are Cd > As > Cu > Pb> Zn >Cr in descending order in Table 1. All heavy metals show high variation characteristics, indicating that the spatial distribution and sources of heavy metals are different and disturbed by anthropogenic activities.

3.3. Assessment of Heavy Metals Pollution in Surface Sediments

Single factor pollution index (PI) refers to the method of evaluating the pollution grade of a certain heavy metals in soil or sediment [39]. The PI results for Pb, Cr, Cu, As, Cd, and Zn are shown in Figure 3. The PI. varied considerably across the different metals. In this work, the PI values were classified as low (PI < 1), moderate (PI between 1 and 3), high ($3 < PI \leq 5$), and extremely high (PI > 5). The PI values for Cd in sediments varied from 0 to 4.25, with about 16.7% of the analyzed samples presenting high PI for Cd. Zn and As present moderate contamination, with PI values between 0.25 and 1.1 for Zn and between 0.13 and 1.2 for As. Moderate PI values were found in 25% of the samples for Zn and 42% for As. These results indicated that Pb, As, and Zn pollution might be widespread in Hulun Lake. Cu, Pb, and Cr exhibited low pollution, with PI values below 1.

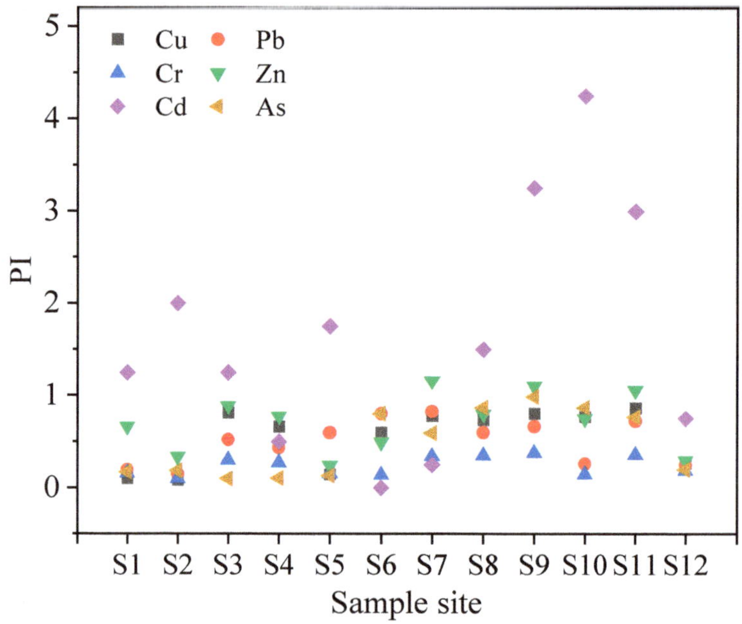

Figure 3. Single factor index (PI) distribution of heavy metals in surface sediments of Hulun Lake.

The Igeo is a widespread tool to assess the contamination degree of sediment. The essence is to remove the corresponding natural or background content from the present heavy metals content to obtain the total enrichment degree of heavy metals caused by anthropogenic activities. The Igeo variation characteristics of heavy metals in the sediments were presented in Figure 4. In the sediments of Hulun Lake, the Igeo of Cd was the highest, with 16.7% at the moderate pollution level and 25% between unpolluted to moderately pollution level. The reason might be related to the discarded batteries of seine netting fishing [40]. In addition, Zn reached a moderate pollution level of 8.3%. Cu, Pb and Cr showed unpollution level of 100%, indicating that these three heavy metals do not significantly contaminate the sediments. The results of the Igeo and PI are basically consistent. The study of heavy metal pollution in Hulun Lake had important research value. In the evaluation of heavy metal pollution, a combination of multiple evaluation methods was used to make the evaluation results more accurate and reference value.

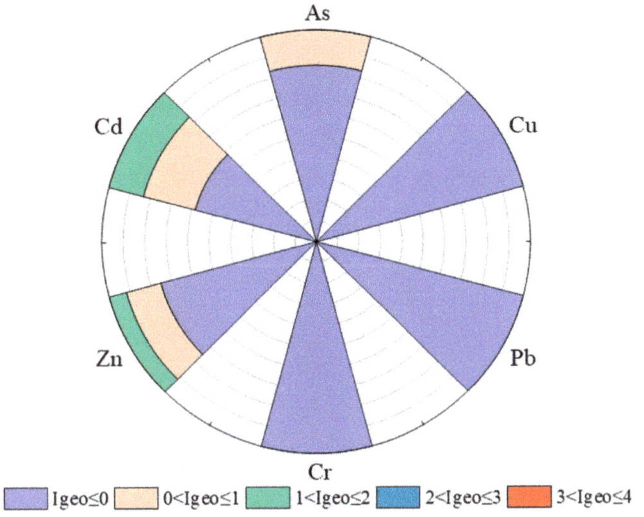

Figure 4. Geo-accumulation rose chart of respective heavy metals in the sediment of Hulun Lake.

3.4. Source Identification of Heavy Metals in Surface Sediments by PMF

Correlation and PMF analyses have been extensively applied to study the sources of heavy metals [41]. The correlation between heavy metals can qualitatively explain their sources and transport [42]. The correlation coefficients of heavy metals concentrations was exhibited in Figure 5. There were stronger correlations among Zn, Cr, and Cu. Cd showed a significant positive correlation with As. Pb exhibited a positive correlation with Cr and Cu. This result reflects the homology of heavy metals in sediments and is consistent with the spatial distribution characteristics of heavy metals mentioned above. The highly correlated results for these heavy metals suggested that they might belong to the same sources with similar levels of contamination and release [43].

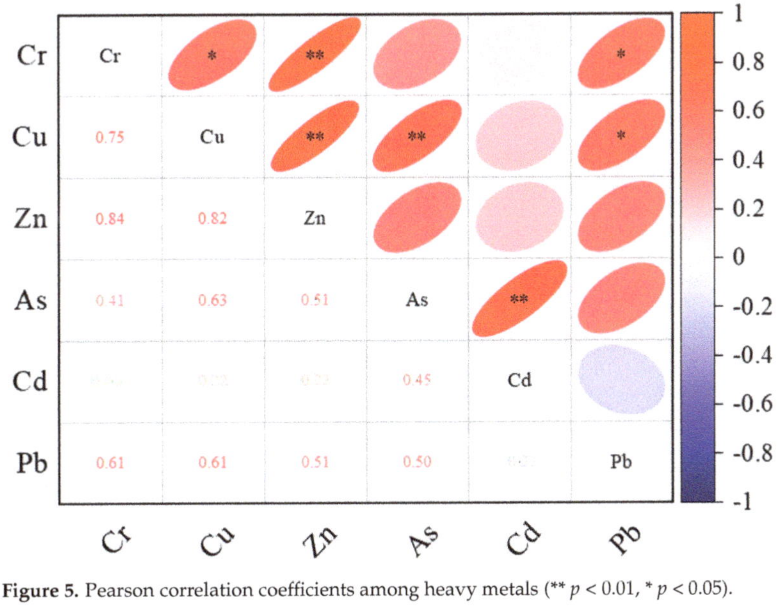

Figure 5. Pearson correlation coefficients among heavy metals (** $p < 0.01$, * $p < 0.05$).

The PMF model was intensively applied to the source identification and concentration quantification of heavy metals in sediments. EPA PMF 5.0 software was used to examine the amounts and unknown concentrations of six heavy metals in surface sediment samples. The analysis results are shown in Figures 6 and 7a. Factor 1 contributed 17.03% of the heavy metals sources in the sediment, among which Pb (52.10%) has a higher loading. The maximum concentration of Pb was below the local background value, reflecting Pb might not be externally contaminated. In the pollution assessment of PI and Igeo, Pb showed a pollution-free level and could be considered to come from the natural background. Therefore, factor 1 was defined as the natural parent material source.

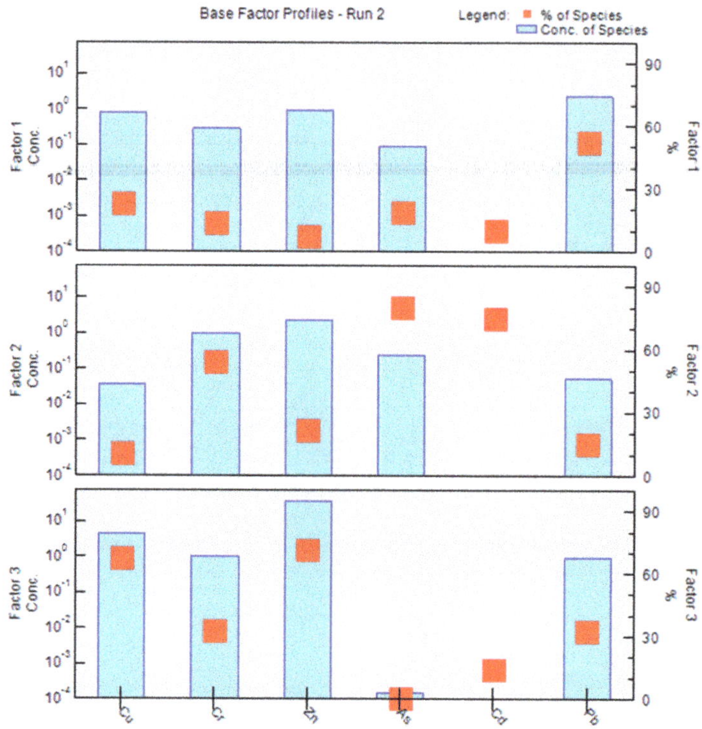

Figure 6. Source profiles and contributions of heavy metals in the studied Hulun Lake sediments obtained with the PMF model.

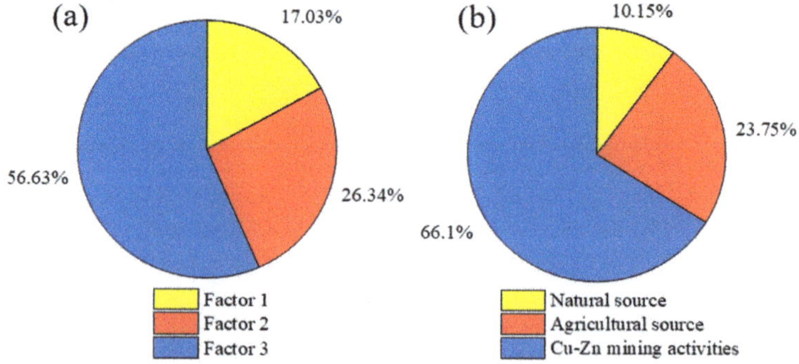

Figure 7. Contributions of various sources to the sediment heavy metal concentrations (**a**) and the associated ecological risks (**b**) in the study area.

Factor 2 had major factor loadings for As, Cu, and Cd and accounted 26.34% of the contributions of various heavy metals sources to the heavy metals concentrations in the sediments of Hulun Lake. Sewage agricultural irrigation and some agricultural products such as phosphate fertilizers, pesticides and organic fertilizers emitted large amounts of Cd [44,45]. According to researchs, the accumulation of Cr, As, and Cd was closely related to agricultural production activities, e.g., long-term application of pesticides [45], especially the content of Cd in sediment is significantly related to the content of total phosphorus. Animal husbandry is developed around the Kelulun River and Wuerxun River. Meanwhile, the water source of Hulun Lake comes from the runoff input of the Urxun River and Kulun River, and the final pollutants flow into the lake through the surface runoff and accumulate in the sediment. Therefore, factor 2 was associated with agricultural sources.

Factor 3 exhibited significant factor loadings for Zn and Cu and accounted a relatively large amount (56.63%) of the contributions of various sources of heavy metals to the heavy metals concentrations in the sediments of Hulun Lake. Cu and Zn are indicators of the source of mineral dust, which is closely related to human activities. There are many zinc copper mines in the north of Hulun Lake [46]. Mining activities result in the release of metals [47]. Some processing wastewater and metal dust will be discharged during metal smelting and mechanical processing. Correlation study results reveal high positive correlations between Cu and Zn and similar spatial distributions, demonstrating that they may have homology. Besides, the upper part of the Kelulun River is densely populated with industrial parks, with leather processing plants and nitrate plants [48,49], which might be contributed to the higher Cu and Zn concentrations in the southwestern part of the lake. Therefore, factor 3 was associated with industrial sources.

In this work, source apportionment of the ecological risks was assessed by combining PMF and PERI [50], which calculation methods were exhibited in the supplementary material. The source contributions to ecological risk were shown in Figure 7b. Cu-Zn mining and industrial activities were considered to be the the main factor, contributing 66.1%, which exceeded 0.2 times the contribution derived from content. The ecological risk contribution of agricultural sources and natural source were 23.75% and 10.15% respectively, both of which are lower than the contribution values from the content. The change of source contribution was related to the toxic reaction factors of heavy metals [50]. In brief, heavy metals from industrial sources have a higher toxicity coefficient than other sources.

3.5. Heavy Metals Temporal Variation in the Core Sediments

The vertical distribution characteristics of lake sediments are considered the records of historical sedimentation of the geographical environment. The vertical changes of element content in rock cores reflect the sedimentary conditions in different historical periods, which can understand the accumulation and superposition history of heavy metals in the study area and reflect the impact of human activities on heavy metals in the study area in different historical stages [49]. In our previous research, the sedimentation rate of the Hulun Lake was calculated as 0.51 cm/a [51]. The sediment deposition rate obtained is similar to that previously reported [41]. It can be seen from Figure 8 that the content of heavy metals changes obviously in stages. Since the Cd concentration is low and the changing trend is not apparent, vertical analysis is not conducted. Heavy metals were detectable in the core sediment, attributed to their constant use and continuous lake input during the study period. The heavy metals concentrations of core sediments at 0–2 cm depth (corresponding to the years 2020–2016) were the lowest, which might be attributed to the remarkable achievements in ecological environment protection and control in the Hulun Lake basin in recent years [51]. This result is in approximate agreement with the results for heavy metals in the surface sediments (Table 1). The maximum peak occured at 6–8 cm, which corresponds to 2008–2004. From 2003 to 2008, Inner Mongolia's GDP growth rate of 17.5% was the highest since the last 20 years and is considered to be the fastest growing economy in China in this time [51]. Heavy metals content in this stage was low and in a stable fluctuation condition between 1978 and 1992, which might respond the background concentration of heavy metals. The economy was still in its early stages of development, moving slowly, and there were fewer heavy metal emissions due to human activity.

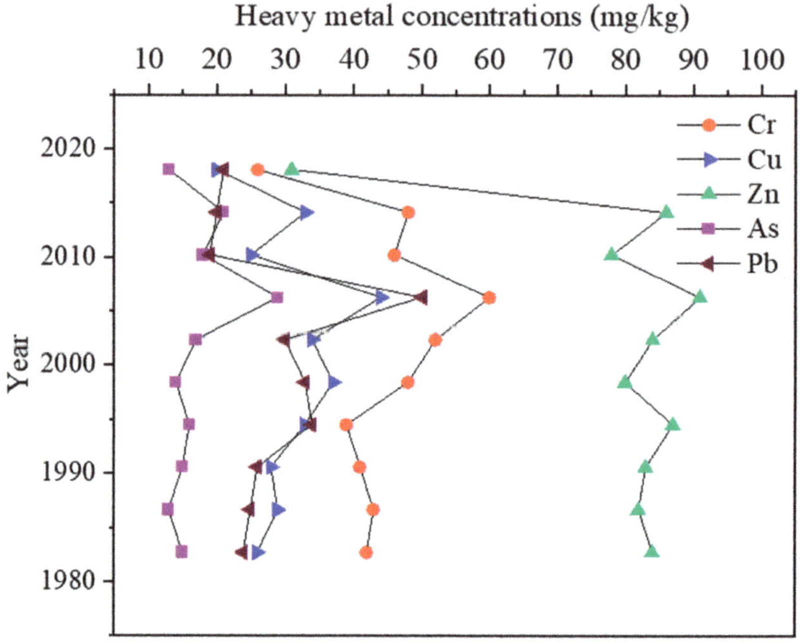

Figure 8. Temporal distribution profiles of heavy metals in the core sediments of Hulun Lake.

4. Conclusions

Heavy metals in lake sediments threaten aquatic ecological environments and public health. In summary, Cu, Pb, Cr, Zn, Cd, and As were widespread in all surface sediments ranging from 2.96–30.21 mg/kg, 5.39–28.93 mg/kg, 9.20–34.11 mg/kg, 24.50–116.10 mg/kg, ND-0.85 mg/kg and 1.50–14.8 mg/kg, respectively. The average concentration of Cd was below the corresponding background value. The distribution of heavy metals depends on population density and intensity of human activity, and heavy metals were terrestrial pollutants. Surface runoff and suspended particulate matter might be the pathways that carry pollutants from the catchment source to the lake. Three primary sources of heavy metals were identified in sediments: the natural parent material source (relevant indicator is Pb), agricultural sources (relevant indicators are As, Cu and Cd), and mining-related industrial sources (relevant indicators are Cu and Zn), whose contributions to heavy metals' pollution was 17.03%, 26.34% and 56.63%, respectively.The findings of the source apportionment of ecological risks showed that industrial sources were the major ecological risk sources (66.1%), followed by agricultural sources (23.75%) and natural sources (10.15%). The PI and Ieo results indicated that the surface sediment were mainly polluted by Cd, followed by Zn and As. Heavy metal concentrations fluctuate with the year, reflecting a gradual increase in heavy metal use and lake inputs as time passes. The highest peak in heavy metals might be related to the high growth rate of GDP during the same period. In this study, the spatial and temporal distribution of heavy metals in Hulun Lake sediments and the analysis of pollution sources were investigated. The study provides data to support future environmental protection in related fields. The results of the study can provide basic data for the health of the aquatic ecosystem of Hulun Lake, as well as to better protect the fishery resources and environment of Hulun Lake waters and maintain the aquatic environment.

Supplementary Materials: The following supporting information can be downloaded at: https://www.mdpi.com/article/10.3390/w15071329/s1, Table S1: The evaluation standard of Geo-accumulation index (Igeo).

Author Contributions: T.L.: Writing—original draft, Investigation, Visualization. D.Z.: Methodology, Formal analysis, Data curation. W.Y.: Data curation, Investigation, Funding acquisition. B.W.: Software. L.H.: Visualization. K.L.: Formal analysis. B.-T.Z.: Supervision, Conceptualization, Resources. All authors have read and agreed to the published version of the manuscript.

Funding: This work was supported by the National Key R & D Program of China (2021YFC3200101, 2018YFD0900805), Key R & D projects in Hebei Province (21373904D) and Experimental Project of Seasonal River Ecological Flow Technology.

Data Availability Statement: Not applicable.

Conflicts of Interest: The authors declare that they have no known competing financial interest or personal relationships that could have appeared to influence the work reported in this paper.

References

1. Irabien, M.J.; Cearreta, A.; Leorri, E.; Gómez, J.; Viguri, J. A 130 year record of pollution in the Suances estuary (southern Bay of Biscay): Implications for environmental management. *Mar. Pollut. Bull.* **2008**, *56*, 1719–1727. [CrossRef]
2. Aung, P.P.; Mao, Y.; Hu, T.; Qi, S.; Tian, Q.; Chen, Z.; Xing, X. Metal concentrations and pollution assessment in bottom sediments from Inle Lake, Myanmar. *J. Geochem. Explor.* **2019**, *207*, 106357. [CrossRef]
3. Hu, J.; Zhou, S.; Wu, P.; Qu, K. Assessment of the distribution, bioavailability and ecological risks of heavy metals in the lake water and surface sediments of the Caohai plateau wetland, China. *PLoS ONE* **2017**, *12*, e0189295. [CrossRef]
4. Li, S.; Hu, X.; Tang, Y.; Huang, C.; Xiao, W. Changes in lacustrine environment due to anthropogenic activities over 240 years in Jiuzhaigou National Nature Reserve, southwest China. *Quat. Int.* **2014**, *349*, 367–375. [CrossRef]
5. Sojka, M.; Jaskuła, J.; Barabach, J.; Ptak, M.; Zhu, S. Heavy metals in lake surface sediments in protected areas in Poland: Concentration, pollution, ecological risk, sources and spatial distribution. *Sci. Rep.* **2022**, *12*, 15006. [CrossRef] [PubMed]
6. Niu, Y.; Chen, F.; Li, Y.; Ren, B. Trends and Sources of Heavy Metal Pollution in Global River and Lake Sediments from 1970 to 2018. In *Reviews of Environmental Contamination and Toxicology*; Springer: Cham, Switzerland, 2021; Volume 257. [CrossRef]
7. Song, Y.; Man, S.C.; Ji, Y.L.; Dong, J.J. Regional background concentrations of heavy metals (Cr, Co, Ni, Cu, Zn, Pb) in coastal sediments of the South Sea of Korea. *Sci. Total Environ.* **2014**, *482–483*, 80–91. [CrossRef] [PubMed]
8. Liu, Q.; Wang, F.; Meng, F.; Jiang, L.; Li, G.; Zhou, R. Assessment of metal contamination in estuarine surface sediments from Dongying City, China: Use of a modified ecological risk index. *Mar. Pollut. Bull.* **2018**, *126*, 293–303. [CrossRef]
9. Goretti, E.; Pallottini, M.; Ricciarini, M.I.; Selvaggi, R.; Cappelletti, D. Heavy metals bioaccumulation in selected tissues of red swamp crayfish: An easy tool for monitoring environmental contamination levels. *Sci. Total Environ.* **2016**, 559. [CrossRef] [PubMed]
10. Ma, Z.; Chen, K.; Yuan, Z.; Bi, J.; Huang, L. Ecological Risk Assessment of Heavy Metals in Surface Sediments of Six Major Chinese Freshwater Lakes. *J. Environ. Qual.* **2013**, *42*, 341–350. [CrossRef]
11. Hu, Y.; Cheng, H.; Tao, S. The Challenges and Solutions for Cadmium-contaminated Rice in China: A Critical Review. *Environ. Int.* **2016**, *92–93*, 515–532. [CrossRef] [PubMed]
12. Li, D.; Yu, R.; Chen, J.; Leng, X.; Zhao, D.; Jia, H.; An, S. Ecological risk of heavy metals in lake sediments of China: A national-scale integrated analysis. *J. Clean. Prod.* **2022**, *334*, 130206. [CrossRef]
13. Fei, X.; Lou, Z.; Xiao, R.; Ren, Z.; Lv, X. Contamination assessment and source apportionment of heavy metals in agricultural soil through the synthesis of PMF and GeogDetector models. *Sci. Total Environ.* **2020**, *747*, 141293. [CrossRef] [PubMed]
14. Ma, Y.; Liu, T.; Zhang, B.-T.; Liu, Y.; Shao, P.; Sun, C.; Zhang, Y.; Bi, J.; Dong, Y.; Wang, S. Spatial-temporal distributions and influential factors of phthalate acid esters in sediments of three lakes in Inner Mongolia. *Environ. Sci. Pollut. Res.* **2022**, *29*, 32800–32812. [CrossRef] [PubMed]
15. Zhang, X.; Li, C.; Zhang, S. Distribution characteristics and ecological risk assessment of heavy metals in Hulun Lake sediments. *J. Agric. Environ. Sci.* **2010**, *29*, 157–162.
16. Mao, P.; Zhang, J.; Li, M.; Liu, Y.; Wang, X.; Yan, R.; Shen, B.; Zhang, X.; Shen, J.; Zhu, X.; et al. Spatial and temporal variations in fractional vegetation cover and its driving factors in the Hulun Lake region. *Ecol. Indic.* **2022**, *135*, 108490. [CrossRef]
17. Karim, Z.; Qureshi, B.A.; Mumtaz, M. Geochemical baseline determination and pollution assessment of heavy metals in urban soils of Karachi, Pakistan. *Ecol. Indic.* **2015**, *48*, 358–364. [CrossRef]
18. Muller, G. Index of Geoaccumulation in Sediments of the Rhine River. *GeoJournal* **1969**, *2*, 109–118.
19. Mandour, A.; El-Sayed, M.K.; El-Gamal, A.A.; Khadr, A.M.; Elshazly, A. Temporal distribution of trace metals pollution load index in the Nile Delta coastal surface sediments. *Mar. Pollut. Bull.* **2021**, *167*, 112290. [CrossRef]
20. Muller, G. Schwermetalle in den sedimenten des Rheins-Vernderungen seit 1971. *Umschav* **1979**, *79*, 778–783.
21. Persaud, D.; Jaagumagi, R.; Hayton, A. *Guidelines for the Protection and Management of Aquatic Sediment. Quality in Ontario*; Ministry of Environment and Energy: Toronto, ON, Canada, 1993.

22. Huang, J.; Guo, S.; Zeng, G.M.; Li, F.; Gu, Y.; Shi, Y.; Shi, L.; Liu, W.; Peng, S. A new exploration of health risk assessment quantification from sources of soil heavy metals under different land use. *Environ. Pollut.* **2018**, *243*, 49–58. [CrossRef]
23. Wang, S.; Cai, L.M.; Wen, H.H.; Luo, J.; Wang, Q.S.; Liu, X. Spatial distribution and source apportionment of heavy metals in soil from a typical county-level city of Guangdong Province, China. *Sci. Total Environ.* **2019**, *655*, 92–101. [CrossRef] [PubMed]
24. Amp, O.O. *EPA Positive Matrix Factorization (PMF) 5.0 Fundamentals and User Guide*; U.S. Environmental Protection Agency Office of Research and Development: Washington, DC, USA, 2015.
25. Zhao, X.; Li, X.; Lu, H. Heavy metal pollution characteristics and potential ecological risk assessment of surface sediments in Dongjiang Lake. *Environ. Sci.* **2022**, *43*, 3048–3057.
26. Kim, I.; Kim, Y.; Kim, R.; Hyon, T. Spatial distribution, origin and contamination assessment of heavy metals in surface sediments from Jangsong tidal flat, Kangryong river estuary, DPR Korea. *Mar. Pollut. Bull.* **2021**, *168*, 112414. [CrossRef] [PubMed]
27. Zi, X.; Zhang, M.; Gu, X. Impact of enclosure culture on heavy metal content in surface sediments of Hongze Lake and ecological risk assessment. *Environ. Sci.* **2021**, *42*, 5355–5363.
28. Tang, C.; Qian, B.; Li, W. Pollution characteristics and risk assessment of heavy metals in surface sediments of dongting lake area. *People's Yangtze River* **2020**, *51*, 49–56.
29. Zhang, J.; Guo, X.; Zeng, Y.; Deng, J. Distribution and pollution assessment of heavy metals in river sediments in taihu lake basin. *Environ. Sci.* **2019**, *40*, 2202–2210. [CrossRef]
30. Jahromi, F.A.; Keshavarzi, B.; Moore, F.; Abbasi, S.; Busquets, R.; Hooda, P.S.; Jaafarzadeh, N. Source and risk assessment of heavy metals and microplastics in bivalves and coastal sediments of the Northern Persian Gulf, Hormogzan Province. *Environ. Res.* **2021**, *196*, 110963. [CrossRef]
31. Karthikeyan, P.; Vennila, G.; Nanthakumar, G.; Aswini, M. Dataset for spatial distribution and pollution indices of heavy metals in the surface sediments of Emerald Lake, Tamil Nadu, India. *Data Brief.* **2020**, *28*, 104877. [CrossRef]
32. Griboff, J.; Horacek, M.; Wunderlin, D.A.; Monferran, M.V. Bioaccumulation and trophic transfer of metals, As and Se through a freshwater food web affected by antrophic pollution in Córdoba, Argentina. *Ecotoxicol. Environ. Saf.* **2018**, *148*, 275–284. [CrossRef]
33. Xie, Z.; Jiang, Y.; Zhang, H.; Wang, D.; Qi, S.; Du, Z.; Zhang, H. Assessing heavy metal contamination and ecological risk in Poyang Lake area, China. *Environ. Earth Sci.* **2016**, *75*, 549. [CrossRef]
34. Wang, W.; Li, W.; Yan, Y.; Liu, B.; Wang, T.; Mao, S.; Song, N.; Dou, H.; Ao, W.; Zou, C. Organic Matter Pollution During the Spring Thaw in Hulun Lake Basin: Contribution of Multiform Human Activities. *Bull. Environ. Contam. Toxicol.* **2020**, *105*, 307–316. [CrossRef] [PubMed]
35. Wang, Y.; Zhang, F.; Zou, Y.; Sun, J.; Lin, X.; Liang, T. Oil source and charge in the Wuerxun Depression, Hailar Basin, northeast China: A chemometric study. *Mar. Pet. Geol.* **2018**, *89*, 665–686. [CrossRef]
36. Ji, Z.; Long, Z.; Zhang, Y.; Wang, Y.; Pei, Y. Enrichment differences and source apportionment of nutrients, stable isotopes, and trace metal elements in sediments of complex and fragmented wetland systems. *Environ. Pollut.* **2021**, *289*, 117852. [CrossRef]
37. Jia, Y.; Wang, L.; Qu, Z.; Yang, Z. Distribution, contamination and accumulation of heavy metals in water, sediments, and freshwater shellfish from Liuyang River, Southern China. *Environ. Sci. Pollut. Res.* **2018**, *25*, 7012–7020. [CrossRef]
38. Li, W.; Li, C.; Jia, K.; Zhang, S. Nutrient element distribution and environmental pollution assessment in Hulun Lake sediments, Inner Mongolia. *Agrl. Environ. Sci.* **2010**, *29*, 5.
39. Salem, D.; Khaled, A.; Nemr, A.E.; El-Sikaily, A. Comprehensive risk assessment of heavy metals in surface sediments along the Egyptian Red Sea coast. *Egypt. J. Aquat. Res.* **2014**, *25*, 349–362. [CrossRef]
40. Chen, Y.; SHI, Q.; QU, J.; HE, M.; LIU, Q. A pollution risk assessment and source analysis of heavy metals in sediments: A case study of Lake Gehu, China. *Chin. J. Anal. Chem.* **2022**, *50*, 100077. [CrossRef]
41. Jiang, X.; Teng, A.; Xu, W.; Liu, X. Distribution and pollution assessment of heavy metals in surface sediments in the Yellow Sea. *Mar. Pollut. Bull.* **2014**, *83*, 366–375. [CrossRef]
42. Shao, D.; Zhan, Y.; Zhou, W.; Zhu, L. Current status and temporal trend of heavy metals in farmland soil of the Yangtze River Delta Region: Field survey and meta-analysis. *Environ. Pollut.* **2016**, *219*, 329–336. [CrossRef]
43. Xiao, H.; Shahab, A.; Xi, B.; Chang, Q.; Siddique, J. Heavy metal pollution, ecological risk, spatial distribution, and source identification in sediments of the Lijiang River, China. *Environ. Pollut.* **2020**, *269*, 116189. [CrossRef] [PubMed]
44. Liu, G. Environmental Geochemistry of Heavy Metals in Hulun Lake. Master's thesis, Inner Mongolia University, Hohhot, China, 2011. Available online: http://a.xueshu.baidu.com/usercenter/paper/show?paperid=3f7d292206f46f24f331a491ed6a3d8d (accessed on 23 March 2023).
45. Wu, J.; Ge, Y.; Li, J.; Lai, X.; Chen, R. A PMF-SSD based approach for the source apportionment and source-specific ecological risk assessment of Le'an river in Jiangxi Province, China. *Environ. Res.* **2023**, *219*, 115027. [CrossRef]
46. Yu, H.; Shi, X.; Sun, B. Analysis on the change of water quality and eutrophication of Hulun Lake from 2011 to 2020. *Study Arid. Areas* **2021**, *38*, 1534–1545. [CrossRef]
47. Su, D.; Zang, S.; Ye, H. Pollution characteristics and source identification of heavy metals in sediment cores of Nanshan Lake in Zhalong Wetland. *Environ. Sci.* **2012**, *33*, 7.
48. Li, J.; Song, L.; Chen, H.; Wu, J.; Teng, Y. Source apportionment of potential ecological risk posed by trace metals in the sediment of the Le'an River, China. *J. Soils Sediments* **2020**, *20*, 2460–2470. [CrossRef]
49. Xie, R.; Sun, D.; Sun, L.; Zhang, N.; Zhang, J.; Zang, S. Sediment record of heavy metals over the last 150 years in Northeast China: Implications for regional anthropogenic activities. *Ecotoxicology* **2021**, *30*, 1354–1365. [CrossRef]

50. Chen, J. Grassland Pearl Reappears Moving Brilliance—Remarkable Achievements Have Been Made in the Ecological Environment Protection and Control of Hulun Lake Basin; Hulun Buir Daily; 2022; pp. 1–2. *Hulun Buir Daily*, 2022; 1–2. [CrossRef]
51. Hakanson, L. An ecological risk index for aquatic pollution control.a sedimentological approach. *Water Res.* **1980**, *14*, 975–1001. [CrossRef]

water

MDPI

Article

Distribution and Ecological Risk Assessment of Pharmaceuticals and Personal Care Products in Sediments of North Canal, China

Shasha Pei [1], Binghua Li [2,*], Boxin Wang [1,*], Jingchao Liu [3] and Xuanying Song [4]

1 Hebei Institute of Water Science, Shijiazhuang 050051, China; peishasha187@163.com
2 Beijing Water Science and Technology Institute, Beijing 100089, China
3 Faculty of Architecture, Civil and Transportation Engineering, Beijing University of Technology, Beijing 100124, China; liujingchao721@126.com
4 China Water Resources Bei Fang Investigation, Design & Research Co., Ltd., Tianjin 300222, China; 15201606308@163.com
* Correspondence: libinghua75@163.com (B.L.); wangboxin293@163.com (B.W.); Tel.: +86-137-1785-6370 (B.L.); +86-186-3304-5293 (B.W.)

Abstract: The pollution of water bodies by pharmaceuticals and personal care products (PPCPs) has attracted widespread concern due to their widespread use and pseudo-persistence, but their effects on sediments are less known. In this study, solid-phase extraction-high performance liquid chromatography–tandem mass spectrometry (SPE-LC/MSMS) was used to investigate the occurrence and ecological risks of five typical pharmaceuticals and personal care products (PPCPs) in thirteen key reservoirs, sluices, dams, and estuaries in the Haihe River Basin. At the same time, the PPCP exchanges of surface water, groundwater, and sediments in three typical sections were studied. Finally, the PPCP's environmental risk is evaluated through the environmental risk quotient. The results showed that the five PPCPs were tri-methoprazine (TMP), sinolamine (SMX), ibuprofen (IBU), triclosan (TCS), and caffeine (CAF). The average concentration of these PPCPs ranged from 0 to 481.19 μg/kg, with relatively high concentrations of TCS and CAF. The relationship between PPCPs in the surface sediments was analyzed to reveal correlations between SMX and TMP, CAF and IBU, CAF and TCS. The risk quotients (RQ) method was used to evaluate the ecological risk of the five detected PPCPs. The major contributors of potential environmental risks were IBU, TCS and CAF, among which all the potential environmental risks at the TCS samples were high risk. This study supplemented the research on the ecological risk of PPCPs in sediments of important reaches of the North Canal to reveal the importance of PPCP control in the North Canal and provided a scientific basis for pollution control and risk prevention of PPCPs.

Keywords: risk assessment; surface sediments; PPCPs; North Canal Basin; spatial distributions

Citation: Pei, S.; Li, B.; Wang, B.; Liu, J.; Song, X. Distribution and Ecological Risk Assessment of Pharmaceuticals and Personal Care Products in Sediments of North Canal, China. *Water* **2022**, *14*, 1999. https://doi.org/10.3390/w14131999

Academic Editor: Heiko L. Schoenfuss

Received: 11 May 2022
Accepted: 16 June 2022
Published: 22 June 2022

1. Introduction

In recent years, the existence of pharmaceuticals and personal care products (PPCPs) as emerging pollutants has become a global concern [1]. PPCPs mainly include prescription, over-the-counter, and consumer chemicals, including perfumes and sun-screen. PPCPs have been detected in various water bodies in China and residual PPCPs in the environment can lead to endocrine disorders, reduced reproductive rates, and reduced life expectancy [2,3]. Exposure to even very low concentrations of PPCPs in the environment poses potential risks to the ecosystem and human health [4–7].

Due to the continuous input of PPCPs into the aquatic environment, a large number of PPCPs have been absorbed in river sediments and have been detected in most aquatic sediments in China. Pan et al. [8] showed that the PPCP content in the sediments of the eastern half of Chaohu Lake was higher in the coastal zone than in the open zone, and showed a decreasing trend from west to east along the main lake flow direction. Chen et al. [9] found that the PPCP pollution was more serious in the Luoma Lake Bay area, with caffeine, ofloxacin, and sulfa-methoxazole in surface sediments of Luoma Lake presenting higher ecological risks.

The binding of PPCPs to sediments affects the bioavailability and bioaccumulation of PPCPs [10,11]. At the same time, when sediments are scoured by river water, PPCPs are re-released into the water environment, causing secondary pollution [12,13]. Therefore, toxicological data and ecological risks of PPCPs in sediments may be overestimated or underestimated [14,15]. At present, studies have shown that the contribution of PPCPs in sediments to water pollution cannot be ignored.

Although PPCPs were reported in some river basins in China [16,17], there are few studies on PPCP concentrations in the North Canal sediments. Meanwhile, as one of the largest PPCP consumption areas in China and even the world, Beijing's daily sewage load is about 3.3 million tons, but the sewage treatment capacity of suburbs and urban areas is only 50% and 83%, respectively [8,18,19]. Therefore, the study of PPCP pollution, sources, and ecological risks in sediments is greatly significant for river pollution control and ecological protection. The purpose of this study is as follows: (i) to determine the content and spatial distribution of PPCPs in surface sediments in the North Canal of China; (ii) to analyze the correlation between PPCP concentration and water quality parameters in river sediments; (iii) to assess the potential risks of PPCPs in the sediment.

2. Material and Method

2.1. Study Area

North Canal is located in the north of China, with a latitude of 40°00′–40°50′ north and longitude of 115°50′–116°25′ east (Figure 1). It originates at the south foot of the Jundu Mountain in Beijing, flows through the Hebei province, before merging into the Haihe River in Tianjin. The region has a temperate monsoon climate, with hot and rainy summers, followed by cold and dry winters. The annual average rainfall is 642.5 mm, with most rainfall concentrated in June to August. The mainstream of the North Canal has a total length of 260 km, a total area of 6166 km^2, and an average annual runoff of 572 million m^3 [19]. As the main drainage channel of Beijing, the North Canal flows through densely populated and highly industrialized areas, resulting in 93% of the water source in the upper reaches of the North Canal belonging to wastewater discharged from sewage treatment plants and 4% to untreated wastewater, while the middle and lower reaches of the North Canal are affected by pesticides, fertilizers, and domestic sewage [20].

2.2. Sampling and Analysis

Sediment samples were collected from 13 typical sections of the North Canal by a rigid plexiglas tube gravity sampler in July 2016. The collected columnar sediment samples were divided into three groups by 0–20 cm, 20–40 cm, 40–60 cm and filtered through a 1 mm screen. Water samples were collected 0–50 cm from the surface using a water harvester at the same location as the sediment samples. Groundwater samples were collected from 13 monitoring wells along the north Canal river, and collected at a depth of 50 m by QED low-flow sampling equipment. Sediment river water and groundwater samples were collected and sent to the laboratory for testing. Details of the analytical procedure are provided elsewhere [13,18]. In short, all samples were stored in pre-cleaned cryogenic containers and immediately transported to the laboratory for processing. In the laboratory, the water samples were concentrated by pre-treated solid phase extraction and sediment samples were extracted by ultrasonic extraction, followed by high performance liquid chromatography–tandem mass spectrometry analysis of target antibiotics, following appropriate quality assurance and quality control procedures, usually including solvent blank procedures and independent inspection standards. Concentrations of 5 PPCPS were determined for all samples. The 5 PPCPS were classified and abbreviated as follows: sulfamethoxazole (SMX), trimethoprimethidine (TMP), caffeine (CAF), ibuprofen (IBU);

triclosan (TCS). The main physicochemical properties of the five target PPCPS are shown in Supplementary Materials Table S1. Limits of detection (limit of detection, LOD) and limits of quantification (limit of quantification, LOQ) are generally determined as the minimum detectable amount of analyte with a signal-to-noise ratio. The LOD and LOQ of each PPCPs are shown in Table S2.

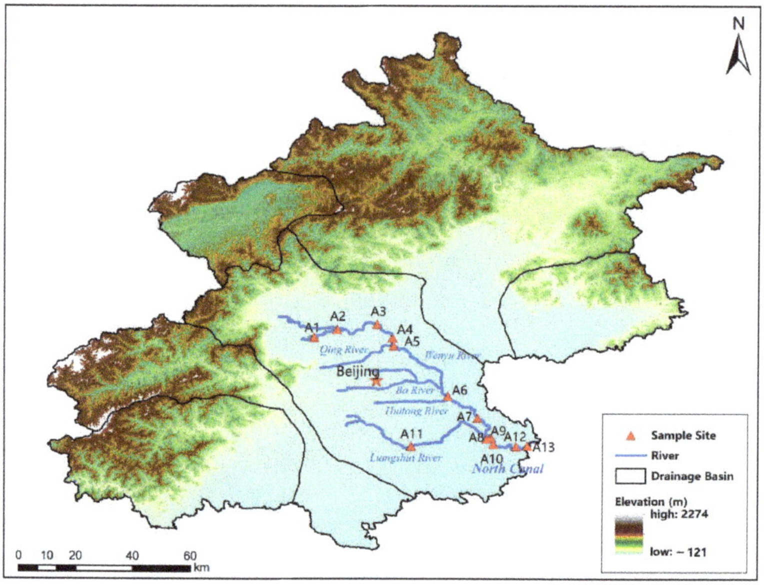

Figure 1. Schematic diagram of the study area and sampling point location.

2.3. *Leaching Potential Assessment*

In this study, the groundwater ubiquity score (GUS) method was used to evaluate the leaching potential of PPCPs, and the parameters were provided by the leaching potential evaluation mode [21,22]. Its calculation formula is as follows:

$$GUS = \log t_{\frac{1}{2}} (4 - \log K_{OC}) \quad (1)$$

K_{OC} is the organic carbon-water allocation coefficient, and $t_{1/2}$ is the soil degradation half-life (days). GUS value classification criteria were as follows: low leaching potential (GUS < 1.8), medium leaching potential ($1.8 \leq GUS \leq 2.8$), and the high leaching potential (GUS > 2.8).

2.4. *Ecological Risk Assessment*

In this study, the environmental risk quotients (RQ) method was selected to assess the potential risk of PPCPs in sediments to aquatic organisms. This entropy is based on the risk assessment method in the European Technical Guidance document [23]. The RQ value is calculated from the measured concentration (MEC) and the predicted no-effect concentration (PNEC) by the following formula:

$$RQ = \frac{MEC}{PNEC} \quad (2)$$

$$PNEC = \frac{EC_{50}}{AF} \text{ or } PNEC = \frac{NOEC}{AF} \quad (3)$$

RQ represents the risk quotient calculated by EC_{50} or NOEC, and MEC represents the measured environmental concentration. PNEC is the predicted no-effect concentration, which is the maximum concentration of a drug known to have no adverse effects on microorganisms or ecosystems in the environment. PNEC is obtained by dividing EC_{50} or NOEC by the assessment factor (AF, 1000 for acute toxicity, 100 for NOEC). The sources of EC50 and NOEC are shown in Table S3. The RQ were divided into the following four categories: RQ < 0.01 (no significant risk), 0.01 < RQ < 0.1 (low risk), 0.1 < RQ < 1 (medium risk), RQ > 1 (high risk) [23].

2.5. Pseudo-Partitioning (P-PC)

To obtain a quantitative understanding of the relationship of the antibiotics between the sediment and water phases, the pseudo-partitioning coefficient ($k_{d,s}$) was used to describe the system and was calculated according to the following equation:

$$k_{d,s} = C_s/C_w\left(L\ kg^{-1}\right) \quad (4)$$

where C_s is the concentration in the sediment, and C_w is the corresponding concentration in the water phase [24].

3. Results and Discussion

3.1. Occurrence of PPCPs in Surface Sediments

The detection results showed that the five PPCPs were detected in the North Canal. The detection rates of TMP, SMX, IBU, TCS, and CAF were 100.00%, 61.53%, 76.92%, 100.00% and 69.00%, respectively. The average concentrations were ND–1.55 μg/kg, ND–0.46 μg/kg, ND–7.47 μg/kg, 1.45–697.63 μg/kg and ND–246.59 μg/kg, respectively. The total concentration of the five PPCPs in the North Canal was 5.53–641.93 μg/kg, and the main pollutants were TCS and CAF. The proportion of the average concentration of the five PPCPs to the total concentration was TCS (65.26%), CAF (32.32%), IBU (2.25%), TMP (0.11%), SMX (0.05%), respectively. The average concentrations of TCS and CAF were the highest among the five PPCPs, which were 57.77 μg/kg and 28.618 μg/kg, 14 to 585 times higher than the other PPCPs. TCS is widely used as an antibacterial agent in a variety of soaps, different kinds of toothpaste, and health care products [25]. CAF, as a stimulant, is ubiquitous in our daily life and has been detected in waters around the world [26].

Compared to previous reports, TCS concentrations in Liuxi River and Zhujiang River sediments were lower than those in the North Canal, while TCS concentrations in Shijing River sediments were higher than those in the North Canal, and CAF concentrations in the North Canal sediments were higher than those in Baiyang Lake. Additionally, CAF concentrations in the North Canal sediments exceeded concentrations in the Songhua River. The contents of the TMP, SMX, and IBU in the sediments of the North Canal were lower than those in other regions of China (Table 1).

3.2. Spatial Variation in PPCPs in Surface Sediments

As shown in Figure 2, the five PPCPs in the North Canal showed no obvious increase or decrease in the upstream and downstream regions but fluctuated constantly from upstream to the downstream. In particular, PPCPs suddenly increased in some stations. The levels of SMX, IBU, and TCS all fluctuate at the A5 site. The A5 site is located at the confluence of the Qinghe River and Wenyu River, where there are three sewage treatment plants. The discharge of reclaimed water may affect the concentration level of PPCPs in the sediment. At the confluence points A8 and A9 of the Liangshui River and North Canal, the PPCP concentration also showed the same level of fluctuations. At the sampling points A3, A8, and A9, the distribution of CAF fluctuates more obviously than the other PPCPs. In summary, the concentration fluctuation range was large in the upstream population aggregation area, while the concentration was low in the downstream villages, towns, and the fluctuation range was small.

Table 1. Occurrence and concentration of PPCPs in sediments in different study areas (μg/kg).

Chemical	Location	Range	Reference
TMP	Nanjing	ND–1.07	Xue et al. 2013 [27]
	Taihu Lake	ND–39.3	Xu et al., 2014 [28]
	Beijing	ND–5.02	Zhang et al. 2017 [29]
	Hanjiang River	ND–10	Hu et al., 2018 [30]
	Baiyang Lake	ND–7.26	Zhang et al., 2018 [31]
	Pearl River	0.1–0.2	Xie et al., 2019 [32]
	This study	ND–1.55	
SMX	Jiulong River	1.2–3.4	Zhang et al., 2011 [33]
	Baiyang Lake	ND–7.9	Li et al., 2012 [34]
	Taihu Lake	ND–49.3	Xu et al., 2014 [28]
	Beijing	ND–0.35	Zhang et al., 2017 [29]
	Hanjiang River	ND–1.2	Hu et al., 2018 [30]
	Haihe River	1.2–2.54	Chen et al., 2018 [35]
	Pearl River	ND	Xie et al., 2019 [32]
	This study	ND–0.46	
IBU	Taihu Lake	ND–21	Xie et al., 2015 [36]
	Guangzhou	ND–3.19	Peng et al., 2017 [37]
	Songhua River	25.2–95.0	He et al., 2018 [38]
	Pearl River	ND–0.02	Xie et al., 2019 [32]
	This study	ND–7.47	
TCS	Liuxi River	ND–116	Zhao et al., 2010 [39]
	Zhujiang River	12.2–196	Zhao et al., 2010 [39]
	Shijing River	345–1329	Zhao et al., 2010 [39]
	Yangtze River	0.18–0.63	Liu et al., 2015 [40]
	Guangzhou	0.84–689	Peng et al., 2017 [37]
	Hanjiang River	0–7.73	Gao et al., 2018 [41]
	Pearl River	ND–0.1	Xie et al., 2019 [32]
	This study	1.45–697.63	
CAF	Taihu Lake	25.4–482	Zhang et al., 2016 [42]
	Chaohu Lake	1.87–3.27	Pan et al., 2016 [14]
	Beijing	ND–1.74	Zhang et al., 2017 [29]
	Songhua River	ND–63.7	He et al., 2018 [38]
	Baiyang Lake	1.37–30.51	Zhang et al., 2018 [31]
	This study	ND–246.59	

ND: not detected.

In the vertical direction, PPCPs were detected in different layers of each point. Although it is currently difficult to prove the exact rate and timing of PPCP infiltration, PPCPs in surface water may migrate vertically and laterally through hydraulic exchange between surface water and groundwater [43,44]. The downward migration of surface PPCPs contaminates sediments as a whole and increases the ecological risk level of sediments. The detection results of PPCPs in sediments show certain spatial and species differences, which are caused by the comprehensive influence of factors, such as input source differences, dilution effects, location differences, sediment properties, and drug properties. It has been reported that photodegradation greatly reduces CAF in rivers [45,46]. It can be observed that the content of PPCPs in sediments is higher than that in water bodies. Meanwhile, it is worth noting that SMX, IBU and TCS are only detected in water bodies at some sites, but not in sediments.

3.3. Pearson Correlation Analysis of PPCP Concentration and Hydrochemical Parameters in Surface Sediments

The hydrochemical parameters of river sediments from 13 stations in the Yunhe River Basin were tested (Figure 3); TP concentrations of A6, A8 and A9 were higher than those of other samples, NH_4-N concentrations of A2, A4 and A9 were higher than those of other samples, and NH_2-N concentrations of A6, A8 and A12 were slightly higher than those of other samples. The concentrations of TN, NH_3-N and OM were stable, and there was no difference among the different stations. From the analysis of the spatial location of the stations, it can be found that the sudden change in concentration at stations A6 and A9 may be related to the intersection of the two rivers [47,48].

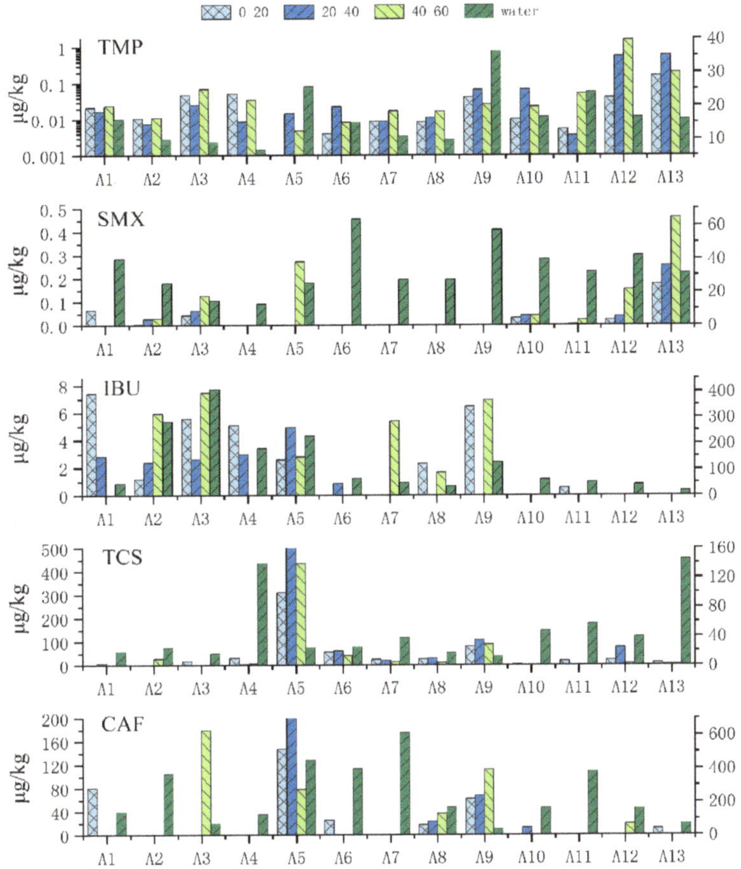

Figure 2. Spatial distribution of PPCP concentration in sediments of the North Canal.

The correlation between the PPCP concentration and hydrochemical parameters in the sediments of the North Canal Basin is shown in Figure 4. The river water quality parameters analyzed include TP (4.11–33.8 mg/kg) and TN (0.16–0.26 g/100 g), NH_4-N (0.03–0.07 mg/g), NH_2-N (0.01–0.02 mg/g), NH_3-N (0.02–0.57 mg/g), OM (113.04–131.59 g/kg). According to Pearson's correlation analysis, TCS and NH_4-N ($p < 0.05$), CAF and NH_4-N ($p < 0.01$), SMX and TMP ($p < 0.05$), CAF and IBU ($p < 0.01$), CAF and TCS ($p < 0.01$), indicating that SMX, TMP, CAF, IBU, and TCS may come from similar sources. However, the correlation coefficient between PPCPs is not high, and the similarity of its sources still needs to be further explored. Meanwhile, NH_4-N is negatively correlated with PPCPs, which may be related to the influence of inorganic conditions in sediments (NH_4-N) on the distribution of PPCPs in sediments [49].

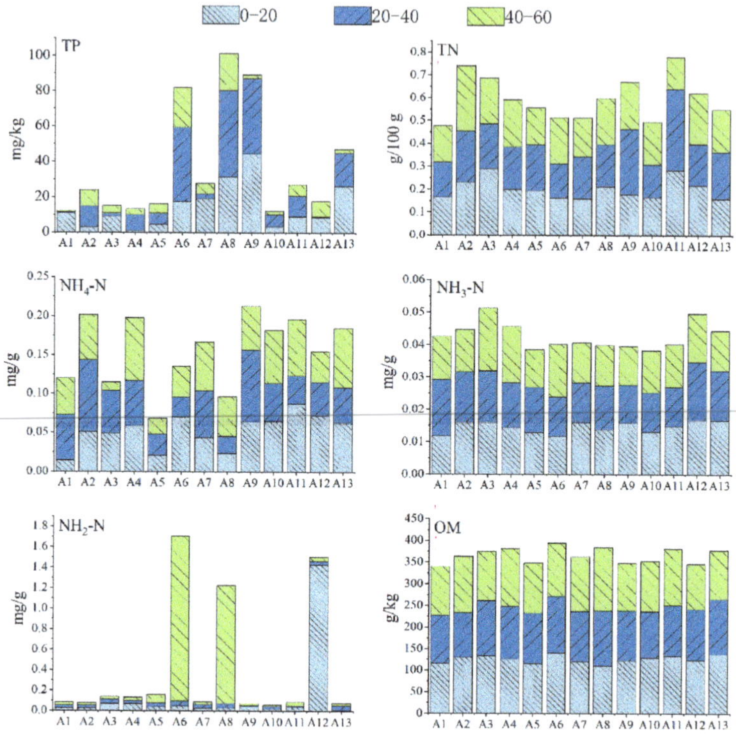

Figure 3. Basic environmental parameters in sediments of the North Canal.

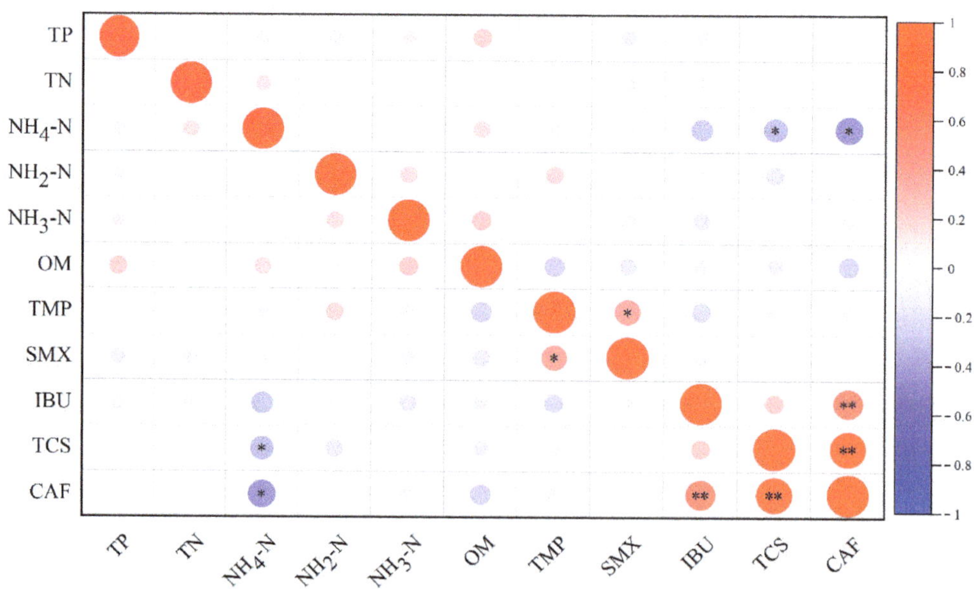

Figure 4. Pearson correlation analysis of PPCPs and environmental parameters. * Correlation is significant at the 0.05 level. ** Correlation is significant at the 0.01 level.

3.4. The Relationship between PPCP Concentration in Surface Sediments and Water

Pseudo-partitioning is used to better understand the relationship between the solid phase and the water phase of PPCPs [50]. The P-PC value is calculated by dividing the concentration in the sampled sediment by its concentration in the water phase. It can be observed from Table 2 that the P-PC value of the TMP, SMX, IBU are at a low level, while the relative P-PC value of TCS and CAF are at a high level, indicating that it is easier for TCS and CAF to accumulate in the North Canal sediments.

Table 2. Pseudo-partitioning of 5 PPCPs in the North Canal.

Site	TMP	SMX	IBU	TCS	CAF
A1	1.21	1.12	14.58	ND	22.69
A2	1.13	1.17	20.99	11.46×10^2	ND
A3	7.64	8.52	18.30	74.50	26.02×10^2
A4	5.34	ND	ND	45.38	ND
A5	0.18	10.76	12.20	18.01×10^3	17.40×10
A6	0.56	ND	ND	16.67×10^2	ND
A7	1.60	ND	11.44×10	34.93×10	ND
A8	0.47	ND	12.91	80.59×10	11.59×10^2
A9	2.71	ND	20.18×10	51.08×10^2	66.5×10
A10	1.36	1.06	ND	7.31	ND
A11	2.22	1.78	ND	7.39	ND
A12	93.74	3.66	ND	18.96×10	11.58×10
A13	12.49	14.80	ND	29.10	ND

ND: not detected.

The leaching capacity of the five PPCPs was shown in Table 3, among which CAF was the most easily leaching due to its high $T_{1/2}$ and low Koc, followed by SMX. TMP and IBU have medium leaching capacity, and TCS has low leaching potential. Infiltration is mainly affected by adsorption and decreases with the increase in PPCP adsorption. The GUS value of PPCPs can better predict the risk of PPCP pollution to groundwater. Although TCS has a low GUS value, it has a relatively high concentration in the surface sediments, which may be related to its extensive and massive use and discharge. Therefore, more attention should be paid to the risk of groundwater pollution. Although the concentration of CAF in surface sediments is lower than that of TCS, its risk of groundwater pollution cannot be ignored because of its strong permeability [22].

Table 3. The $\log T_{1/2}$, $\log K_{oc}$ and GUS of PPCPs in sediment.

Name	$\log T_{1/2}$	$\log K_{OC}$	GUS	Leaching Potential
TMP	1.89	2.857	2.16	Middle
SMX	1.88	2.412	2.98	High
IBU	1.48	2.626	2.03	Middle
TCS	2.08	4.369	−0.77	Low
CAF	1.48	1	4.43	High

Although the PPCP migration rule between surface water sediment and groundwater cannot be analyzed as a typical seasonal river, the hydraulic connection between river water and groundwater is mainly that surface water supplies groundwater [51]. According to the P-PC and GUS values, river sediments are the key channel for PPCPs in river water to enter groundwater. PPCPs in the river water will accumulate in sediments first and then spread further into groundwater through the three typical sections in Table 4.

Table 4. Concentrations of PPCPs in three media in typical sections.

Site	Type	TMP	SMX	IBU	TCS	CAF
A12	surface water (ng/L)	0.45	10.02	176.79	133.67	304.02
	sediment (0–20) (μg/kg)	0.04	0.02	0.00	22.64	0.00
	sediment (20–40) (μg/kg)	0.55	0.04	0.00	75.75	0.00
	sediment (40–60) (μg/kg)	1.56	0.15	0.00	7.41	18.73
	groundwater (ng/L)	0.17	7.32	189.07	6.49	612.30
A10	surface water (ng/L)	0.64	14.10	172.18	171.16	280.75
	sediment (0–20) (μg/kg)	0.01	0.03	0.00	2.83	0.00
	sediment (20–40) (μg/kg)	0.07	0.04	0.00	1.16	12.04
	sediment (40–60) (μg/kg)	0.02	0.04	0.00	0.34	0.00
	groundwater (ng/L)	0.01	0.74	144.70	13.05	170.67
A3	surface water (ng/L)	0.12	4.49	369.06	79.13	444.68
	sediment (0–20) (μg/kg)	0.05	0.04	5.54	16.63	0.00
	sediment (20–40) (μg/kg)	0.03	0.06	2.62	0.74	0.00
	sediment (40–60) (μg/kg)	0.07	0.13	7.48	1.18	178.51
	groundwater (ng/L)	ND	ND	10.93	2.67	153.37

Compared with the concentration of sediment, it can be observed that part of the PPCPs are enriched in the sediment during the infiltration of surface water, except IBU, into groundwater. Therefore, PPCPs in sediment have the risk of water diffusion to groundwater, and is a key indicator affecting the ecological risk to water environments [52–54].

3.5. Environmental Risks of PPCPs in Surface Sediments

To evaluate the possible environmental risks caused by five PPCPs in the water sediments of the North Canal, the potential environmental risks of PPCPs in the sediments of thirteen sampling points in the North Canal basin were evaluated by the RQ values. RQ values corresponding to TMP, SMX, IBU, TCS, and CAF are in the range of 0.02 to 2.47, 0 to 1.03, 0 to 2.52×10, 4.63 to 1.24×10^3, and 0 to 5.07×10^2, respectively (Figure 5).

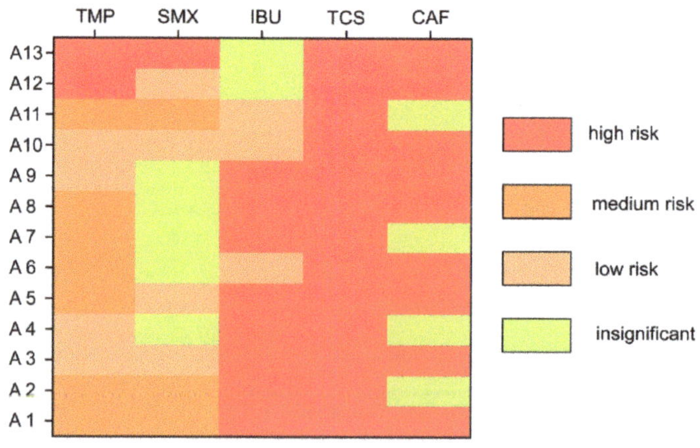

Figure 5. Ecological risks of PPCPs in the sediment samples of the North Canal Basin.

TMP and SMX are at medium and low risk upstream, TMP is at a high risk at A12 and A13, and SMX is at high risk at A13. IBU, CAF, and TCS are highly toxic to the surface sediments in the study area and are the main potential ecological risk factors for the surface sediments in the study area. IBU, TCS, and CAF are the main contributors to the potential environmental risks of the five PPCPs in the North Canal, and the potential environmental risks of the TCS samples are all high risks. The potential ecological risks of TCS and CAF of A5 and A9 are extremely high. At the same time, the combined effect of multiple PPCPs may further increase the level of ecological risk, and the impact and persecution of the aquatic environment cannot be ignored.

PPCPs in most areas pose a threat to the ecosystem and may have adverse effects on aquatic organisms [52]; therefore, it is worth paying attention to the control and elimination of PPCPs in sediments. It is difficult to completely remove PPCPs in sediments of urban rivers, and their accumulation in sediments will affect the survival and reproduction of benthic animals as recently reported; CAF is a psychoactive compound with high ecotoxicological relevance in many other natural water domains. TCS has been shown to produce cytotoxic genotoxicity and endocrine disruptor effects, while TCS in the environment can increase bacterial resistance at the same time. PPCPs in sediments will further spread into groundwater, thus threatening the safety of drinking water. For example, IBU entering the body for a long time will lead to renal failure [22,34].

4. Conclusions

As an important tributary of the Haihe River system, the North Canal flows through the most densely populated area with the highest urbanization intensity in China and plays a major role in urban drainage and landscape greening. With the development of the regional economy and society and the need for water environment protection, the ecological risk caused by PPCP pollution is widely concerning.

In this study, the distribution and migration of five PPCPs in sediments of the North Canal were investigated, and the ecological risk of PPCPs was evaluated by using environmental risk quotients. The relationship between sediment PPCPs and the river ecological environment was analyzed from the perspective of aquatic organisms. According to the analysis results, the concentrations of trimethopretin (TMP), sinolamine (SMX), and ibuprofen (IBU) in the sediments of the North Canal were low, while the concentrations of triclosan (TCS) and caffeine (CAF) were relatively high, and TCS and CAF have been enriched in some reaches. IBU, TCS, and CAF in the sediments have high ecological risk levels, which may affect the survival of regional organisms. Therefore, the existence of new pollutants, such as PPCPs, should not be ignored in order to maintain the stability of the river ecosystem. This paper has practical guiding significance for river water quality management and integrated river management.

Supplementary Materials: The following supporting information can be downloaded at: https://www.mdpi.com/article/10.3390/w14131999/s1, Table S1: Physiochemical properties of the antibiotics considered in the study; Table S2: Limit of detection (LOD) and limit of quantification (LOQ) for the the target compounds; Table S3: Toxicity data used to derivate the predicted no effect concentrations (PNECs) in this study. Bold is the lowest NOEC. References [55–67] are cited in the supplementary material.

Author Contributions: S.P., methodology, data curation and software; B.L., conceptualization, formal analysis and supervision; B.W., methodology and supervision; J.L., data curation and software; X.S., data curation. All authors have read and agreed to the published version of the manuscript.

Funding: This study was funded by the Key R&D Program of Hebei Province (Grant No: 21374201D), the National Natural Science Foundation of China (Grant No: 41730749) and the Sino-French Cooperation Project—Ecological Sewage Treatment System (Grant No: 13000021V8G4VKXBKLUMX).

Institutional Review Board Statement: Not applicable.

Informed Consent Statement: Not applicable.

Acknowledgments: The authors would like to express their gratitude to the editors and anonymous experts for their constructive comments.

Conflicts of Interest: The authors declare no conflict of interest.

References

1. Arpin-Pont, L.; Bueno, M.; Gomez, E.; Fenet, H. Occurrence of PPCPs in the marine environment: A review. *Environ. Sci. Pollut. Res.* **2016**, *23*, 4978–4991. [CrossRef] [PubMed]
2. Cizmas, L.; Sharma, V.K.; Gray, C.M.; McDonald, T.J. Pharmaceuticals and personal care products in waters: Occurrence, toxicity, and risk. *Environ. Chem. Lett.* **2015**, *13*, 381–394. [CrossRef]
3. Caliman, F.A.; Gavrilescu, M. Pharmaceuticals, personal care products and endocrine disrupting agents in the environment—A review. *Clean–Soil Air Water* **2009**, *37*, 277–303. [CrossRef]
4. Ashfaq, M.; Nawaz Khan, K.; Saif Ur Rehman, M.; Mustafa, G.; Faizan Nazar, M.; Sun, Q.; Iqbal, J.; Mulla, S.I.; Yu, C.P. Ecological risk assessment of pharmaceuticals in the receiving environment of pharmaceutical wastewater in Pakistan. *Ecotoxicol. Environ. Saf.* **2017**, *136*, 31–39. [CrossRef] [PubMed]
5. Chaves, M.J.S.; Barbosa, S.C.; Malinowski, M.M.; Volpato, D.; Castro, I.B.; Franco, T.; Primel, E.G. Pharmaceuticals and personal care products in a Brazilian wetland of international importance: Occurrence and environmental risk assessment. *Sci. Total Environ.* **2020**, *734*, 139374. [CrossRef]
6. Fonseca, E.; Hernández, F.; Ibáñez, M.; Rico, A.; Pitarch, E.; Bijlsma, L. Occurrence and ecological risks of pharmaceuticals in a Mediterranean river in Eastern Spain. *Environ. Int.* **2020**, *144*, 106004. [CrossRef] [PubMed]
7. Madikizela, L.M.; Ncube, S.; Chimuka, L. Analysis, occurrence and removal of pharmaceuticals in African water resources: A current status. *J. Environ. Manag.* **2020**, *253*, 109741. [CrossRef]
8. Pan, X.; Qiang, Z.; Wang, W. Distribution and ecological risk of sedimentary PPCPs in the eastern drinking water source area of Chaohu Lake. *Environ. Chem.* **2016**, *35*, 2234–2244.
9. Chen, Y.; Xu, Y.N.; Xiang, S.; Chen, S.Q.; Huang, T.Y.; Pang, Y.; Huang, J.H. Occurrence characteristics and ecological risk assessment of PPCPs in surface sediments of Luoma Lake. *Environ. Sci. Res.* **2021**, *34*, 9. [CrossRef]
10. Dai, G.; Wang, B.; Huang, J.; Dong, R.; Deng, S.; Yu, G. Occurrence and source apportionment of pharmaceuticals and personal care products in the Beiyun River of Beijing, China. *Chemosphere* **2015**, *119*, 1033–1039. [CrossRef]
11. Tamura, I.; Kimura, K.; Kameda, Y.; Nakada, N.; Yamamoto, H. Ecological risk assessment of urban creek sediments contaminated by untreated domestic wastewater: Potential contribution of antimicrobials and a musk fragrance. *Environ. Technol.* **2013**, *34*, 1567–1575. [CrossRef] [PubMed]
12. Yang, L.; Wang, T.; Zhou, Y.; Shi, B.; Meng, J. Contamination, source and potential risks of pharmaceuticals and personal products (PPCPs) in Baiyangdian Basin, an intensive human intervention area, China. *Sci. Total Environ.* **2020**, *760*, 144080. [CrossRef] [PubMed]
13. Wu, J.; Li, J.; Teng, Y.; Chen, H.; Wang, Y. A partition computing-based positive matrix factorization (PC-PMF) approach for the source apportionment of agricultural soil heavy metal contents and associated health risks. *J. Hazard. Mater.* **2019**, *388*, 121766. [CrossRef] [PubMed]
14. Martínez-Hernández, V.; Meffe, R.; Herrera, S.; Arranz, E.; de Bustamante, I. Sorption/desorption of non-hydrophobic and ionisable pharmaceutical and personal care products from reclaimed water onto/from a natural sediment. *Sci. Total Environ.* **2014**, *472*, 273–281. [CrossRef]
15. Long, E.R.; Dutch, M.; Weakland, S.; Chandramouli, B.; Benskin, J.P. Quantification of pharmaceuticals, personal care products, and perfluoroalkyl substances in the marine sediments of Puget Sound, Washington, USA. *Environ. Toxicol. Chem.* **2013**, *32*, 1701–1710. [CrossRef]
16. Wu, D.; Sui, Q.; Yu, X.; Zhao, W.; Li, Q.; Fatta-Kassinos, D.; Lyu, S. Identification of indicator PPCPs in landfill leachates and livestock wastewaters using multi-residue analysis of 70 PPCPs: Analytical method development and application in Yangtze River Delta, China. *Sci. Total Environ.* **2021**, *753*, 141653. [CrossRef]
17. Chaturvedi, P.; Shukla, P.; Giri, B.S.; Chowdhary, P.; Chandra, R.; Gupta, P.; Pandey, A. Prevalence and hazardous impact of pharmaceutical and personal care products and antibiotics in environment: A review on emerging contaminants. *Environ. Res.* **2021**, *194*, 110664. [CrossRef]
18. Dai, G.; Wang, B.; Fu, C.; Dong, R.; Huang, J.; Deng, S.; Wang, Y.; Yu, G. Pharmaceuticals and personal care products (PPCPs) in urban and suburban rivers of Beijing, China: Occurrence, source apportionment and potential ecological risk. *Environ. Sci. Process. Impacts* **2016**, *18*, 445–455. [CrossRef]
19. Zhang, C.; Wan, Z.; Jing, Z.; Zhang, S.; Zhao, Y. Calculation of ecological water requirements of urban rivers using a hydrological model: A case study of Beiyun River. *J. Clean. Prod.* **2020**, *262*, 121368. [CrossRef]
20. Ren, J.; Liang, J.; Ren, B.; Zheng, X.; Guo, C. New patterns of temporal and spatial variation in water quality of a highly artificialized urban river-course—A case study in the Tongzhou Section of the Beiyun River. *Water* **2018**, *10*, 1446. [CrossRef]
21. US EPA. *Method 1694: Pharmaceuticals and Personal Care Products in Water, Soil, Sediment, and Biosolids by HPLC/MS/MS: EPA-821-R-08-002*; US EPA: Washington, DC, USA, 2007.
22. Wu, J.; Liu, J.; Pan, Z.; Wang, B.; Zhang, D. Spatiotemporal distributions and ecological risk assessment of pharmaceuticals and personal care products in groundwater in north china. *Hydrol. Res.* **2020**, *51*, 911–924. [CrossRef]
23. Harada, A.; Komori, K.; Nakada, N.; Kitamura, K.; Suzuki, Y. Biological effects of PPCPs on aquatic lives and evaluation of river waters affected by different wastewater treatment levels. *Water Sci. Technol.* **2008**, *58*, 1541–1546. [CrossRef] [PubMed]
24. Kim, S.C.; Carlson, K. Temporal and spatial trends in the occurrence of human and veterinary antibiotics in aqueous and river sediment matrices. *Environ. Sci. Technol.* **2007**, *41*, 50–57. [CrossRef] [PubMed]

25. Bedoux, G.; Roig, B.; Thomas, O.; Dupont, V.; Le Bot, B. Occurrence and toxicity of antimicrobial triclosan and by-products in the environment. *Environ. Sci. Pollut. Res.* **2012**, *19*, 1044–1065. [CrossRef] [PubMed]
26. Moore, M.T.; Greenway, S.L.; Farris, J.L.; Guerra, B. Assessing caffeine as an emerging environmental concern using conventional approaches. *Arch. Environ. Contam. Toxicol.* **2008**, *54*, 31–35. [CrossRef] [PubMed]
27. Xue, B.; Zhang, R.; Wang, Y.; Xiang, L.; Li, J.; Gan, Z. Antibiotic contamination in a typical developing city in South China: Occurrence and ecological risks in the Yong jiang river impacted by tributary discharge and anthropogenic activities. *Ecotoxicol. Environ. Saf.* **2013**, *92*, 229–236. [CrossRef]
28. Xu, J.; Zhang, Y.; Zhou, C.; Guo, C.; Wang, D.; Du, P.; Luo, Y.; Wan, J.; Meng, W. Distribution, sources and composition of antibiotics in sediment, overlying water and pore water from Taihu Lake, China. *Sci. Total Environ.* **2014**, *497*, 267–273. [CrossRef]
29. Zhang, P.W.; Zhou, H.D.; Zhao, G.F.; Li, K.; Zhao, X.H.; Liu, Q.N.; Ren, M.; Zhao, D.D.; Li, D.J. Potential risk and distribution characteristics of PPCPs in surface water and sediment from rivers and lakes in Beijing, China. *Environ. Sci.* **2017**, *38*, 1852–1862. [CrossRef]
30. Hu, Y.; Yan, X.; Shen, Y.; Di, M.; Wang, J. Antibiotics in surface water and sediments from Hanjiang River, Central China: Occurrence, behavior and risk assessment. *Ecotoxicol. Environ. Saf.* **2018**, *157*, 150–158. [CrossRef]
31. Zhang, P.; Zhou, H.; Li, K.; Zhao, X.; Liu, Q.; Li, D.; Zhao, G. Occurrence of pharmaceuticals and personal care products, and their associated environmental risks in a large shallow lake in north China. *Environ. Geochem. Health* **2018**, *40*, 1525–1539. [CrossRef]
32. Xie, H.W.; Hao, H.S.; Xu, N.; Liang, X.X.; Gao, D.X.; Xu, Y.; Gao, Y.; Tao, H.C.; Wong, M.H. Pharmaceuticals and personal care products in water, sediments, aquatic organisms, and fish feeds in the pearl river delta: Occurrence, distribution, potential sources, and health risk assessment. *Sci. Total Environ.* **2019**, *659*, 230–239. [CrossRef]
33. Zhang, D.; Lin, L.; Luo, Z.; Yan, C.; Zhang, X. Occurrence of selected antibiotics in Jiulongjiang River in various seasons, South China. *J. Environ. Monit.* **2011**, *13*, 1953–1960. [CrossRef] [PubMed]
34. Li, W.; Shi, Y.; Gao, L.; Liu, J.; Cai, Y. Occurrence of antibiotics in water, sediments, aquatic plants, and animals from Baiyangdian Lake in North China. *Chemosphere* **2012**, *89*, 1307–1315. [CrossRef] [PubMed]
35. Chen, H.; Jing, L.; Teng, Y.; Wang, J. Characterization of antibiotics in a large-scale river system of China: Occurrence pattern, spatiotemporal distribution and environmental risks. *Sci. Total Environ.* **2018**, *618*, 409–418. [CrossRef] [PubMed]
36. Xie, Z.; Lu, G.; Liu, J.; Yan, Z.; Ma, B.; Zhang, Z.; Chen, W. Occurrence, bioaccumulation, and trophic magnification of pharmaceutically active compounds in Taihu Lake, China. *Chemosphere* **2015**, *138*, 140–147. [CrossRef] [PubMed]
37. Peng, F.J.; Pan, C.G.; Zhang, M.; Zhang, N.S.; Windfeld, R.; Salvito, D.; Selck, H.; Van den Brink, P.J.; Ying, G.-G. Occurrence and ecological risk assessment of emerging organic chemicals in urban rivers: Guangzhou as a case study in China. *Sci. Total Environ.* **2017**, *589*, 46–55. [CrossRef] [PubMed]
38. He, S.; Dong, D.; Zhang, X.; Sun, C.; Wang, C.; Hua, X.; Zhang, L.; Guo, Z. Occurrence and ecological risk assessment of 22 emerging contaminants in the Jilin Songhua River (Northeast China). *Environ. Sci. Pollut. Res.* **2018**, *25*, 24003–24012. [CrossRef]
39. Zhao, J.L.; Ying, G.G.; Liu, Y.S.; Chen, F.; Yang, J.F.; Wang, L. Occurrence and risks of triclosan and triclocarban in the Pearl River system, South China: From source to the receiving environment. *J. Hazard. Mater.* **2010**, *179*, 215–222. [CrossRef]
40. Liu, W.R.; Zhao, J.L.; Liu, Y.S.; Chen, Z.F.; Yang, Y.Y.; Zhang, Q.Q.; Ying, G.G. Biocides in the Yangtze river of china: Spatiotemporal distribution, mass load and risk assessment. *Environ. Pollut.* **2015**, *200*, 53–63. [CrossRef]
41. Gao, Y.; Jie, L.I.; Nan, X.U.; Jinren, N.I.; University, P. Pollution levels and ecological risks of PPCPs in water and sediment samples of Hanjiang river. *Environ. Chem.* **2018**, *37*, 1706–1719. [CrossRef]
42. Zhang, P.W.; Zhou, H.D.; Zhao, G.F.; Li, K.; Liu, Q.N.; Ren, M.; Zhao, D.D.; Li, D.J. Spatial and temporal distribution characteristics and potential risks of PPCPs in surface sediments of Taihu Lake. *Environ. Sci.* **2016**, *37*, 3348–3355. [CrossRef]
43. Lewandowski, J.; Putschew, A.; Schwesig, D.; Neumann, C.; Radke, M. Fate of organic micropollutants in the hyporheic zone of a eutrophic lowland stream: Results of a preliminary field study. *Sci. Total Environ.* **2011**, *409*, 1824–1835. [CrossRef] [PubMed]
44. Sakakibara, K.; Tsujimura, M.; Song, X.; Zhang, J. Interaction between surface water and groundwater revealed by multi-tracer and statistical approaches in the Baiyangdian lake watershed, north china plain. *Hydrol. Res. Lett.* **2016**, *10*, 74–80. [CrossRef]
45. Marques, R.; Sampaio, M.J.; Carrapic, O.P.M.; Silva, C.G.; Morales-Torres, S.; Drazi, G.D.; Fariaa, J.L.; Silvaa, A.M.T. Photocatalytic degradation of caffeine: Developing solutions for emerging pollutants. *Catal. Today* **2013**, *209*, 108–115. [CrossRef]
46. Ngo, T.H.; Van, D.A.; Tran, H.L.; Nakada, N.; Huynh, T.H. Occurrence of pharmaceutical and personal care products in Cau river, Vietnam. *Environ. Sci. Pollut. Res.* **2021**, *28*, 12082–12091. [CrossRef] [PubMed]
47. Beretta, M.; Britto, V.; Tavares, T.M.; Silva, S.; Pletsch, A.L. Occurrence of pharmaceutical and personal care products (PPCPs) in marine sediments in the Todos os Santos Bay and the north coast of Salvador, Bahia, Brazil. *J. Soils Sediments* **2014**, *14*, 1278–1286. [CrossRef]
48. Lapointe, B.E.; Herren, L.W.; Paule, A.L. Septic systems contribute to nutrient pollution and harmful algal blooms in the St. Lucie Estuary, Southeast Florida, USA. *Harmful Algae* **2017**, *70*, 1–22. [CrossRef]
49. Liang, X.; Chen, B.; Nie, X.; Shi, Z.; Huang, X.; Li, X. The distribution and partitioning of common antibiotics in water and sediment of the Pearl River Estuary, South China. *Chemosphere* **2013**, *92*, 1410–1416. [CrossRef]
50. Zhang, P.W. Environmental Behavior and Potential Risk of PPCPs in Typical Water Bodies of Haihe River Basin. Ph.D. Thesis, China Institute of Water Resources and Hydropower Research, Beijing, China, 2018.
51. Chen, X.Y.; Zhang, K.; Chao, L.J.; Liu, Z.Y.; Du, Y.H.; Xu, Q. Quantifying natural recharge characteristics of shallow aquifers in groundwater overexploitation zone of North China. *Water Sci. Eng.* **2021**, *14*, 184–192. [CrossRef]

52. Morais, S.A.; Delerue-Matos, C.; Gabarrell, X. An uncertainty and sensitivity analysis applied to the prioritisation of pharmaceuticals as surface water contaminants from wastewater treatment plant direct emissions. *Sci. Total Environ.* **2014**, *490*, 342–350. [CrossRef]
53. Katsikaros, A.G.; Chrysikopoulos, C.V. Occurrence and distribution of pharmaceuticals and personal care products (PPCPs) detected in lakes around the world—A review. *Environ. Adv.* **2021**, *6*, 100131. [CrossRef]
54. Barceló, D.; Petrovic, M. Pharmaceuticals and personal care products (PPCPs) in the environment. *Anal. Bioanal. Chem.* **2007**, *387*, 1141–1142. [CrossRef]
55. Quinn, B.; Gagné, F.; Blaise, C. An investigation into the acute and chronic toxicity of eleven pharmaceuticals (and their solvents) found in wastewater effluent on the cnidarian, Hydra attenuata. *Sci. Total Environ.* **2008**, *389*, 306–314. [CrossRef]
56. Ando, T.; Nagase, H.; Eguchi, K.; Hirooka, T.; Nakamura, T.; Miyamoto, K.; Hirata, K. A novel method using cyanobacteria for ecotoxicity test of veterinary antimicrobial agents. *Environ. Toxicol. Chem. Int. J.* **2007**, *26*, 601–606. [CrossRef] [PubMed]
57. Choi, K.; Kim, Y.; Jung, J.; Kim, M.H.; Kim, C.S.; Kim, N.H.; Park, J. Occurrences and ecological risks of roxithromycin, trimethoprim, and chloramphenicol in the Han River, Korea. *Environ. Toxicol. Chem. Int. J.* **2008**, *27*, 711–719. [CrossRef] [PubMed]
58. Parolini, M.; Pedriali, A.; Binelli, A. Application of a biomarker response index for ranking the toxicity of five pharmaceutical and personal care products (PPCPs) to the bivalve Dreissena polymorpha. *Arch. Environ. Contam. Toxicol.* **2013**, *64*, 439–447. [CrossRef] [PubMed]
59. Ding, T.D.; Li, W.; Kan, X.L.; Li, J.Y. Research advances on the pollution of pharmaceutical and personal care products (PPCPs) in natural waters and their toxicity to freshwater algae. *Ying Yong Sheng Tai Xue Bao = J. Appl. Ecol.* **2019**, *30*, 3252–3264.
60. Białk-Bielińska, A.; Stolte, S.; Arning, J.; Uebers, U.; Böschen, A.; Stepnowski, P.; Matzke, M. Ecotoxicity evaluation of selected sulfonamides. *Chemosphere* **2011**, *85*, 928–933. [CrossRef]
61. Park, S.; Choi, K. Hazard assessment of commonly used agricultural antibiotics on aquatic ecosystems. *Ecotoxicology* **2008**, *17*, 526–538. [CrossRef]
62. Brain, R.A.; Ramirez, A.J.; Fulton, B.A.; Chambliss, C.K.; Brooks, B.W. Herbicidal Effects of Sulfamethoxazole in Lemna gibba: Using p-Aminobenzoic Acid As a Biomarker of Effect. *Environ. Sci. Technol.* **2008**, *42*, 8965–8970. [CrossRef]
63. Han, G.H.; Hur, H.G.; Kim, S.D. Ecotoxicological risk of pharmaceuticals from wastewater treatment plants in Korea: Occurrence and toxicity to Daphnia magna. *Environ. Toxicol. Chem.* **2006**, *25*, 265–271. [CrossRef] [PubMed]
64. Han, S.; Choi, K.; Kim, J.; Ji, K.; Kim, S.; Ahn, B.; Yun, J.; Choi, K.; Kjim, J.S.; Zhang, X.; et al. Endocrine disruption and consequences of chronic exposure to ibuprofen in Japanese medaka (Oryzias latipes) and freshwater cladocerans Daphnia magna and Moina macrocopa. *Aquat. Toxicol.* **2010**, *98*, 256–264. [CrossRef] [PubMed]
65. Dussault, È.B.; Balakrishnan, V.K.; Sverko, E.D.; Solomon, K.R.; Sibley, P.K. Toxicity of human pharmaceuticals and personal care products to benthic invertebrates. *Environ. Toxicol. Chem. Int. J.* **2008**, *27*, 425–432. [CrossRef]
66. Yang, L.H.; Ying, G.G.; Su, H.C.; Stauber, J.L.; Adams, M.S.; Binet, M.T. Growth-inhibiting effects of 12 antibacterial agents and their mixtures on the freshwater microalga Pseudokirchneriella subcapitata. *Environ. Toxicol. Chem.* **2010**, *27*, 1201–1208. [CrossRef] [PubMed]
67. Lower, N. The effects of contaminants on various life-cycle stages of Atlantic salmon (Salmo salar L.). Ph.D. Thesis, University of Portsmouth, Portsmouth, UK, 2008.

water MDPI

Article

A Multi-Dimensional Comprehensive Assessment (MDCA) Method for the Prioritization of Water Pollution Treatment Technologies in China

Jiao Li [1,2], Yujiao Wen [1], Jinyuan Jiang [3,*], Wei Tan [3,*] and Tianyi Zhang [4]

1 Technical Centre for Soil, Agricultural and Rural Ecology and Environment, Ministry of Ecology and Environment, Beijing 100012, China
2 Engineering Research Center of Groundwater Pollution Control and Remediation, Ministry of Education, Beijing Normal University, Beijing 100012, China
3 Chinese Research Academy of Environmental Sciences, Beijing 100012, China
4 College of Architecture and Civil Engineering, Beijing University of Technology, Beijing 100124, China
* Correspondence: jiangjy@craes.org.cn (J.J.); weitan_2012@163.com (W.T.); Tel.: +86-13671109299 (J.J.); +86-18810797391 (W.T.)

Abstract: Water pollution treatment technology assessment methods can be used to guide the selection of scientific and reasonable water pollution treatment technologies. At present, China has not yet established a standardized methodological system to scientifically evaluate these technologies, which makes it difficult to effectively screen water pollution treatment technologies suitable for specific watersheds or regions and restricts the sustainable development of local economy and society. In this study, an MDCA framework for water pollution treatment technologies was developed using a sustainable assessment approach. The framework begins with the definition of water pollution treatment technologies' decision-making problems and then proceeds through the following: select potential water pollution treatment technologies; identify indicators; decision making; indicator scoring; indicator weighting; select appropriate assessment model; uncertainty analysis; and other steps to ultimately determine preferred options. To demonstrate the validity and applicability of the framework, typical urban wastewater treatment technologies were selected for case validation. The results showed that the comprehensive assessment results obtained by the multidimensional assessment model based on the ideal point method and weighted method were basically consistent. SBR and TAS can be used as recommended technologies for urban sewage treatment in the study area. However, these two technologies also have shortcomings, such as the unsatisfactory economic benefit of SBR, and the high sludge production and poor resistance to hydraulic shock loading of TAS. Among the six alternative technologies, CWS had the worst environmental benefit, mainly due to the low ammonia removal rate. A2/O has the worst economic and technical performance, mainly due to high investment and operation cost, relatively complex operation management, and poor resistance to hydraulic shock load. The method established in this study can not only select the technology, but also identify the shortcomings of the technology, therefore realizing the systematization and standardization.

Keywords: water pollution treatment technologies; comprehensive assessment; multi-criteria decision analysis; sustainability assessment

Citation: Li, J.; Wen, Y.; Jiang, J.; Tan, W.; Zhang, T. A Multi-Dimensional Comprehensive Assessment (MDCA) Method for the Prioritization of Water Pollution Treatment Technologies in China. *Water* **2023**, *15*, 751. https://doi.org/10.3390/w15040751

Academic Editor: Laura Bulgariu

Received: 8 January 2023
Revised: 8 February 2023
Accepted: 10 February 2023
Published: 14 February 2023

1. Introduction

According to the 'China Ecological Environment Bulletin' [1], in the 2018 state-controlled sections of the ten major river basins, the proportion of sections with water quality of Class IV-V and inferior to Class V is 18.9 % and 6.8 %, respectively; eutrophicated lakes (reservoirs) accounted for 29.0 % of the 111 monitored lakes (reservoirs). With rapid economic growth and a significant increase in resource and energy consumption, China is still facing a prominent conflict between social development and water environmental protection [2–5]. Since the 1970s, China has been committed to water pollution prevention and control, and especially since the Ninth Five-Year Plan, water pollution control has received unprecedented attention from the state and the government. According to the "National Medium and Long-term Science and Technology Development Plan (2006–2020)" [6], the "Water Pollution Control and Treatment" special project has been set as one of the 16 major national science and technology projects. It is the largest investment in water pollution control science and technology projects since the founding of the People's Republic of China, with a total investment of more than 30 billion yuan, focusing on removing major scientific and technological bottlenecks in water pollution that restrict economic and social development. According to incomplete statistics, more than a thousand key technologies of water pollution control have been developed based on this special project, involving point and surface source pollution control, toxic and harmful pollutants control, ecological restoration of water bodies, drinking water purification, etc. Meanwhile, Chinese scholars have also conducted a lot of research on water pollution control technologies [7–11], which has provided strong technical support for water pollution control projects in Chinese watersheds.

However, although a large number of water pollution control technologies have been developed, it is still a major challenge in China's watershed water pollution treatment to identify green and low-carbon water pollution treatment technologies that are suitable for specific watersheds or regions in practical applications. Water pollution treatment technology involves many types and large quantities, and different types of water pollution treatment technologies have their own characteristics. Thus, how to establish a unified standard to realize the systematization and standardization of technology assessment is a difficult point in the field of water pollution treatment technology assessment. At present, there have been some studies on environmental technology at home and abroad. Decision making regarding the pollution source assessment in the USA have typically been driven by practical factors such as time and money, which rarely considers technology [12]. In the EU, the direct costs, time, and technology are the assessment criteria involved in the decision making [13]. Research on technology assessment in China started late. Scholars such as Huang Qingming first proposed a technology assessment system around 1980 and conducted detailed research on the methodology and practice of technology assessment [14,15]. After entering the 21st century, technology assessment began to be carried out on a larger scale in the field of water pollution prevention and treatment in China, and many scholars adopted different assessment methods to evaluate various water pollution treatment technologies from multiple perspectives. For example, Li Cong [16] selected six typical water pollution treatment technologies used in the heavy chemical industry as research objects and gave preference to water treatment technologies based on the environmental cost-benefit approach. Liang Jingfang [17] constructed a water pollution prevention and treatment technology assessment model for the pharmaceutical industry based on hierarchical analysis and the fuzzy comprehensive assessment method and conducted a technical assessment of wastewater treatment technologies of 16 pharmaceutical enterprises.

In general, there are relatively few studies on the assessment of water pollution treatment technologies in China, and the existing studies have many problems. For example, a majority of studies lack a systematic assessment system, most of the assessment objects are only for a certain industry, and the assessment methods and evaluation criteria are relatively chaotic. These problems make the assessment results more subjective, it is difficult to make horizontal comparisons between technologies, and we are not able to derive practical guidance for water pollution control and management in a specific river basin or region. Therefore, how to comprehensively assess these water pollution treatment technologies from the perspective of sustainable development is an urgent need. Multi-criteria decision analysis (MCDA) is a technique commonly used to facilitate decision-making when processing and aggregating numerous and sometimes conflicting attributes [18,19]. The main advantages come from the robustness, consistency, transparency, and repeatability of the decision-making process [20]. The method considers not only direct costs and time, but also other factors such as overall environmental impact and social impact [21] and is suitable for systematic and comprehensive assessment of a large number of water pollution treatment technologies. Currently, there are no cases in China where they have been applied to the screening of water pollution treatment technologies in specific watersheds or regions. Additionally, many of the available assessment methods lack uncertainty analysis.

Therefore, the main objective of this study is to develop a green and sustainable assessment method to realize the assessment of existing water pollution treatment technologies in China under a unified framework. The detailed objectives are to (1) propose an MDCA framework method for the prioritization of water pollution treatment technologies, considering key attributes in environmental, economic, and technical fields; (2) determine the indicator screening method, assessment model, uncertainty analysis method, etc.; (3) verify the validity and applicability of this proposed method, taking typical urban sewage treatment technology assessment our focus.

2. Materials and Methods

2.1. Framework of MDCA

With the existing water pollution treatment technologies in China as the assessment object, and screening the technologies suitable for the specific river basin or region as the assessment purpose, a framework for comprehensive assessment of water pollution treatment technologies is proposed by multi-attributes decision making technique (Figure 1). The framework involves eight steps: (1) define decision problem; (2) select potential water pollution treatment technologies; (3) identify indicators for decision making; (4) indicator scoring; (5) indicator weighting; (6) select appropriate assessment model; (7) uncertainty analysis; (8) determine preferred options. The framework only provides principles for each step and is not absolutely specific or definitive. In practical application, it is necessary to carry out technology assessment and selection in accordance with the relevant provisions of the framework according to the specific water environment problems and water pollution control objectives faced by a particular river basin or region.

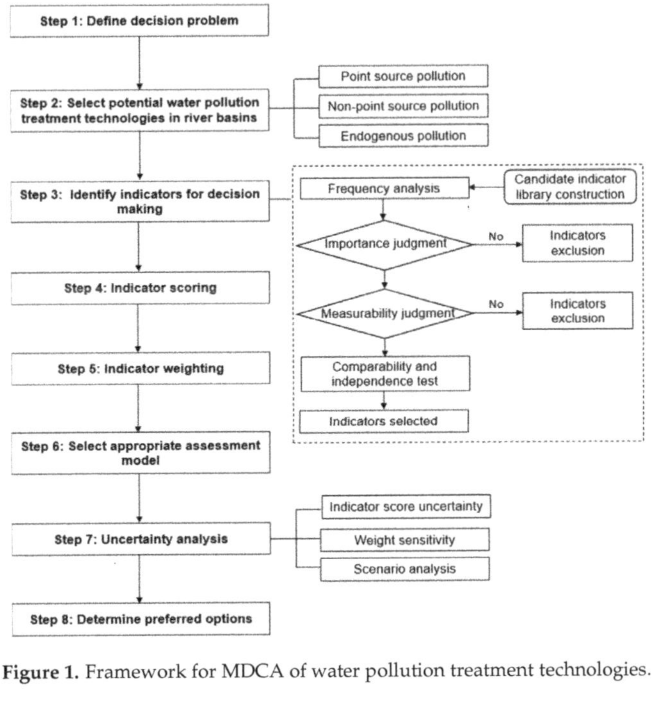

Figure 1. Framework for MDCA of water pollution treatment technologies.

2.2. Define Decision Problem

The first step was to define the decision problem. Due to the different treatment effects of different water pollution treatment technologies, the water purified by different technologies will have varying degrees of impact on the water environment of the basin after being discharged into the river and have different impacts on related social and economic values. In order to select the technologies to better protect the water environment of the river basin or region, comprehensive decision making and evaluation of water pollution treatment technologies are required.

2.3. Select Potential Water Pollution Treatment Technologies

The second step was to select the potential water pollution treatment technologies suitable for a specific pollution type in the river basin or region for the final decision. The types of pollution sources can be roughly divided into point source pollution, non-point source pollution, and endogenous pollution. For one pollution type, there are usually multiple water pollution treatment technologies that can be applied. For example, urban sewage point source pollution requires treatment technologies such as physical treatment, chemical coagulation and sedimentation, biofilm, activated sludge, and constructed wetland [22,23]. Non-point source treatment of rural domestic pollution involves treatment technologies such as constructed wetland, biogas purification tank, septic tank, land law, and stabilization pond [24,25]. The treatment of river sediment pollution (endogenous) includes treatment technologies such as excavation and dredging, sediment oxygenation, biological treatment, and ecological restoration [26]. It should be noted that if the water pollution control goal in the study area cannot be achieved by a single technology type, the control goal should be decomposed into several sub-goals, and each sub-goal should be screened according to the proposed MDCA framework.

2.4. Identify Indicators for Decision Making

The next step was to identify the indicators to be used for technology assessment. There is a broad consensus in favor of tiered approaches to sustainability assessment as it can minimize the cost and complexity of decision making [27]. The MDCA indicator system of water pollution control technology is constructed by using the four levels of "target layer—criteria layer—element layer—indicator layer". The target layer is the comprehensive assessment of water pollution treatment technology. The criteria layer covers three dimensions: environmental, economic, and technical performance. The element layer includes environmental benefits, secondary pollution, costs, benefits, technical reliability, and technical applicability. Due to the many types and large number of the technologies involved, it is not possible to construct a unified indicator set to cover all types of water pollution treatment technologies. The indicator layer selection can be completed by considering the demand of stakeholders and following the principles of scientificity and objectivity.

After establishing a library of alternative indicators, the collected data for these indicators is subjected to frequency analysis, importance judgment, measurability judgment, comparability test, and independence test. Those that meet a series of tests can be used as the final selected indicators. While ensuring that the assessment indicator system comprehensively covers the characteristics of the evaluated technologies, the number of indicators should be selected as concisely as possible. Therefore, this study focuses mainly on the top ranked indicators in terms of frequency, while the low frequency indicators are judged by experts to determine their importance with only those of "high" importance are considered. The selected indicators provided a holistic assessment of treatment technology for a certain pollution type in the river basin or region, encompassing environmental, economic, and technological attributes. It should be noted that the indicators are screened in the same way, while the selected indicators of the indicator layer depend on the characteristics of different pollution types.

Here, frequency analysis refers to the collection and sorting of indicators related to water pollution treatment technology assessment and indicator data in practical application cases to form a library of alternative indicators, so as to conduct statistical analysis of the frequency of indicators in all cases. Importance judgment is to determine the importance level of an indicator by expert consultation, so as to avoid missing the less frequent but valuable indicators. The importance of an indicator is divided into three levels: high, medium, and low, denoted by "+++", "++" and, "+", respectively. Measurability judgment is to analyze whether a single indicator can be obtained accurately and in a timely manner, and to discard indicators that are not available based on the current level and not important. Comparability test is to test the discrimination degree of the values taken by a single indicator. For the indicators lacking data support, the comparability test is conducted by the expert research method. For the indicators supported by basic data, the comparability test is conducted by the coefficient of variation method. The coefficient of variation ≤ 0.1 indicates that the indicator is not comparable. Independence test is to quantitatively analyze the degree of cross-over between the composite indicators to avoid the duplicate assessments caused by the correlation between indicators. The correlation coefficient test is used to indicate the degree of strength of linear correlation between the two composite indicators. The closer the absolute value of the correlation coefficient is to 1, the higher the correlation between the variables; the closer the absolute value of the correlation coefficient is to 0, the lower the correlation between the variables; when the correlation coefficient is 0, it indicates that there is no correlation between the variables. In this study, Spearman's correlation coefficient method was used to assess the correlation between indices at 0.05 and 0.01 levels, respectively.

2.5. Indicator Scoring

Indicators were divided into qualitative and quantitative indicators. For qualitative indicators, the expert judgment method was used to score according to the characteristics of each water pollution treatment technology on a scale of 0 to 100, and the qualitative indicators were converted into quantitative indicators to facilitate the subsequent model calculation [28]. Quantitative indicators are assigned values by using measured data collected from surveys. Indicator values were normalized to provide a common numerical scale that would enable comparison of indicator attributes.

2.6. Indicator Weighting

Weights were assigned to each indicator attribute to indicate the degree of importance [29]. The simplest way is the equal method, which distributes weights equally among all selected indicators. The unequal weight method that considers the relative importance of indicators can be roughly divided into two categories: objective methods and subjective methods. The objective method is to perform mathematical analysis to assign weights based only on the initial measured data; the subjective method determines the weights based on stakeholder preferences taken from previous stakeholder engagement studies and from the literature. The sum of the weights of all indicators is usually equal to 1.

2.7. Select Appropriate Assessment Model

It is the task of this step to calculate the attribute values for each decision criterion by analyzing the multidimensional indicator data above mentioned and then calculating the total preference score for each alternative water pollution treatment technology. The multi-criteria decision analysis is a technique commonly applied to handle multidimensional information and enable decision makers to evaluate options. Assessment model selection is arbitrary, depending primarily on purpose, needs, and stakeholder preferences. The solution methods include sequential optimization, simple additive weight, distance-based method, domination method, ε-control method, etc.

2.8. Uncertainty Analysis

The process of assigning indicator values is usually accompanied by uncertainty [30]. The measured data of the quantitative indicators for the same technology would vary due to the influence of various factors such as region and operating period, while the score values of the qualitative indicators will also be different subject to the subjectivity of different experts, indicating that the indicator value for a certain technology is not a fixed value. In addition, different weighting methods would cause great differences in weight assignment, which would directly affect the evaluation results. The uncertainty of decision making is mainly controlled by indicator scores and weight assignment. Therefore, it is necessary to conduct uncertainty analysis to assess the impact that different indicator scores and weightings may have on the final output. Some commonly used uncertainty analysis methods include stochastic simulation, sensitivity analysis, and scenario analysis. The aim of this analysis is to identify crossover points where the priority of water pollution control technologies in the river basin or region may change.

2.9. Determine Preferred Options

The preferred option of water pollution treatment technologies in a river basin or region is guided by the total score derived by the assessment process. The final output is not a certain result, but a suggestion; the uncertainty evaluation of indicator scoring and weighting is considered in this study. A comprehensive assessment of water pollution treatment technologies is conducted based on three dimensions: environmental, economic and technological, which can find out the strengths and weaknesses of each technology in each dimension and select the suitable water pollution treatment technology according to the actual technological needs of a specific river basin or region. This study proposes to use the three-dimensional coordinate method to represent the assessment results; that is, the values of each technology calculated in the three dimensions of environment, economy, and technology are expressed in the form of (x,y,z) coordinates to realize the visualization of the assessment results of each dimension in three-dimensional space.

3. Case Study

3.1. Study Area

Liaohe river basin is located in the southwest of Northeast China, with a river length of 1340 km and a basin area of 221,400 km^2 (Figure S1). It is composed of two major river systems: Liaohe river and Daliaohe river. This region belongs to the temperate and warm temperate semi-humid continental monsoon climate, with an average annual temperature of 4–9 °C and an average annual precipitation of 350–1000 mm. The topography in the basin is complex, dominated by low mountains, hills, and plains. It spans the four provinces of Hebei, Inner Mongolia, Jilin, and Liaoning. With the rapid expansion of cities, the natural and original Liaohe river basin has gradually transformed into an urban inland river. The main body of the basin is the central city cluster of Liaoning Province. Liaoning province is located in the lower reaches of the Liaohe river basin, which mainly includes the main stream of Liaohe river, Hun river, and Taizi river. The main streams and their tributaries are faced with the dual effects of the original agricultural non-point source pollution and the surging urban point source pollution. Urban domestic sewage discharge was one of the most important sources of urban point pollution in the Liaohe river basin in Liaoning Province. The proposed MDCA method for water pollution treatment technology is applied to the urban sewage treatment in Liaoning Province as a typical case in this study.

3.2. Potential Options

According to the survey, the main technologies for urban domestic sewage treatment in the region included Anoxic-Oxic biological treatment technology (A/O), Anaerobic-Anoxic-Oxic biological treatment technology (A2/O), sequencing batch reactor activated sludge wastewater treatment technology (SBR), traditional activated sludge treatment technology (TAS), sewage treatment technology of constructed wetland system (CWS), biofilm sewage treatment technology (BT), and other technologies. The first six urban domestic sewage treatment technologies were selected as potential options. A total of 77 urban sewage treatment plants involving these six technologies in the region were collected as a case set. The number of cases for each technique and a brief description of the corresponding techniques are shown in Table S1.

3.3. MDCA Indicators of Urban Sewage Treatment Technology

The indicators screening method above was applied to identify the indicators for the comprehensive assessment of urban sewage treatment technologies. Twelve indicators were finally obtained from fifty-five alternative indicators according to the indicator selection criteria, including chemical oxygen demand (COD) removal rate (H1), biological oxygen demand (BOD) removal rate (H2), suspended solid (SS) removal rate (H3), ammonia nitrogen removal rate (H4), total nitrogen (TN) removal rate (H5), total phosphorus (TP) removal rate (H6), sludge production (H7), investment cost (E1), operating cost (E2), resistance to hydraulic shock load (J1), operational stability (J2), ease of operation management (J3), etc. Thus, the urban sewage treatment technology assessment indicator system based on "target layer—criterion layer—element layer—indicator layer" was constructed. The assessment indicator system and the calculation formula of each indicator in the indicator layer are shown in Table S2.

3.4. Assessment Model

(1) MDCA model based on ideal point method

One of the methods to solve discrete multi-criteria decision analysis problems is based on the distance method. The method can be divided into two types [31]: (a) an ideal point is set, which is an N-dimensional vector (when the distance between the alternative and the ideal point reaches the minimum, this is the best scheme); (b) a lowest value is set, which is also an N-dimensional vector. The scheme with the greatest distance from the alternative is identified as the best. To avoid the trouble caused by dimensional differences, all indicators need to be standardized. The distance between the alternative and the ideal point is expressed by Equation (1), and the distance between the alternative and the lowest point is expressed by Equation (2). Indicators are divided into positive and negative indicators. A positive indicator means that the larger the indicator value, the better; a negative indicator means that the smaller the indicator value, the better. After calculation by Equations (1) and (2), all indicators become positive indicators.

$$S_i = \begin{cases} \frac{a_i^{max} - a_i}{a_i^{max} - a_i^{min}} \\ 1 - \frac{b_i^{max} - b_i}{b_i^{max} - b_i^{min}} \end{cases} \tag{1}$$

$$s_i = \begin{cases} \frac{a_i - a_i^{min}}{a_i^{max} - a_i^{min}} \\ 1 - \frac{b_i - b_i^{min}}{b_i^{max} - b_i^{min}} \end{cases} \tag{2}$$

where S_i is the distance between indicator i and the ideal point after normalization; s_i is the distance between indicator i and the lowest point after normalization; a_i^{max} and a_i^{min} are the ideal point value and the lowest point value for the positive indicator i, respectively; b_i^{max} and b_i^{min} are the ideal point value and the lowest point value for the negative indicator i, respectively; a_i and b_i are the original data values for positive and negative indicators, respectively.

The ideal point method is the most commonly used and effective method based on distance method. This method synthesizes the ideal distance and the lowest distance to construct a comprehensive equilibrium formula. This study not only considers the distance between the comprehensive assessment results and the ideal point, but also considers the distance between the assessment results and the ideal point from the three dimensions of environmental, economic, and technical performance, respectively, so as to construct the assessment model based on the multi-dimension ideal point method. The constructed equilibrium formulae are showed below.

$$F_{en} = \frac{\sum_{i=1}^{n}(w_{i,en} \times s_{i,en})}{\sum_{i=1}^{n}(w_{i,en} \times S_{i,en}) + \sum_{i=1}^{n}(w_{i,en} \times s_{i,en})} \quad (3)$$

$$F_{ec} = \frac{\sum_{i=1}^{n}(w_{i,ec} \times s_{i,ec})}{\sum_{i=1}^{n}(w_{i,ec} \times S_{i,ec}) + \sum_{i=1}^{n}(w_{i,ec} \times s_{i,ec})} \quad (4)$$

$$F_{te} = \frac{\sum_{i=1}^{n}(w_{i,te} \times s_{i,te})}{\sum_{i=1}^{n}(w_{i,te} \times S_{i,te}) + \sum_{i=1}^{n}(w_{i,te} \times s_{i,te})} \quad (5)$$

$$F_{c} = \frac{\sum_{i=1}^{n}(w_{i,c} \times s_{i,c})}{\sum_{i=1}^{n}(w_{i,c} \times S_{i,c}) + \sum_{i=1}^{n}(w_{i,c} \times s_{i,c})} \quad (6)$$

where F_{en}, F_{ec}, F_{te}, F_c are the results of the evaluated water pollution treatment technologies in the three dimensions of environment, economy, technical performance, and comprehensive assessment, respectively; $w_{i,en}$, $w_{i,ec}$, $w_{i,te}$, $w_{i,c}$ are the weights of indicator i in the three dimensions of environmental, economic, and technical performance as well as the comprehensive situation, respectively; $s_{i,en}$, $s_{i,ec}$, $s_{i,te}$, $s_{i,c}$ are the distances from the lowest point value after normalization of indicator i in the three dimensions of environmental, economic, and technical performance as well as the comprehensive situation; $S_{i,en}$, $S_{i,ec}$, $S_{i,te}$, $S_{i,c}$ are the distances from the ideal point value after normalization of indicator i in the three dimensions of environmental, economic, and technical performance as well as the comprehensive situation; n indicates the number of indicators. The larger the value of Fen, Fec, Fte, and Fc, the better the alternative scheme, and the maximum scheme is the best choice.

(2) MDCA model based on weighted method

Another commonly used method for discrete multi-criteria decision analysis is simple weighting method. The weighting method is similar to the ideal point method in that the data are first normalized. The calculation formulae are shown in Equation (7):

$$V_i = \begin{cases} \frac{a'_i - a'^{min}_i}{a^{max}_i - a^{min}_i} \\ 1 - \frac{b'_i - b'^{min}_i}{b'^{max}_i - b'^{min}_i} \end{cases} \quad (7)$$

where V_i is the indicator value after normalization; $a\prime^{max}_i$ and $a\prime^{min}_i$ are the maximum value and the minimum for the positive indicator i, respectively; $b\prime^{max}_i$ and $b\prime^{min}_i$ are the maximum value and the minimum value for the negative indicator i, respectively; $a\prime_i$ and $b\prime_i$ are the original data values for positive and negative indicators, respectively. The calculation formula for multi-dimensional assessment model based on weighting method is shown as follows.

$$F_c = F_{en} + F_{ec} + F_{te} = \sum_{i=1}^{n}(w_{i,en} \times V_{i,en}) + \sum_{i=1}^{n}(w_{i,ec} \times V_{i,ec}) + \sum_{i=1}^{n}(w_{i,te} \times V_{i,te}) \quad (8)$$

where V_{en}, V_{ec}, V_{te} are the normalized values after indicators in the three dimensions of environmental, economic, and technical performance. Others have the same meaning as above.

3.5. *Uncertainty Analysis*

3.5.1. Indicator Score Uncertainty

The assignment of indicator is accompanied by many uncertainties. The assessment of water pollution treatment technology involves a variety of environmental, economic, and technical indicators, the value of which is subject to uncertainty due to various factors. For quantitative indicators, the indicator data obtained by a certain water pollution treatment technology in different engineering cases, different regions, and different operating periods are different, so the indicator value is not a fixed value. For qualitative indicators, the scores given by different stakeholders also vary.

In this study, mathematical statistics is used to investigate the influence of indicator assignment uncertainty on the assessment results of water pollution treatment technologies. The specific steps are: (1) assigning values to the indicators; (2) conducting descriptive statistics on the indicators and analyzing the distribution, mean, and standard deviation of the indicator data; (3) determining the uncertainty level of the indicators, which is divided into three levels: high, medium, and low; (4) obtaining the assessment scores under the three dimensions of environmental, economic, and technical performance and the comprehensive assessment score based on the multi-dimensional weighting method assessment model; (5) analyzing the distribution of the assessment results and studying the influence of the indicator assignment uncertainty on the assessment results.

3.5.2. Weight Sensitivity

The different weights of indicators directly affect the assessment results. From the perspective of sustainable development, the assessment of water pollution treatment technology is carried out by considering the three influencing factors of environmental, economic, and technical performance. Different stakeholders hold two views on the importance of environmental, economic, and technological influences in the comprehensive assessment of water pollution treatment technologies: (1) environmental, economic, and technological factors are equally important, i.e., equal weigh mode; (2) environmental factors are the most important while economic factors are the least important, i.e., unequal weight mode [32].These two weighting scenarios are set to analyze the variability of the assessment results. Under the equal weight scenario, the weights of environmental, economic, and technical factors are each 1/3. The weight coefficient of 0.5 is set for environmental factor, followed by the technical factor with the weight coefficient set at 0.3, then the weight coefficient of 0.2 is set for the economic factor in the non-equal weight scenario.

In order to further study the influence of indicator weights on the assessment results, the scenario analysis method is used to carry out the weight sensitivity of different indicators. Firstly, one of the indicators is set as the variable indicator A, and the weights of other indicators B and C are set to be equal. Then, the weight of the variable indicator A is changed in turn, and the weights of other indicators are changed accordingly. Eight scenario modes are set [32]. The weight of variable indicator A in scenario 1 is set to 1, and the weight of indicators B and C are both 0. The weights of variable indicator A in scenario 2–7 are set to 3/4, 3/5, 2/5, 1/2, 1/3, 2/5, and 1/4, respectively, and indicators B and C share the remaining weights equally. The weight of variable indicator A in scenario 8 is set to 0, and the weights of indicators B and C are 0.5, respectively. Here, variable indicator A can be environmental, economic, or technical performance, and indicators B and C are the remaining two indicators besides the variable indicator.

4. Results

4.1. MDCA Based on Ideal Point Method

The score distribution of environmental, economic, and technical dimensions based on the uncertainty of indicator assignment is shown in Figure 2. The uncertainty between the scores of the six alternative technologies on the environmental and technological dimensions did not differ significantly, but the uncertainty of the BT score on the economic dimension differed significantly from the other five technologies. Comparing the indicators in the economic dimension of the six alternative technologies, it was found that the operating cost assignment interval of BT was significantly larger than that of the other five technologies. The confidence intervals for scores on the environmental, economic, and technical dimensions differed significantly. In general, the confidence interval for the scores on the technical dimension was the smallest, while the uncertainty is the largest for the environmental dimension. The variability of confidence intervals across dimensions is determined by a combination of the number and the uncertainty for indicators.

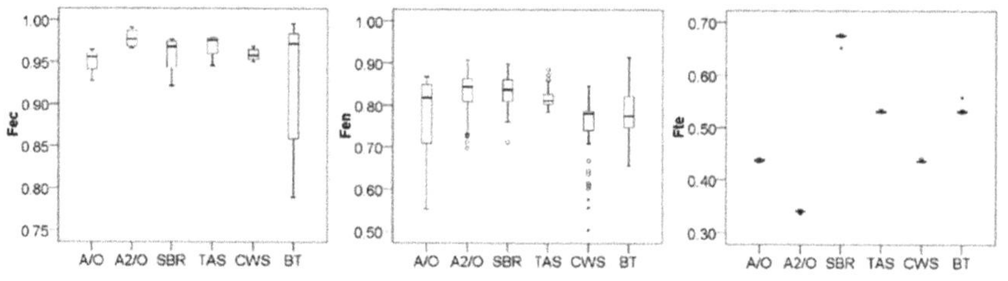

Figure 2. Score distribution of environmental, economic, and technical dimensions based on the uncertainty of indicator assignment.

The three-dimensional assessment results of environmental, economic, and technical performance did not change regardless of equal weight or unequal weight mode. The value of F_{en} is lower than that of F_{ec} and F_{te}. The technical performance discrimination of the six alternative technologies was the most obvious, and the F_{te} values were SBR > TAS ≈ BT > T1 ≈ A/O > A2/O. The assessment results of environmental, economic, and technological dimensions were graded according to 0–0.6, 0.6–0.7, 0.7–0.8, 0.8–0.9, 0.9–1. It was found that A2/O and TAS had relatively good environmental benefits, while CWS had the worst environmental benefits. The assessment results differentiation of economic dimension is not very high. All the Fec values of A/O, SBR, and CWS were above 0.9, and A2O was less economical than other technologies. The highest Fte value is SBR, and the worst is A2/O. The comprehensive effects of the six technologies in environmental, economic, and technological aspects were also evaluated from the total score. Table 1 showed the proportion of comprehensive scores classification of the six alternative techniques in equal weight and unequal weight modes. It can be found that whether it is an equal or an unequal weight mode, the water pollution treatment technology with the best comprehensive assessment effect was SBR, followed by TAS and BT for urban domestic sewage treatment in the study area.

Table 1. Proportion of comprehensive scores classification of six alternative techniques in equal weight and unequal weight modes.

Alternative Technologies	Modes	Proportion (%)				
		[0.8–1]	[0.7–0.8)	[0.6–0.7)	[0.5–0.6)	[0–0.5)
A/O	Equal	1.0	88.2	10.8	0.0	0.0
	Non-equal	0.0	67.7	30.9	1.4	0.0
A2/O	Equal	0.0	71.4	26.8	1.8	0.0
	Non-equal	0.0	47.6	49.7	2.6	0.1
SBR	Equal	92.2	2.0	0.0	1.5	4.4
	Non-equal	55.4	38.7	3.4	2.5	0.0
TAS	Equal	5.9	94.1	0.0	0.0	0.0
	Non-equal	0.0	99.5	0.5	0.0	0.0
CWS	Equal	0.0	75.0	25.0	0.0	0.0
	Non-equal	0.0	22.9	70.8	6.3	0.0
BT	Equal	0.0	94.6	5.4	0.0	0.0
	Non-equal	0.0	74.8	24.5	0.7	0.0

4.2. MDCA Based on Weighted Method

The impact of indicator assignment uncertainty by using the weighted assessment model on the scores of environmental, economic, and technical dimensions was consistent with that of the ideal point assessment model, which was not be repeated here. Three-dimensional assessment results of environmental, economic, and technical performance indicated that the degree of differentiation for Fen and Fec values were far less than the Fte values for the six alternative techniques under eight weighted scenarios (Figure 3). Clearly, the best technology in the technical dimension was SBR, and the worst was TAS. Figure 4 showed the comprehensive score distribution of the six alternative technologies under eight weighted scenarios by multi-dimensional assessment model based on th weighted method. It can be seen that the selection of weights directly affected the assessment result and then the final decision-making. Special attention should be paid to soliciting the views of various stakeholders to determine the weights in the comprehensive assessment of water pollution treatment technology. In this case, with the decrease in the weight of environmental factors, the comprehensive scores discrimination of the six alternative technologies became more and more obvious. Compared with environmental factors, economic and technical factors are more sensitive to the impact of comprehensive assessment results. When only environmental factors were considered (scenario 1), the differences in the comprehensive scores of the six alternative technologies were not particularly obvious. The highest comprehensive score went to SBR, followed by TAS, with the decrease in the weights of environmental factor and the increase in the weights of economic and technical factors.

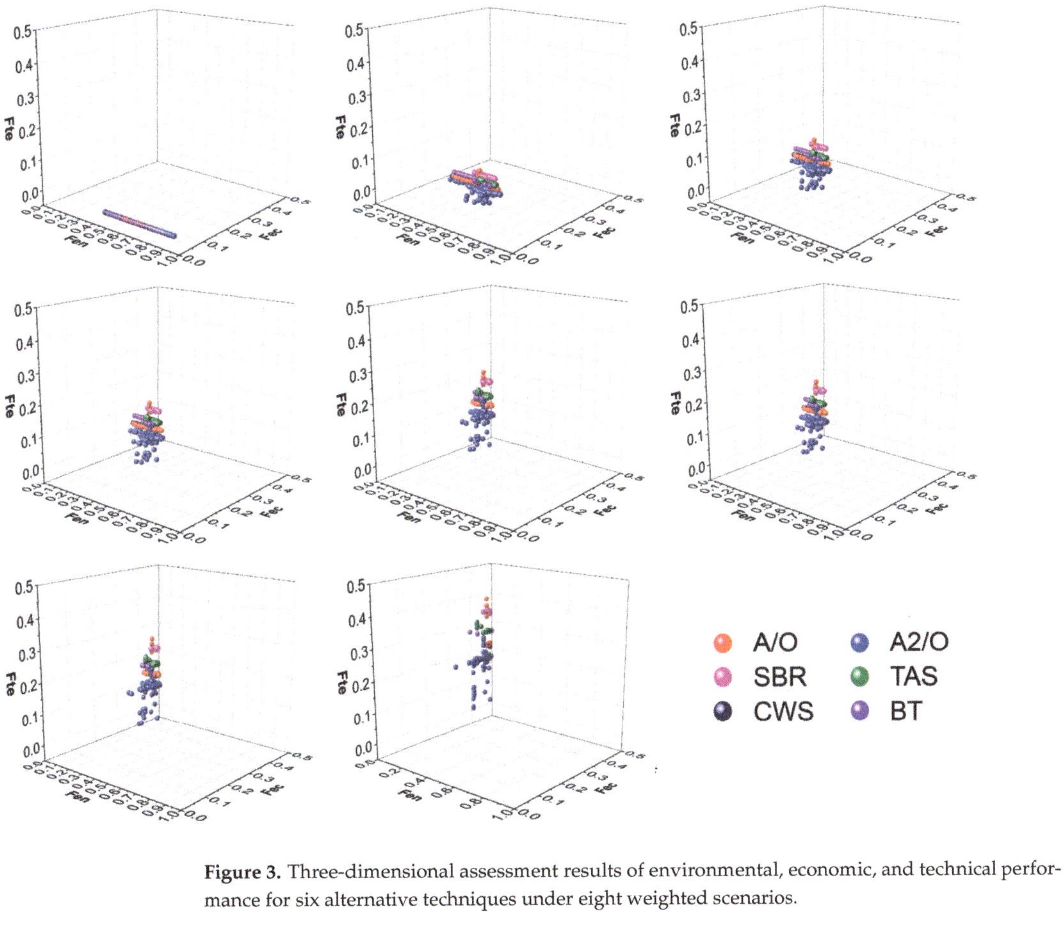

Figure 3. Three-dimensional assessment results of environmental, economic, and technical performance for six alternative techniques under eight weighted scenarios.

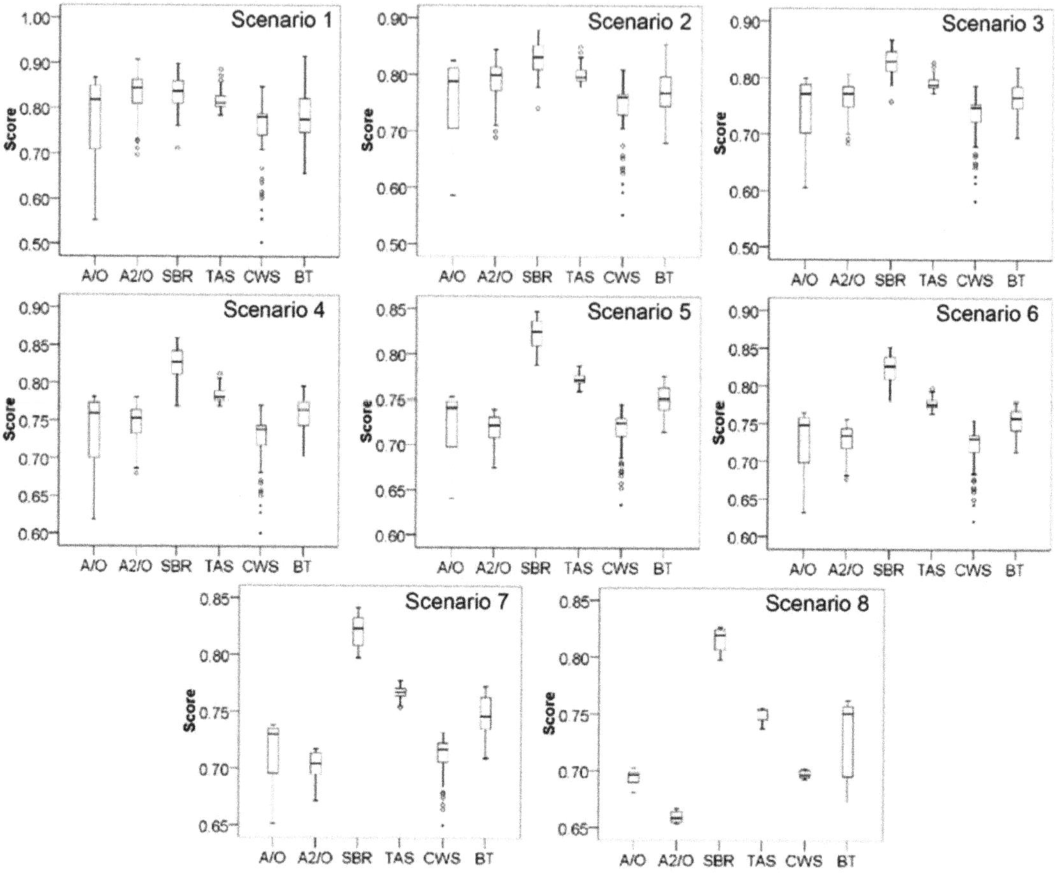

Figure 4. Comprehensive score distribution of six alternative technologies under eight weighted scenarios.

4.3. Comparison Results of Two MDCA Models

The three-dimensional score distribution and evaluation results of environmental, economic, and technical performance calculated by the two assessment models based on the ideal point method and the weighted method showed that although the scores of each dimension obtained by the two models were different, the uncertainty law of the evaluation results was similar. The reason is that the uncertainty of the evaluation results depends on the uncertainty of the indicator assignment. From the evaluation results of the three dimensions of environmental, economic, and technical performance, the assessment model based on the multi-dimensional ideal point method was not affected by the weight change of the criterion layer, and the three-dimensional evaluation results were unique. However, the multi-dimensional assessment model based on the weighted method changed with the change of the weighted scenario mode of the criterion layer, and the three-dimensional evaluation results change accordingly. The three-dimensional assessment results of the two models were consistent except that the technologies with the worst technical performance were A2O by the ideal point method and TAS by the weighted method respectively, primarily affected by weight sensitivity. The ranking of comprehensive assessment results of alternative technologies by the two models was basically the same, and the higher Fc values was SBR, followed by TAS.

4.4. Influencing Factors for MDCA Results

The three-dimensional assessment results showed that TAS and A2O had the better environmental benefits, while CWS had the worst environmental benefits. A/O, SBR, and CWS had high economic benefits, and A2O had the worst economic and technical performance. In order to identify the shortcomings of alternative technologies, comparative analysis of the indicators in environmental, economic, and technological dimensions was carried out with the results shown in Figure 5. It indicated that the poor assessment result of environmental dimension for CWS was mainly due to the low ammonia removal rate. A2/O had high investment cost and operation cost, relatively complex operation management, and poor resistance to hydraulic shock load, resulting in low Fec value and Fte value. The Fc values of SBR and TAS were relatively high, which can be used as the recommended technologies for urban sewage treatment in the study area. However, it can be seen from Figure 5 that the economic benefit for SBR was not advantageous, and TAS had a high sludge production quantity and poor resistance to hydraulic shock load. Therefore, the assessment results can not only select the technology, but also identify the shortcomings of the technology, thus providing guidance for improving the shortcomings of the technology and improving the applicability of the technology in the future.

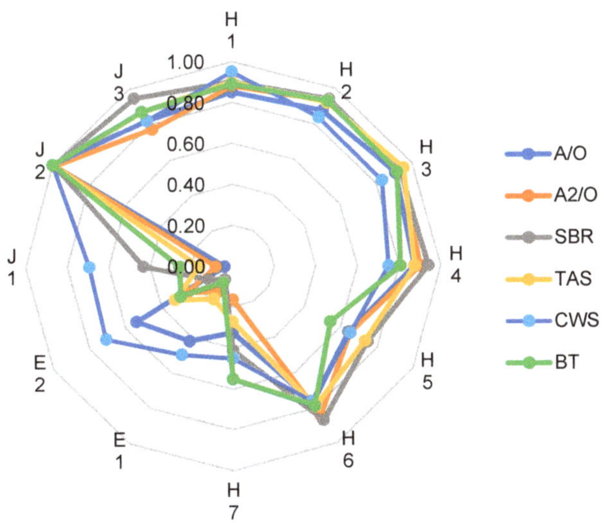

Figure 5. Comparison of indicators in environmental, economic, and technical dimensions for six alternative technologies.

5. Discussion

On the basis of fully considering the general and individual characteristics of different types of water pollution treatment technologies, the comprehensive assessment indicator system of water pollution treatment technologies is constructed based on the three dimensions of environmental, economic, and technical performance, and the assessment model and the expression form of evaluation results are proposed in this study. The assessment methodology established has the advantages and characteristics as follows:

(1) A new method of indicator screening for the indicator layer is proposed. It is found that the final indicators are very different due to scholars' different understanding of the characteristics of assessment objects and the meaning of the indicators. On the premise of following the construction principle of the assessment system, this study establishes a library of alternative indicators and selects the indicators after a series of screenings, such as frequency analysis, importance judgment, measurability judgment, and comparability

and independence tests, to ensure the scientificity, objectivity, and representativeness of the selected indicators as much as possible;

(2) The expression of the assessment results can objectively reflect the advantages and shortcomings of the water pollution treatment technology. In practical applications, according to the characteristics of water environmental pollution and the demand targets, the technology with a high score is not necessarily the most suitable water pollution treatment technology for a specific river basin or region. Therefore, this study proposes to visually reflect the score of each assessment indicator in a certain dimension through a radar chart, to express the assessment results of environmental, economic, and technical performance through three-dimensional coordinates, and to calculate the comprehensive assessment score so that the technology suitable for the region can be selected more accurately according to actual needs. In addition, based on the assessment results, the shortcomings of a potential technology can be identified and further improved.

(3) The assessment system is systematic and standardized. A four-level framework structure is adopted in the design of the comprehensive assessment indicator system framework of water pollution treatment technology. The common characteristics of the technologies are reflected at the criterion layer and the element layer; that is, the criterion layer and the element layer are unified when different types of water pollution treatment technologies are evaluated, while the indicator layer needs to put forward specific assessment indicators according to the characteristics of different types of technology. In this way, the comprehensive assessment can not only be unified in the overall framework; the respective characteristics of different technologies can also be reflected. The determination of assessment models and indicator weights is also defined in the study as a unified methodology and principle.

6. Conclusions

In this study, an MDCA framework of water pollution treatment technologies was developed by using the sustainable assessment method. The framework begins with the definition of water pollution treatment technologies' decision making problems, then proceeds through the following: selecting potential water pollution treatment technologies; identifying indicators for decision making; indicator scoring; indicator weighting; selecting appropriate assessment model; uncertainty analysis; and other steps to ultimately determine preferred options. The established comprehensive assessment indicator system for water pollution treatment technologies unifies the indicators of the criterion layer and the element layer; while for the indicators of the indicator layer, new indicator screening methods that meet the characteristics of different technology types are proposed. This indicator system realizes the integration of common and individual characteristics of water pollution treatment technologies under the same framework. The representation of assessment results based on radar charts and the three-dimensional coordinate method is helpful in selecting suitable water pollution treatment technologies and identifying their potential shortcomings.

The applicability of the MDCA method is verified by combination with typical cases. Taking urban wastewater treatment technology as an example, a comprehensive assessment indicator system for urban wastewater treatment technology was established and a multi-dimensional assessment model combined with uncertainty analysis was adopted to screen technologies applicable to urban wastewater treatment in the Liaohe river basin in Liaoning Province. The results showed that SBR had the highest comprehensive assessment score, followed by TAS, which can be used as recommended technologies. However, these two technologies also have shortcomings, such as the unsatisfactory economic benefit of SBR, and the high sludge production and poor resistance to hydraulic shock loading of TAS. The assessment results not only select the technology but also identify the shortcomings. It is noted that indicator assignment and weighting are key factors affecting uncertainty in assessment results.

Supplementary Materials: The following supporting information can be downloaded at: https://www.mdpi.com/article/10.3390/w15040751/s1, Figure S1: Location of study area; Table S1: Number of cases and brief description of potential urban sewage treatment technologies; Table S2: Indicator system of urban sewage treatment technology assessment and the calculation formula of each indicator in the indicator layer; Table S3: Descriptive statistics of environmental, economic and technical indicator data for the assessment o six alternative wastewater treatment technologies.

Author Contributions: J.L.: Conceptualization, methodology, software, writing the manuscript and manuscript revision; Y.W.: data validation, statistical analysis; J.J.: supervision, funding acquisition; W.T.: literature screening, review, and revision of the manuscript; T.Z.: data curation, visualization. All authors have read and agreed to the published version of the manuscript.

Funding: The study was supported by the National Major Science and Technology Project for Water Pollution Control and Treatment (funder: J.J.; No. 2017ZX07401), the Open Project Program of Engineering Research Center of Groundwater Pollution Control and Remediation, Ministry of Education of China (funder: J.L.; No.GW202205), and Joint Research Project on Ecological Protection and Quality Development in the Yellow River Basin (funder: J.J.; No. 2022-YRUC-01-0203).

Data Availability Statement: Not applicable.

Conflicts of Interest: The authors declare that they have no known competing financial interest or personal relationships that could have appeared to influence the work reported in this paper.

References

1. *China Ecological Environment Bulletin in 2018*; Ministry of Ecology and Environment of the People's Republic of China: Beijing, China, 2019. Available online: https://www.mee.gov.cn/hjzl/sthjzk/zghjzkgb/201905/P020190619587632630618.pdf (accessed on 7 January 2023). (In Chinese)
2. Han, D.M.; Matthew, J.C.; Cao, G.L. Deep challenges for China's war on water pollution. *Environ. Pollut.* **2016**, *218*, 1222–1233. [CrossRef]
3. Wang, Q.; Yang, Z.M. Industrial water pollution, water environment treatment, and health risks in China. *Environ. Pollut.* **2016**, *218*, 358–365. [CrossRef]
4. Li, J.L.; Yang, J.X.; Liu, M.M.; Ma, Z.W.; Fang, W.; Bi, J. Quality matters: Pollution exacerbates water scarcity and sectoral output risks in China. *Water Res.* **2022**, *224*, 119059. [CrossRef]
5. Tang, W.Z.; Pei, Y.S.; Zheng, H.; Zhao, Y.; Shu, L.M.; Zhang, H. Twenty years of China's water pollution control: Experiences and challenges. *Chemosphere* **2022**, *295*, 133875. [CrossRef]
6. State Council of the People's Republic of the China. National Medium and Long-Term Science and Technology Development Plan (2006–2020). 2016. Available online: http://www.gov.cn/gongbao/content/2006/content_240244.htm (accessed on 7 January 2023). (In Chinese)
7. Zhou, L.; Wu, Y.L.; Zhang, S.Q.; Li, Y.F.; Gao, Y.; Zhang, W.; Tian, L.; Li, T.; Du, Q.; Sun, S.Q. Recent development in microbial electrochemical technologies: Biofilm formation, regulation, and application in water pollution prevention and control. *J. Water Process Eng.* **2022**, *49*, 103135. [CrossRef]
8. Wang, H.H.; Xiong, J.X. Governance on water pollution: Evidence from a new river regulatory system of China. *Econ. Model.* **2022**, *113*, 105878. [CrossRef]
9. Ding, M.; Zeng, H. A bibliometric analysis of research progress in sulfate-rich wastewater pollution control technology. *Ecotox. Environ. Safe.* **2022**, *238*, 113626. [CrossRef]
10. Wang, X.; Zhang, M.; Liu, L.L.; Wang, Z.P.; Lin, K.F. Using EEM-PARAFAC to identify and trace the pollution sources of surface water with receptor models in Taihu Lake Basin, China. *J. Environ. Manag.* **2022**, *321*, 115925. [CrossRef]
11. Li, M.Y.; Li, L.; Sun, Y.; Ma, H.Y.; Zhang, H.; Li, F. Facile synthesis of dual-hydrolase encapsulated magnetic ZIF-8 composite for efficient removal of multi-pesticides induced pollution in water. *Chemosphere* **2023**, *314*, 137673. [CrossRef]
12. Lu, Y. Experience of American environmental technology management. *Environ. Edu.* **2007**, *5*, 59. (In Chinese)
13. Wang, K.J. The enlightenment of foreign environmental technology management to China. *Environ. Prot.* **2007**, *8*, 32. (In Chinese)
14. Huang, Q.M.; Xiang Wang, X. Technology selection and evaluation in technology transfer. *J. Mang. Eng.* **1986**, *2*, 41. (In Chinese)
15. Huang, Q.M. Theory and method of comprehensive evaluation of technical economy. *Tech. Eco.* **1989**, *Z1*, 33–38. (In Chinese)
16. Li, C. Evaluation and Optimization of Water Pollution Control Technology in Liaohe River Basin. Master's Thesis, Northeastern University, Liaoning, China, 2011. (In Chinese).
17. Liang, J.F. Study on Evaluation Method of Water Pollution Control Technology in Pharmaceutical Industry. Master's Thesis, Hebei University of Science and Technology, Shijiazhuang, China, 2010. (In Chinese)
18. Abrishamchi, A.; Ebrahimian, A.; Tajrishi, M.; Marino, M.A. Case study: Application of multicriteria decision making to urban water supply. *J. Water Resour. Plann. Manag.* **2005**, *131*, 326–335. [CrossRef]

19. Hajkowicz, S.; Collins, K. A review of multiple criteria analysis for water resources planning and management. *Water Resour. Manag.* **2007**, *21*, 1553–1566. [CrossRef]
20. Chen, S.J.; Hwang, C.L. *Fuzzy Multiple Attribute Decision-Making: Methods and Applications*; Springer-Verlag: Berlin, Germany, 1992.
21. Kiker, G.A.; Bridges, T.S.; Varghese, A.; Seager, T.P.; Linkov, I. Application of multicriteria decision analysis in environmental decision making. *Integr. Environ. Assess. Manag.* **2005**, *1*, 95–108. [CrossRef]
22. Hyrycz, M.; Ochowiak, M.; Krupińska, A.; Włodarczak, S.; Matuszak, M. A review of flocculants as an efficient method for increasing the efficiency of municipal sludge dewatering: Mechanisms, performances, influencing factors and perspectives. *Sci. Total Environ.* **2022**, *820*, 153328. [CrossRef]
23. Zhao, Q.; Peng, Y.; Li, J.; Gao, R.; Jia, T.; Deng, L.; Du, R. Sustainable upgrading of biological municipal wastewater treatment based on anammox: From microbial understanding to engineering application. *Sci. Total Environ.* **2022**, *813*, 152468. [CrossRef]
24. Xue, J.; Wang, Q.; Zhang, M. A review of non-point source water pollution modeling for the urban–rural transitional areas of China: Research status and prospect. *Sci. Total Environ.* **2022**, *826*, 154146. [CrossRef]
25. Shen, Z.; Liao, Q.; Hong, Q.; Gong, Y. An overview of research on agricultural non-point source pollution modelling in China. *Sep. Purif. Technol.* **2012**, *84*, 104–111. [CrossRef]
26. Yin, Z. The sustainable management of microplastics pollution in sediments from China: Promulgating relevant laws and implementing targeted management. *Sci. Total Environ.* **2022**, *856*, 159262. [CrossRef] [PubMed]
27. Saaty, T.L. *The Analytical Hierarchy Process: Planning, Priority Setting, Resource Allocation*; McGraw Hill: New York, NY, USA, 1980.
28. Sorvari, J.; Antikainen, R.; Pyy, O. Environmental contamination at Finnish shooting ranges—The scope of the problem and management options. *Sci. Total Environ.* **2006**, *366*, 21–31. [CrossRef]
29. Sam, K.; Coulo, F.; George, P. A multi-attribute methodology for the prioritisation of oil contaminated sites in the Niger Delta. *Sci. Total Environ.* **2017**, *579*, 1323–1332. [CrossRef]
30. Butler, J.; Jia, J.; Dyer, J. Simulation techniques for the sensitivity analysis of multi-criteria decision models. *Eur. J. Oper. Res.* **1997**, *103*, 531–546. [CrossRef]
31. Zarghami, M.; Szidarovszky, F. *Multicriteria Analysis: Applications to Water and Environment Management*; Springer Berlin Heid: Berlin, Germany, 2014.
32. Wu, J.; Xiong, Y.N.; Ge, Y.X.; Yuan, W.C. A sustainability assessment-based methodology for the prioritization of contaminated site risk management options. *Environ. Sci. Pollut. Res.* **2022**, *29*, 7503–7513. [CrossRef]

water

Article

Effect of Ecological Water Supplement on Groundwater Restoration in the Yongding River Based on Multi-Model Linkage

Tian Nan [1,2] and Wengeng Cao [1,2,*]

1 The Institute of Hydrogeology and Environmental Geology, Chinese Academy of Geosciences, Shijiazhuang 050061, China
2 Key Laboratory of Groundwater Sciences and Engineering, Ministry of Natural Resources, Shijiazhuang 050061, China
* Correspondence: caowengeng@mail.cgs.gov.cn

Abstract: Evaluating the effect of ecological water supplement on groundwater restoration quantitatively could produce positive contributions to both water cycle theory and surface–groundwater conjunctive management. Therefore, in this paper, a groundwater flow numerical model has been established after calculating the river section seepage rate using a fuzzy mathematical method in the Yongding River channel. The simulated results show that the model could accurately reflect the real groundwater dynamic features. Then, a data-driven random forest(RF) model has been established to quantitatively evaluate the contributions of the factors which influence the groundwater level variation. The Nash-Sutcliffe efficiency coefficient(NSE) of the RF model is 0.93. It shows excellent ability to identify the rising zone of groundwater level. The study shows that the infiltration capacity is strong in the upstream area of the Yongding River, and the seepage rate is over 0.7. The lowest seepage rate is 0.19 at the downstream end, while the seepage rate in the middle area is basically between 0.4 and 0.7. From 2018 to 2019, the ecological water supplement of the Yongding River has played a significant role in raising the groundwater level along the river channel. Additionally, its contribution analyzed by the RF model to the change of groundwater level is 25%. Groundwater exploitation is the most important variable affecting the groundwater level variation. The impact depth of groundwater level fluctuation reaches about 10 m. The impact range where the groundwater level average uplifts 1.86 m is 502.13 km^2. The influence direction gradually changes from around the ecological water supplement section to along the Yongding River channel. The groundwater level variation along the tangential direction of the Yongding River is slowing down. The groundwater level would entirely uplift with 170×10^6 m^3/year ecological water supplement of the Yongding River and 35.77×10^6 m^3/year groundwater mining reduction in the downstream area until 2035.

Keywords: Yongding River; ecological water supplement; fuzzy mathematics; groundwater numerical model; random forest

Citation: Nan, T.; Cao, W. Effect of Ecological Water Supplement on Groundwater Restoration in the Yongding River Based on Multi-Model Linkage. *Water* **2023**, *15*, 374. https://doi.org/10.3390/w15020374

Academic Editors: Yuanzheng Zhai and Alexander Yakirevich

Received: 19 December 2022
Revised: 6 January 2023
Accepted: 10 January 2023
Published: 16 January 2023

1. Introduction

For the sake of the sustainable development and utilization of groundwater resources, the MAR (managed aquifer recharge) technology system has been widely promoted and applied in recent years [1–4]. Artificial ecological water supplement for groundwater restoration in the North China Plain is an important practice of MAR technology at the watershed scale. Relative studies have shown that the river recharge could have a positive effect on the recovery of groundwater level within the local scale [5] but could also have a negative impact on groundwater quality [6]. On the one hand, river recharge increases the recharge pathway of groundwater aquifers, and then the amount of water recharged to the aquifers would be increased, and the water table would be uplifted [7,8]. On the other hand, due to the difference between the recharge source and groundwater, the variation of groundwater quality would be present [9,10]. Some research has also shown that seasonal changes in surface–groundwater exchange would lead to ecological risks [11–13]. In summary, river ecological water replenishment can have a significant impact on the surface–groundwater circulation and ecological ambient. The influence of ecological water supplement on the groundwater level variation should be evaluated.

Numerical simulation is one of the most popular approaches for analyzing surface–groundwater interactions. Many kinds of numerical models have been proposed to study the surface–groundwater exchange. The mainstream models include the fully coupled model and semi-coupled model [14]. The fully coupled model requires detailed investigations of surface hydrological processes, water migration in the unsaturated zone and saturated zone, and then a reasonable conceptual model could be established [15–17]. The fully coupled model is advantageous for small-scale studies. However, for a medium to large regional scale study, the dimension explosion of parameters and the computing cost would create crucial problems [18–20]. The semi-coupled model based on MODFLOW can deal with surface water transport process as boundary conditions. Additionally, the modeling process is simpler and more flexible than the fully coupled model. It is often used in the study of surface–groundwater interactions at the regional scale [21]. However, whether it is fully coupled or semi-coupled, the hydrogeological parameters in the surface–groundwater conversion process are the key factors affecting the model accuracy [22–24]. In addition, with the development of computer science, because of the high simulation precision and the flexible data processing flow, data-driven models have been widely used in groundwater hydrogeological research. Various machine learning models have been widely applied to predict the variation trend of groundwater level and budget [25–27]. Within these machine learning algorithms, the random forest algorithm has flexible parameter adjustment, high prediction accuracy, strong generalization ability, and is not likely to produce an over-fitting problem. It has fewer requirements on data quality and has strong robustness to feature variable selection. At present, it is commonly used for solving both classification and regression problems. Moreover, the random forest model is also friendly for the explanation of feature variables [28–30].

The Yongding River flows through Beijing, Hebei, and Tianjin and into the Bohai Sea. It has begun to dry up since the 1980s. By 2010, the Yongding River had actually been cut off because of the overexploitation in the upstream areas [31,32]. In an effort to recover the ecological environment and the socio-economic vitality around the Yongding River, various water supplement projects have been launched, including reservoir water replenishment, reclaimed water replenishment, the Yellow River Diversion Project, and the South-to-North Water Diversion Project [33,34]. Since 2017, the amount of water recharged from these projects mentioned above has gradually increased [35]. The treatment effect of ecological water supplement on groundwater aquifers along the Yongding River region needs to be evaluated. However, the research on water supplement of the Yongding River is mainly focused on the influence on groundwater quality in the river channel [36–38], but the systematic and comprehensive research on the relationship between ecological water supplement and groundwater level variation has not been formed, and the contribution degree to water level recovery and the influence range of water supplement are not clearly evaluated [39,40].

Therefore, this study has proposed a systematic method based on multi-models for effective evaluation of the ecological water recharge effect on groundwater restoration. First, a fuzzy mathematical method has been used to figure out the seepage rate along the Yongding River channel. Additionally, the ecological water infiltration amount of each section along the Yongding River has been calculated. Then, a numerical simulation model under ecological water supplement conditions has been established for analyzing the regional groundwater level and budget variation. Additionally, the affected area of ecological recharge has been determined. Finally, the contribution of the ecological water recharge for the recovery of groundwater level has been quantitatively evaluated by using the random forest model.

2. Materials and Methods

2.1. Geological and Hydrogeological Conditions in the Study Area

The study area is from Mentougou to Beichen along the main Yongding River channel. The total study area is 3842.05 km^2 (Figure 1a). In addition to urban areas, there are farmlands, forestlands, and grasslands in the study area [41,42]. The length of the Yongding River channel in this study is 162.5 km.

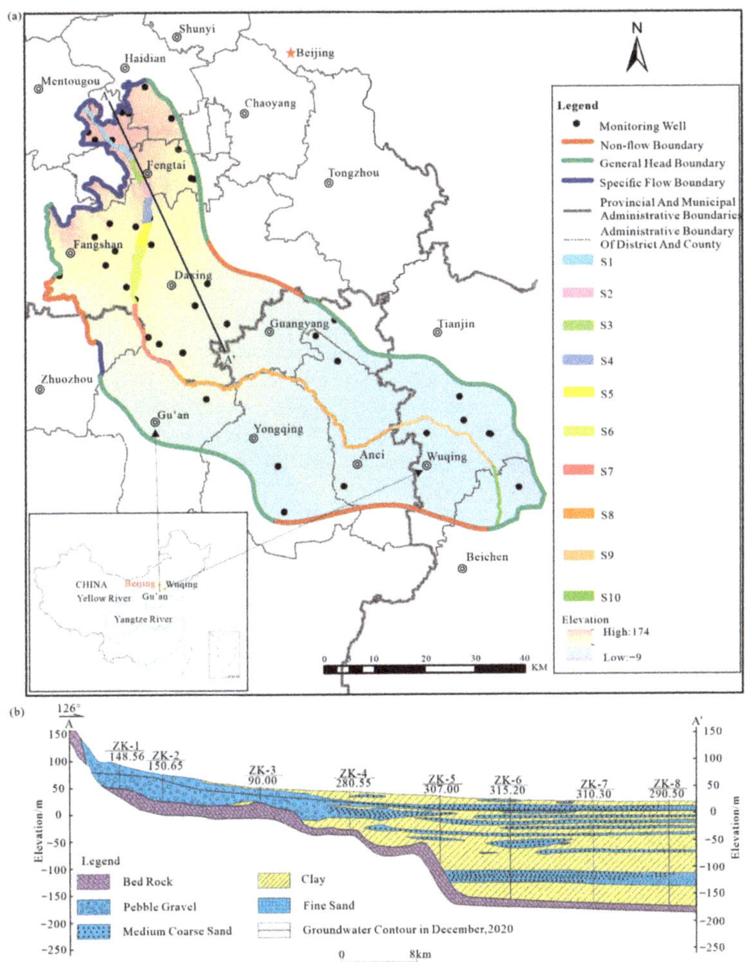

Figure 1. (**a**) location of study area; (**b**) hydrogeological profile in study area (modified after Hao et al. (2014)) [5].

The study area has a typical monsoon climate. The average intensity of precipitation is 595 mm/year, and the average intensity of water surface evaporation is 1120 mm/year (1959–2019). Precipitation is mostly concentrated from June to September each year [43]. The Yongding River Comprehensive Management and Ecological Restoration Project has been in operation for more than 2 years [44,45], and the impact on the groundwater level fluctuation has been gradually revealed. The regional groundwater flows from the northwest to the southeast.

The aquifers mainly consist of coarse-grained sediments in the piedmont area, and gradually changed into finer-grained sediments in the downstream plain (Figure 1b) [46]. The piedmont area is mainly from Mentougou to Fengtai. In the piedmont area, the aquifer is mainly a single phreatic aquifer which is composed of relatively coarse gravel. The thickness of the phreatic aquifer is about 50 to 200 m. The range of hydraulic conductivity changes from 10 to 200 m/d. In the middle reaches of the Yongding River, the aquifers have changed into a multi-layer structure. The aquifers are mainly composed of middle and fine sand, with a few gravels. The thickness of aquifers is about 100 to 150 m. Additionally, the hydraulic conductivity in the middle reaches is usually within 30 m/d. The downstream area is roughly from the Yongqing River channel to the very southeast part of the study area. The aquifers are composed of fine sand. Each aquifer is relatively thin, and the thickness is usually less than 10 m. Additionally, between aquifers, the area is filled by a large number of clay layers. Therefore, the hydraulic conductivity in the downstream aquifers is generally less than 10 m/d [47,48].

The main recharge items in this area include mountain lateral runoff, precipitation infiltration, and water infiltration of the Yongding River; the main discharge items include artificial exploitation, evaporation, and lateral outflow [49]. The bottom boundary of the Quaternary aquifer is impermeable sediment [50].

2.2. *Workflow and Data Collection*

The flowchart of this research is shown in Figure 2. Firstly, the relevant data will be collected and preprocessed. After that, the seepage rate of each section of the Yongding River channel will be calculated using the fuzzy mathematics method. Then, the amount of infiltration can be estimated. Next, using the calculation result of ecological water supplement infiltration, a loosely coupled numerical model will be established. Groundwater restoration effect will be analyzed by using the numerical model. Finally, a random forest model will be used to evaluate the contributions of source and sink items on the groundwater level variation. The specific data and methodology descriptions have been shown as follows.

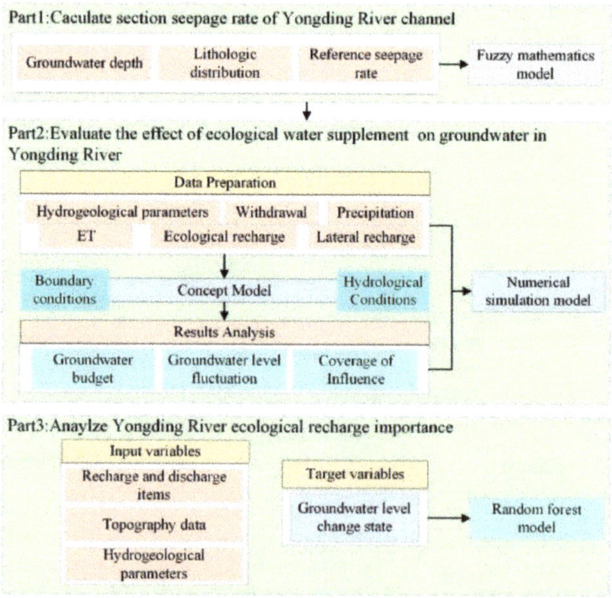

Figure 2. Flowchart of this study.

The data used for the study consist mainly of precipitation, evaporation, groundwater extraction, topography, water levels, and river recharge. Precipitation and evaporation data come from the monthly observation data of the National Meteorological

Administration. Topographic data come from Alos satellite remote sensing interpretation results [51]. Groundwater exploitation data are collected from Beijing, Tianjin, and Hebei hydrological reports [52–57]. The Yongding River ecological water supplement data from 2018 to 2019 are collected from Haihe River Water Conservancy Commission [58,59]. The water level data are from monthly observations from national groundwater monitoring wells in the area from 2016 to 2019 [60]. The drilling data of the sections are derived from the hydrogeological survey of the Institute of Hydrogeology and Environmental Geology, Chinese Academy of Geological Sciences, in 2019 [61].

2.3. Models

2.3.1. Fuzzy Mathematics Model

Fuzzy mathematical methods are used to solve for the river section seepage rates. The method mainly satisfies the mathematical calculation by constructing an affiliation function to quantify the qualitative indicators [62]. Since the Yongding River channel recharges the shallow groundwater along its route in the form of seepage, and the seepage rate is mainly determined by the lithology of unsaturated zone [63], this study establishes an affiliation function between the lithology distribution of the unsaturated zone and the seepage rate. The infiltration test data of the Yongding River channel, provided by the Haihe Water Conservancy Commission, are used as the reference value to calculate the permeability of each section of the Yongding River channel. Firstly, the river channel is divided into four types of lithology: clay and silt, fine sand, medium coarse sand, and pebble gravel (Figure 1b). The empirical values of the seepage rate of the four types of lithology are given, and the permeability evaluation values of the four types of lithology are obtained after standardization S_i $(i = 1, 2, 3, 4;\ S_i \in A)$; the Yongding River is divided into n sections according to the recharge conditions. According to the area A_{ni} of four types of lithology in the profile and the total area A_n of the section, the proportion P_{ni} of four types of rock strata in the nth section is calculated:

$$P_{ni} = \frac{A_{ni}}{\sum A_n} \tag{1}$$

Then, the combined permeability evaluation value of the nth rock section is calculated as S_n :

$$S_n = \sum P_{ni}\, S_i \tag{2}$$

Based on the maximum and minimum seepage rates λ_{max}, λ_{min} gained from the infiltration test, a linear equation between the seepage rate of the section and the permeability evaluation value S_n is developed:

$$\lambda_n = \frac{S_n(\lambda_{max} - \lambda_{min})}{S_{nmax} - S_{nmin}} + \lambda_{min} \tag{3}$$

Solving the equation can obtain the seepage rate of each section.

2.3.2. Numerical Simulation Model

The numerical simulation model is the main method applied to estimate the effect of ecological water supplement on the groundwater level and budget variation.

The groundwater aquifers in the Yongding River region were generalized as a three-dimensional heterogeneous transient groundwater flow system. The model was composed of two conceptual aquifers. The upper layer was the mainly concerned unconfined aquifer group, and the lower layer was the confined aquifer group. The western boundary receiving the piedmont lateral recharge was set as a specific flow boundary, and the southern and northern parts were set as a no-flow boundary. The remaining boundaries had the variation water exchange with the outside area, and then they were set to the general head boundary. Precipitation, river recharge, and irrigation infiltration were designed as recharge boundaries. The model bottom boundary had been regarded as a no-flow boundary. The main hydrogeological parameters used in the model were hydraulic conductivity, specific yield, and storage coefficient. The distribution of the initial hydrogeological parameters were assigned based on the interpretation of the hydrogeological test, drilling work, and the previous research results in the same study area [64,65]. These parameter values were finally determined during the model calibration process.

The model was developed with MODFLOW. The grid size was 500 m × 500 m. The first layer was discretized into 15,850 effective cells, and the second layer was discretized into 11,653 effective cells. The model was discretized into 48 stress periods from January 2016 to December 2019. In the simulation period, ecological water supplement was allocated from April to July.

The main sources of ecological water supplement include the discharge of water from the Guanting reservoir and the water introduced from the South-to-North Water Diversion Project. Because of the shallow groundwater level, the surface water generally has no direct contacts in the study area. The water first enters the Yongding River, and then infiltrates into the unsaturated zone. After partial interception, the rest of it is finally recharged to the shallow aquifers. Additionally, the amount of infiltration has been calculated with the fuzzy mathematics method proposed in this paper.

The distribution of precipitation recharge rates was figured out by the amount of precipitation and infiltration coefficient. Evapotranspiration was calculated by the Evapotranspiration Package. Because the mathematical relationship between evaporation and groundwater level was clear, the groundwater evaporation distribution was the same as the area where the groundwater depth was less than 4 m [66]. Similarly, the evaporation intensity was verified by model fitting. Artificial exploitation of groundwater was put into the model in the form of mining well and calculated by well module.

The trial-and-error method was applied for model identification and verification. The groundwater flow field measured on 31 December 2019 and the observation groundwater level process line were used to identify and verify the model simulation results.

2.3.3. Machine Learning Model

The random forest selected in this study is a popular machine learning strategy (generated in Bell Labs in the 1990s), which belongs to the ensemble learning method in machine learning algorithms. It works by generating multiple classifiers, each of which has an independent learning and predicting ability. These predictions of the classifiers are finally combined into an integrated result. Additionally, the accuracy of the final result is much better than any single prediction result. At present, it is commonly used to simulate and analyze the evolution of groundwater level and water quality [67,68].

In this study, whether the water level rises was taken as the target label (rising to 1, not rising to 0), and the groundwater source and sink items, hydrogeological parameters, and topographic features were taken as the feature variables. The 15378 samples' data in the study area were randomly divided into 80% training data and 20% testing data. The Random Forest Classifier package in the Python ensemble library was used to establish the model.

2.3.4. Model Performance Evaluation Metrics

The Nash–Sutcliffe efficiency coefficient (NSE), the model sensitivity, and the specificity are used as the evaluation criteria. NSE is used to evaluate the accuracy of the regression models. The sensitivity evaluates the accuracy of positive samples prediction. The specificity reflects the accuracy of negative samples prediction. The criterion is close to 1, illustrating that the model is more reliable.

$$NSE = 1 - \frac{\sum_{i=1}^{n}(o_i - y_i)^2}{\sum_{i=1}^{n}(o_i - \overline{o}_i)^2} \tag{4}$$

$$\text{Sensitivity} = \text{TP}/(\text{TP} + \text{FN}) \tag{5}$$

$$\text{Specificity} = \text{TN}/(\text{TN} + \text{FP}) \tag{6}$$

where y_i is the simulated value, o_i is the observed value, and $\overline{o}_i$ is the average value of the observed values. i marks the number of observation wells used in the model performance evaluation and n is the total number of observation wells. TP is the positive sample that the model predicts correctly, FN is the negative sample that the model predicts incorrectly, and the sum of TP and FN is the total number of correct positive samples. TN is the negative sample which has been correctly predicted by the model, FP is the positive sample which has not been correctly predicted by the model, and the sum of TN and FP is the total number of correct negative samples. TP, FN, TN, and FP values are obtained from the simulation results.

The influence of feature variables on water level change is evaluated by feature importance value distribution and Shapley value distribution [69]. The feature importance distribution is applied to assess the contribution of feature variables to water level variation by using the Gini impurity reduction method. The Gini impurity is used to decide the splitting direction of the decision trees. Disturb the order of feature variables separately, and the change of the orders will influence the accuracy of the model. The more important the feature variable is, the more decreases of the accuracy of the model will be produced. Shapley value is used to calculate the marginal contribution of every feature variable. Corresponding to each sample, the random forest model will generate a prediction value, which is the Shapley value. The Shapley value not only provides the degree of the feature variable influence, but also reflects the positive and negative influence of the feature variables. In this study, the greater the Shapley value is, the more the feature variable promotes the rise of water level, and vice versa.

3. Results

3.1. Evaluation of the Infiltration Volume of the Yongding River Channel

The seepage rate of each water supplement section of the Yongding River (Figure 1a) calculated by fuzzy mathematics method is shown in Table 1. Then, as the ecological water replenishment volume data in each section of the Yongding River channel have been determined, the infiltration amount of each section is calculated.

Table 1. Ecological water supplement infiltration volume and seepage rate for each section of the Yongding River.

Section	Seepage Rate	Ecological Water Supplement ($\times 10^6$ m^3)	Infiltration Volume ($\times 10^6$ m^3)
S1	0.72	4.90	3.54
S2	0.72	44.07	31.82
S3	0.76	3.66	2.78
S4	0.62	40.52	25.02
S5	0.43	10.52	4.50
S6	0.38	22.83	8.68
S7	0.48	12.05	5.72
S8	0.48	0	0
S9	0.22	0	0
S10	0.19	0	0
Total	0.59	138.55	82.06

The lithology of section S3 in the upstream of the Yongding River is mainly composed of medium coarse sand and coarser grained sediment, and at the same time, the proportion of aquifer is much higher than the aquitard. Therefore, the seepage rate in this section is the highest in all sections. It reaches 0.76 in section S3. Moreover, because the aquifer of S1 and S2 is slightly thinner than S3, the seepage rates of S1 and S2 are also kind of lower than S3, but the values are also over 0.7. In the lower reaches of S10, due to the high proportion of clay, and the aquifers mainly composed of fine sand, the seepage rate is the lowest in S10 of 0.19. In other sections, the clay content gradually rises from upstream to downstream areas. Additionally, the sediment constructing the aquifers has changed from coarse to fine. Then, the seepage rate distribution shows a synchronous and gradually decreasing trend. The results of this study are highly in agreement with Ji's study which used the numerical simulation method in 2021. However, the method proposed in this paper is more flexible and convenient.

The infiltration amount in each section of the Yongding River is calculated with the seepage rate. The water recharge from S2 is the largest in all sections, which is up to 31.82×10^6 m^3, accounting for 38.8% of the total ecological water supplement. The ecological water recharge from S3 is the smallest one, about 2.78×10^6 m^3, accounting for only 3.4% of the total ecological water supplement. About 60% of the ecological water supplement volume in the study area could infiltrate and recharge the shallow aquifer. The loss of the supplement water is partly intercepted by the vadose zone, and partly consumed in the process of water transport through evapotranspiration. Substituting the calculation results of the ecological water supplement infiltration volume into the numerical model, the coupled model could be used to analyze the effect of the ecological water supplement of the Yongding River on groundwater flow system.

3.2. Numerical Model Performance

The numerical model has been fitted and calibrated by comparing the observation data of 64 typical observation wells in the study area from 2016 to 2019 and the measured groundwater flow field on 31 December 2019 with the simulation value.

The identification and verification results show (Figure 3) that the simulated flow field fits well with the measured flow field. There is a slight difference between the southern and eastern boundaries. At the same time, the simulated groundwater levels have a similar variation trend with that in the observation wells. It proves that the model could precisely describe the groundwater variation features through the ecological water supplement period. The mean absolute error between the measured data and the simulated data of groundwater level of all the observation wells in the calibration period is 0.91 m. As shown in Figure 3b, all the simulated and observed fitting points are very close to the 1:1 line, and the *NSE* is 0.82. This means that the simulation result has an overall good performance. Local error mainly comes from insufficient knowledge on the hydraulic parameters and regional boundary conditions. On the whole, the model established in this paper could correctly present the dynamic features of groundwater flow in the Yongding River channel and could be used to quantitatively analyze the spatial and temporal effects of ecological water supplement on groundwater flow system.

With the model corrected by fitting the groundwater flow field and typical observation wells, the location of flow boundary has been determined. The lateral groundwater flow mainly comes from the southern part of Haidian and Mentougou. It is the lateral recharge of the study area. The total amount of the lateral recharge was 14×10^6 m^3/year before 2017, and it increased to 100×10^6 m^3/year after 2017. Then, the no-flow boundary and general head boundary have been calibrated (Figure 1a). Finally, the amount of outflow through the boundary has been figured out. According to the change of groundwater flow field in 2018 and 2019, the groundwater mainly flows out from the eastern and southern parts of the study area. The flow-out boundary is mainly from east Fengtai. Then, the northern part of Wuqing and the southern part of Beichen are also the flow-out boundary. The calibrated model has captured the boundary features. The out flow calculated by the model was 263×10^6 m^3/year before 2017, and it decreased to 40×10^6 m^3/year after 2017.

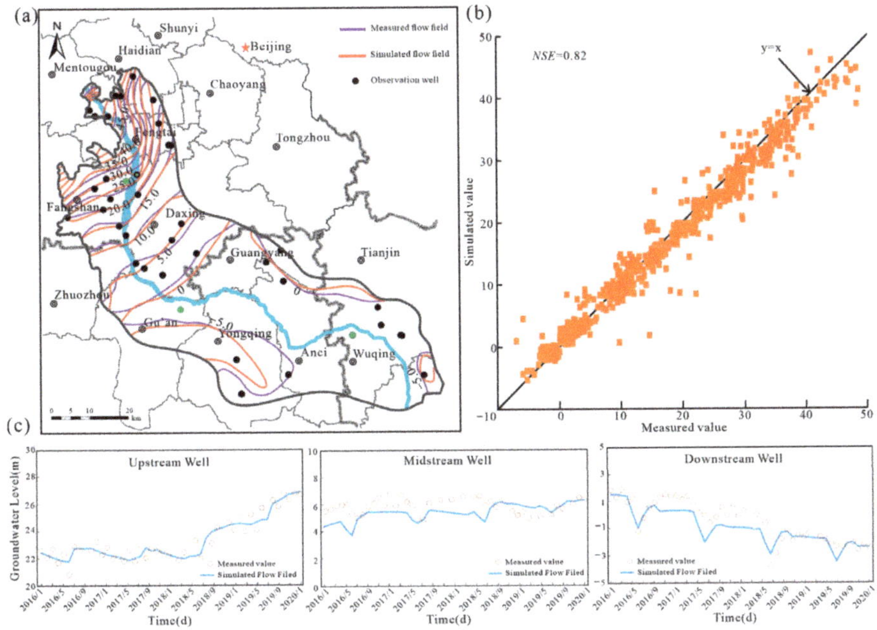

Figure 3. (**a**) the fitting plot at the end of the simulation period; (**b**) the fitting curve of all simulated and measured groundwater levels; (**c**) the fitting curve of typical observation well in the ecological water supplement zone.

The hydraulic conductivity and specific yield are the most important parameters to control groundwater flow. The hydraulic conductivity presents the capacity of aquifers to transport the groundwater based on the difference of hydraulic gradient. The specific yield reflects the water release ability of the unconfined aquifer. The smaller the specific yield is, the more obvious the groundwater level fluctuation is. Figure 4 shows the calibrated distributions of hydraulic conductivity and specific yield. The maximum calibrated hydraulic conductivity value is about 150 m/d, and it is presented in the southern part of the Mentougou area which is close to the mountain area. The aquifer is a single structure composed of gravel. Because the aquifers change into multi-layer structure and the sediments change into middle to fine sand, the calibrated hydraulic conductivity decreases. To the east of the high values area, the hydraulic conductivity values drop to below 50 m/d. The hydraulic conductivity calibrated in the middle part of the study area is usually around 20 to 30 m/d. In the downstream of the Yongding River, the hydraulic conductivity calibrated in this model changes from 15 to 5 m/d. The hydraulic conductivity distribution is in accord with the sedimentary law in the study area. Additionally, the calibrated specific yield distribution has a similar variation trend with the hydraulic conductivity distribution. The value of specific yield is relatively high in the piedmont area, which can reach 0.18. Then, the calibrated specific yield value reduces to 0.15 in the south of the Daxing area which is the middle part of the study area. The smallest calibrated value of specific yield is 0.07 in the downstream of the Yonding River. All of the calibrated parameter results in this paper are highly consistent with those of previous studies [5,70], but the study area has been further extended compared with the former studies.

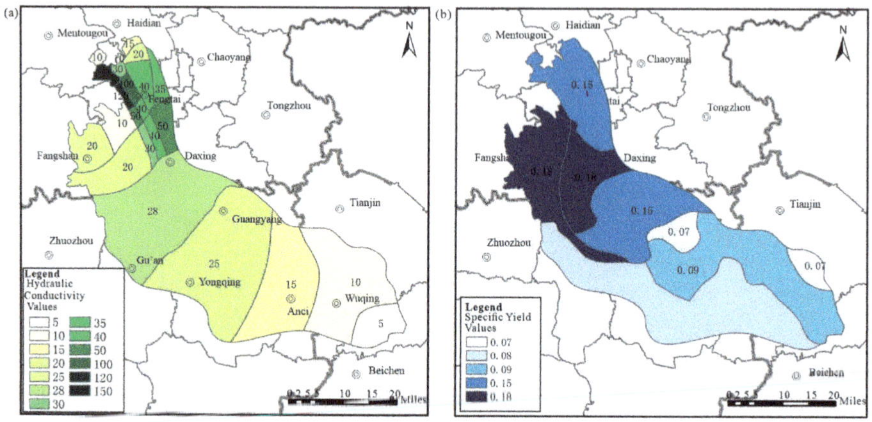

Figure 4. (**a**) the identification verified hydraulic conductivity distribution; (**b**) the identification verified distribution.

3.3. Influence of the Ecological Water Supplement of the Yongding River on the Groundwater Flow Systems

3.3.1. Actual Groundwater Level Variation in the Study Area from 2017 to 2019

Comparing the groundwater flow field in the study area between 2017 and 2019, the groundwater level was raised in the upstream area where the ecological water supplement had arrived (Figure 5). The groundwater level recovery range was generally between 2 and 8 m. The groundwater level in the piedmont generally increased by more than 8 m. In the area between Fangshan and Daxing, the river channel section could only get a few infiltrations of ecological water supplement. Thus, the groundwater level showed a small variation. The groundwater changed commonly between −2 m and 2 m. The groundwater level decrease mainly occurs in the southeastern parts of the study area. The groundwater level in these areas generally decreased by about 2 m. One reason for the groundwater level decline was that the ecological water supplement has not arrived at the downstream area from 2018 to 2019. The other reason was that the groundwater exploitation in the downstream area exceeded the natural recharge.

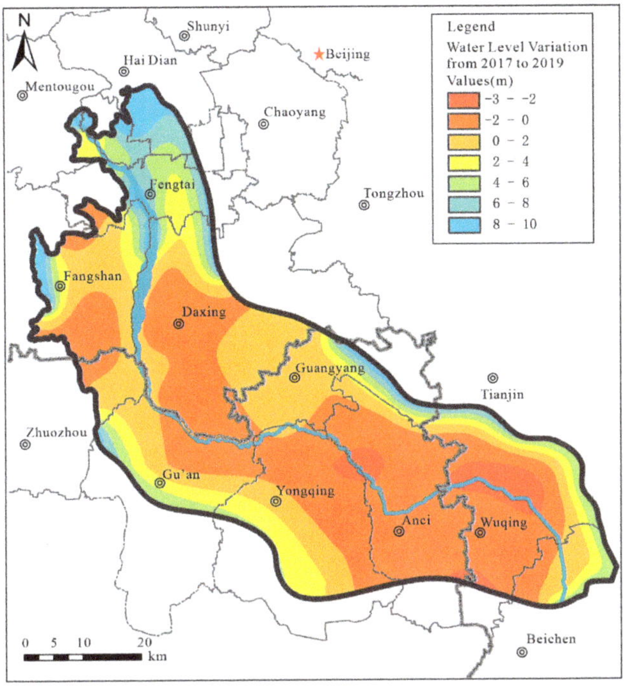

Figure 5. Differences in groundwater flow fields in the study area before and after ecological water supplement.

3.3.2. Groundwater Budget Variation before and after Ecological Water Supplement

It can be found from the change of groundwater budget (Table 2) that the ecological water recharge of the Yongding River could increase the recharge of 82×10^6 m^3/year for the study area, accounting for 9.47% of the total recharge. After ecological water supplement, the shallow groundwater changed from negative equilibrium state to positive equilibrium state. The equilibrium difference was 383×10^6 m^3/year, and the contribution proportion of ecological water supplement infiltration was 21.41%. The effect of ecological water supplement on groundwater quantity change was significant in the study area.

Table 2. Groundwater budget before and after ecological water supplement in the study area from 2016 to 2019.

	Before Ecological Water Recharge			After Ecological Water Recharge	
	Budget Tems	**Volume ($\times 10^6$ m^3/year)**	**Percentage**	**Volume ($\times 10^6$ m^3/year)**	**Percentage**
Recharge	Precipitation infiltration	578	68.65%	507	58.55%
	Mountain front recharge	141	16.75%	98	11.32%
	Lateral inflow	14	1.66%	100	11.55%
	Ecological water supplement infiltration	0	0.00%	82	9.47%
	Regression quantity of well irrigation	109	12.95%	79	9.12%
	Total recharge	842	100%	866	100.00%

Table 2. *Cont.*

	Before Ecological Water Recharge			After Ecological Water Recharge	
	Budget Tems	**Volume (×10⁶ m³/year)**	**Percentage**	**Volume (×10⁶ m³/year)**	**Percentage**
Discharge	Evaporation	18	1.80%	3	0.46%
	Lateral outflow	263	26.10%	40	6.19%
	Exploitation	725	72.10%	603	93.34%
	Total discharge	1006	100.00%	646	100.00%
	Recharge and discharge difference	−163		220	

3.3.3. The Effect of the Ecological Water Supplement of the Yongding River on Groundwater Level Variation

For further explaining the effect of ecological water supplement on groundwater level variation in the study area, the single factor variation scenario has been simulated. The simulation results of the model with ecological water supplement conditions and the model without ecological water supplement conditions were compared. The simulated flow fields on 31 December 2019 of the two models have been compared with the uplift variation variogram (Figure 6a). It can be found that the groundwater level in the upstream area which received sufficient ecological water recharge has obviously increased by over 4 m. Additionally, the maximum increase differences in the groundwater level variation with the two different conditions are located in the upstream area of the Yongding River nearby Fengtai and Daxing. The maximum increase in groundwater level is 8 to 10 m. The groundwater level rising zone could extend to the northern part of Gu'an. Additionally, the average groundwater level increase is 1.86 m.

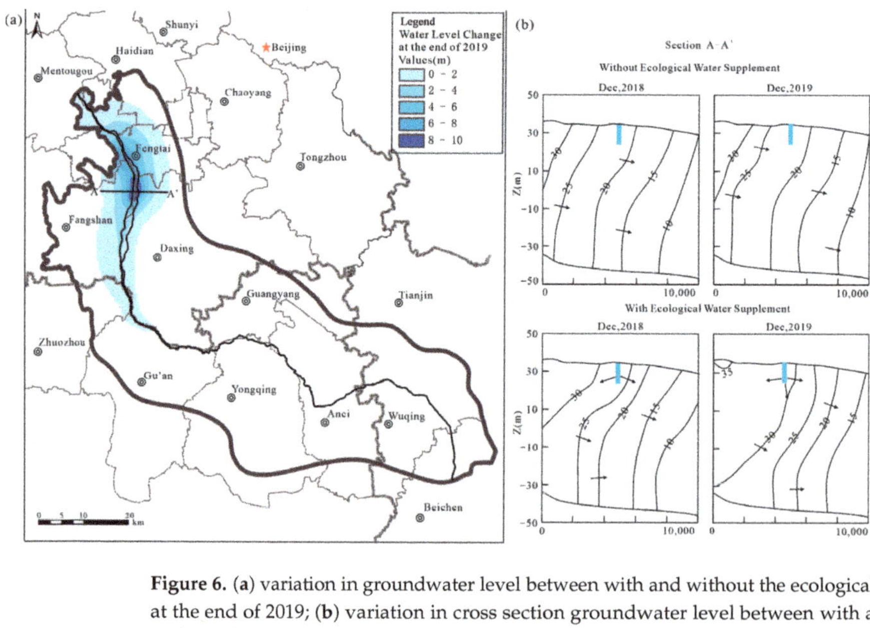

Figure 6. (**a**) variation in groundwater level between with and without the ecological water recharge at the end of 2019; (**b**) variation in cross section groundwater level between with and without the ecological water scenarios from 2018 to 2019.

A typical cross section has been selected to demonstrate the differences in flow field after the ecological water supplement. The selected cross section is located in the middle of S4 where the groundwater level changes the most obviously. Through the comparison of the cross-section flow fields, it can be seen (Figure 6b) that the groundwater level and hydraulic gradient decreased from west to east along with decreasing land surface elevations under the without ecological water recharge condition. Additionally, the flow trend has hardly changed from 2018 to 2019. Under the ecological water supplement condition, the groundwater field is different from that under the without ecological water supplement condition, and the groundwater level obviously increased. The ecological water recharge caused the groundwater level near the river channel to rise significantly in 2018, and the average groundwater level increase could be about 5 m. As the ecological water supplement progresses, the groundwater level gradually increased up to around 10 m. The influenced range from the center of the Yongding River channel had gradually spread to more than 5 km near the Yongding River channel by 2019.

3.3.4. The Impact Range of the Yongding River Ecological Water Supplement

Comparing the flow field between with ecological water supplement and without ecological water supplement (Figure 7), it can be seen that with the supplement water gradually recharged into the aquifers, the groundwater level on both sides of the Yongding River has increased. Taking the interpolation errors of the model and the minor fluctuations in the groundwater level into account, the groundwater level increase greater than 0.5 m has been selected as the standard to determine the impact range.

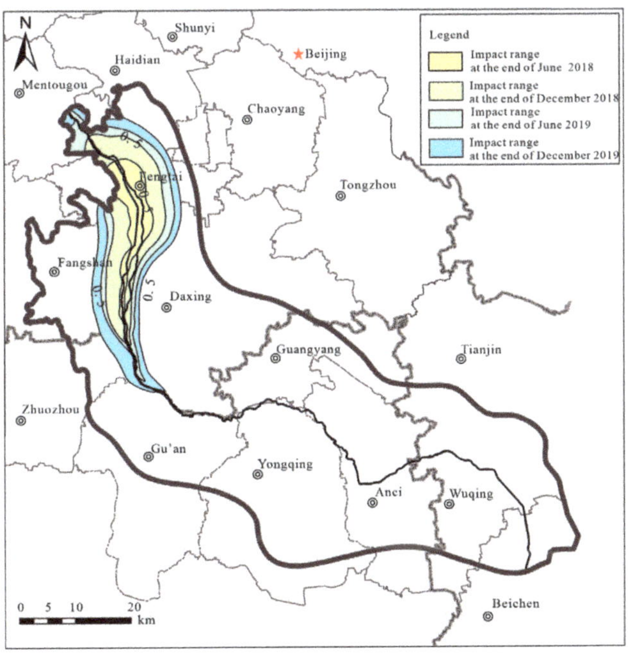

Figure 7. Variogram of impact range of the ecological water supplement of the Yongding River.

By the end of 2018, the farthest impact range of ecological water supplement could reach Zhuozhou. At this time, the farthest influence distance was 6.18 km, and the total impact range was 246.15 km^2. From the end of 2018 to the beginning of 2019, the impact range of ecological water supplement in the study area continued to expand, and the distribution feature of impact range gradually changed from developing along the Yongding River channel to the planar development pattern. The main development direction was northwest–southeast, which was consistent with the direction of groundwater flow. By 2019, the groundwater affected by ecological water supplement had reached Gu'an as far as possible. The areas affected by ecological water recharge include Beijing urban area, Fangshan District,

Daxing District, Zhuozhou City, and Gu'an County. The farthest influence distance at the end of the year reached 8.13 km, and the total impact range was 502.13 km^2. In addition, it can be found that when the maximum influence distance exceeds 6 km, the tangential impact range development rate along the river channel gradually decreases, and the development rate of the impact range along the river channel increases.

4. Discussion

4.1. Contribution of the Ecological Water Supplement in the Yongding River Channel to Groundwater Restoration

The random forest model can be used to identify the groundwater level uplift zones, and then the feature importance value and Shapley value can be used to estimate the contribution of the feature variables to the groundwater level restoration.

The selected feature variables (Table 3) include meteorological indicators, landform indicators, human activity indicators, and hydraulic features.

Table 3. Dataset of the driving factors.

Feature Variables			
Type	**Indicator**	**Source**	**Form**
Meteorological factors	Precipitation	Statistical data	Accumulate
	Evaporation	Statistical data	Accumulate
Topographical factors	Landform	Remote sensing data	Distributed
	Lateral inflow	Numetical model	Distributed
Human factors	Ecological water supplement	Statistics caculating	Accumulate
	Artificial mining	Statistical data	Accumulate
Hydraulic feature	Hydraulic conductivity	Numerical model calibration	Distributed
	Specific yield	Numerical model calibration	Distributed
Labeled data			
Groundwater level rises or not		Yes (1)	
		No (0)	

The *NSE* of the established random forest model is 0.93, the sensitivity is 0.92, and the specificity is 0.94, indicating that the model can well reflect the influence of feature variables on the groundwater level variation in the study area.

As can be seen from the distribution of feature importance value (Figure 8), human activities are the main driving factors affecting groundwater level recovery in the study area, with the contribution of 58%. According to the Shapley (Figure 9) distribution map, it can be seen that artificial groundwater mining mainly restrains the rise of groundwater level, with the contribution of 33%. The greater the amount of exploitation is, the more difficult it is for the water level to rise. The long-time over-exploitation is also the main reason leading to the regional groundwater level drawdown in the last decades [71].

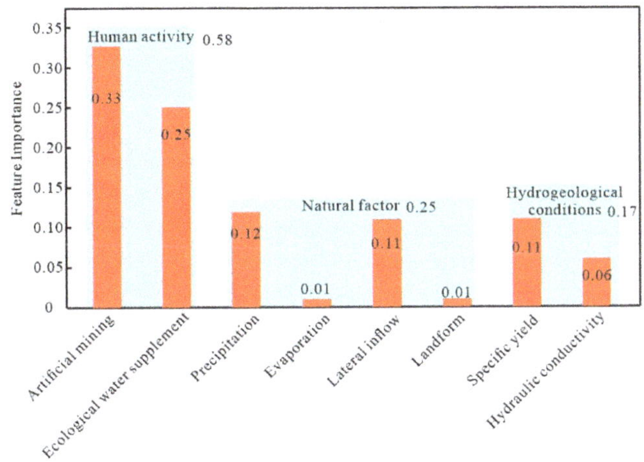

Figure 8. Feature importance map of driving factors.

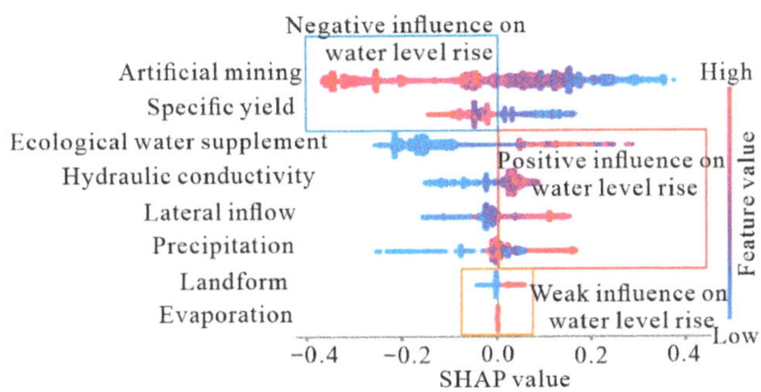

Figure 9. Shapley value distribution of driving factors.

The ecological water supplement infiltration could play a significant role in promoting the recovery of groundwater level, and the contribution is 25%. With the increase in ecological water supplement infiltration, and especially because of the great infiltration capacity in the upstream area of the Yongding River, the groundwater level in the study area has increased obviously [72]. The ecological water recharge in the groundwater cycle will become more and more important with sufficient surface water and reclaimed water sources.

Secondly, precipitation would have a positive effect on the recovery of groundwater level in the study area, with the contribution of 12%. The precipitation recharges the groundwater in the form of areal supply [73]. The precipitation in the study area is the steady recharge source. It is the most important recharge item in the study area.

The lateral recharge and specific yield contributions are more than 10%. The lateral recharge usually depends on the hydraulic gradient and the lithology of calculation section. The greater the hydraulic gradient and permeability are, the greater the lateral recharge is, and the more the groundwater level can be uplifted [74]. In the study area, the piedmont area has the largest lateral recharge, but the recharge section is relatively short, and the total lateral recharge in the study area is small. Therefore, the lateral recharge cannot be the main recharge source in the study area.

The specific yield reflects the groundwater level change ability in the study area [75]. It can be seen that the larger the specific yield is, the less obvious the groundwater level change is. The groundwater aquifers in the upstream area of the Yongding River are mainly composed of coarse sediments. The specific yield in the upstream area is large, and the groundwater level fluctuation is significant. However, in the middle and low reaches of the Yongding River, the groundwater aquifers are composed mainly of fine sand. The groundwater level variation impacted by the specific yield is not obvious.

The contributions of other feature factors to the groundwater level variation in the study area are very small.

4.2. Suggestions of Ecological Water Supplement and Groundwater Exploitation

Using the established groundwater flow numerical model in this study, the prediction study was carried out for the purpose of achieving the groundwater level rise in the entire area in 2035. By adjusting the amount of ecological water supplement and groundwater exploitation, in the study area, the groundwater level could be entirely recovered state. Additionally, the specific scheme proposed in this study is shown as follows:

1. The ecological water supplement should increase by 40×10^6 m^3/year in the downstream area. The potential artificial ecological water supply sources mainly include the water from the mid-route of the South-to-North Water Diversion Project, the water from the Guanting Reservoir, and the reclaimed water collected from the cities along the Yongding River. Then, the total ecological water supplement in the study area is 170×10^6 m^3/year.
2. The groundwater exploitation should be reduced by 35.77×10^6 m^3/year in the downstream area, including the eastern part of Yongqing County, the southern part of Langfang City, the southern part of Wuqing District, and the western part of Beichen District.

Figure 10 shows the groundwater level variation from 2017 to 2035 under the scheme mentioned above. In the upstream area, the groundwater level would increase by over 20 m because the infiltration of ecological water supplement is far more than other sections. By contrast, the filtration of ecological water supplement is very limited in the downstream area, and the groundwater level just increases by about less than 0.3 m/year. Between the upstream area and downstream area, the groundwater level could increase by 2 to 8 m, and the average groundwater level uplift is around 5 m.

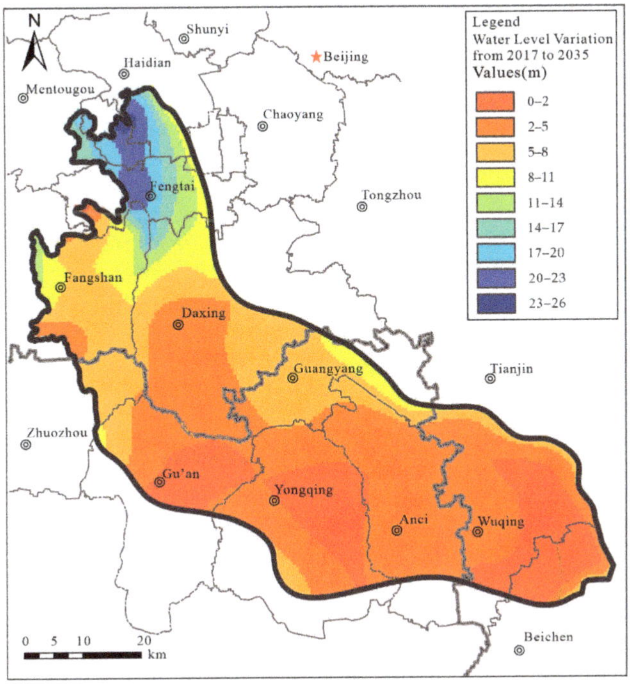

Figure 10. The predicted groundwater level variation of the study area.

5. Conclusions

In this paper, the fuzzy mathematics model, numerical simulation model, and random forest model have been linked to evaluate the effect of ecological water supplement in the Yongding River on groundwater restoration. The models established in this study have good performance and can reflect the features of groundwater flow. The infiltration volume of the ecological water supplement, the groundwater level variation, the impact range of the ecological water supplement, the contribution of ecological water supplement to the groundwater restoration, and the ecological water supplement suggestion have been systematically studied by using the linkage models.

The seepage rate of the upper reaches of the Yongding River could reach more than 0.7, and the downstream is generally about 0.2. The infiltration volume of the Yongding River channel accounts for about 60% of the total ecological water supplement, which can be effectively recharged into the groundwater aquifer. The contribution of ecological water supplement infiltration to the uplift of groundwater level is 25%, indicating that ecological water supplement could significantly promote the groundwater restoration in the study area.

The average groundwater level increase was about 5 m/year from 2018 to 2019. The groundwater recharge and discharge difference was changed from -163×10^6 m^3 to 220×10^6 m^3 before and after ecological water supplement. Additionally, the infiltration of ecological water supplement could account for 21.41% of the contribution to the difference of groundwater budget.

By the end of 2019, the maximum distance affected by ecological groundwater supplement in the Yongding River was 8.13 km, and the impact range was 502.13 km^2. By 2035, under the scheme of increasing the ecological water supplement to 170×10^6 m^3/year and reducing 35.77×10^6 m^3/year groundwater exploitation in the downstream area, the entire groundwater level will be uplifted in the study area.

Author Contributions: Conceptualization, T.N. and W.C.; methodology, T.N.; software, T.N.; validation, T.N. and W.C.; formal analysis, T.N.; investigation, W.C.; resources, W.C.; data curation, W.C.; writing—original draft preparation, T.N.; writing—review and editing, T.N. and W.C.; supervision, W.C.; funding acquisition, W.C. All authors have read and agreed to the published version of the manuscript.

Funding: This research was funded by National Key R&D Program of China (2022YFC3703701), National Natural Science Foundation of China (41972262), and Hebei Natural Science Foundation for Excellent Young Scholars (D2020504032).

Data Availability Statement: Publicly available datasets were analyzed in this study. The precipitation data, evaporation data and ecological water supplement data can be found here: http://data.cma.cn/ and http://www.hwcc.gov.cn/wwgj/slsjk/ (accessed on 18 December 2022). The drilling data presented in this study are available on request from the corresponding author. The data are not publicly available due to the regulations of the Institute of Hydrogeology and Environmental Geology. Restrictions apply to the availability of groundwater level and exploitation data. Data was obtained from Hydrology Bureau of Haihe River Water Conservancy Commission and National Groundwater Monitoring Center. And are available from the authors with the permission of Hydrology Bureau of Haihe River Water Conservancy Commission and National Groundwater Monitoring Center.

Acknowledgments: This research was financially supported by National Key R&D Program of China (2022YFC3703701), National Natural Science Foundation of China (41972262), and Hebei Natural Science Foundation for Excellent Young Scholars (D2020504032). The authors would like to thank the reviewers and editors for their helpful comments and suggestions.

Conflicts of Interest: The authors declare no conflict of interest.

References

1. Escalante, E.F.; Dillon, P.; Stefan, C.; Sapiano, M.; Bonilla, J.; International Association of Hydrogeologists Commission on Managing Aquifer Recharge. Report of Activities Madrid, May, 2019. 2020. Available online: https://recharge.iah.org/ (accessed on 18 December 2022).
2. Gale, I.; Dillon, P. *Strategies for Managed Aquifer Recharge (MAR) in Semi-Arid Areas*; UNESCO Paris: Paris, France, 2005.
3. Yan, Z.; Zhou, Z.; Sang, X.; Wang, H. Water replenishment for ecological flow with an improved water resources allocation model. *Sci. Total Environ.* **2018**, *643*, 1152–1165. [CrossRef] [PubMed]
4. Zheng, Y.; Vanderzalm, J.; Hartog, N.; Escalante, E.F.; Stefan, C. The 21st century water quality challenges for managed aquifer recharge: Towards a risk-based regulatory approach. *Hydrogeol. J.* **2022**. [CrossRef] [PubMed]
5. Hao, Q.; Shao, J.; Cui, Y.; Xie, Z. Applicability of artificial recharge of groundwater in the Yongding River alluvial fan in Beijing through numerical simulation. *J. Earth Sci.* **2014**, *25*, 575–586. [CrossRef]
6. Gan, Y.; Wang, Y.; Duan, Y.; Deng, Y.; Guo, X.; Ding, X. Hydrogeochemistry and arsenic contamination of groundwater in the Jianghan Plain, central China. *J. Geochem. Explor.* **2014**, *138*, 81–93. [CrossRef]
7. Hu, L.; Xu, Z.; Huang, W. Development of a river-groundwater interaction model and its application to a catchment in Northwestern China. *J. Hydrol.* **2016**, *543*, 483–500. [CrossRef]
8. Longcang, S.; Rong, G.; Jiawen, L.; Jianshi, G.; Xiaohu, T.; Guizhang, Z.; Yuezan, T. A integrated method to quantify flow exchanges between surface water and groundwater: Take Suzhou section of the Xinbian River as an example. *Adv. Water Sci.* **2022**, *33*, 57–67.
9. Du, Y.; Ma, T.; Deng, Y.; Shen, S.; Lu, Z. Sources and fate of high levels of ammonium in surface water and shallow groundwater of the Jianghan Plain, Central China. *Environ. Sci. Process. Impacts* **2017**, *19*, 161–172. [CrossRef]
10. He, J.; Yao, J.; Li, A.; Tan, Z.; Xie, G.; Shi, H.; Zhang, X.; Sun, W.; Du, P. Potential impact of water transfer policy implementation on lake eutrophication on the Shandong Peninsula: A difference-in-differences approach. *Hydrol. Res.* **2020**, *51*, 1063–1076. [CrossRef]
11. Liang, Y.; Ma, R.; Wang, Y.; Wang, S.; Qu, L.; Wei, W.; Gan, Y. Hydrogeological controls on ammonium enrichment in shallow groundwater in the central Yangtze River Basin. *Sci. Total Environ.* **2020**, *741*, 140350. [CrossRef]
12. Koch, J.C.; Runkel, R.L.; Striegl, R.; McKnight, D.M. Hydrologic controls on the transport and cycling of carbon and nitrogen in a boreal catchment underlain by continuous permafrost. *J. Geophys. Res. Biogeosci.* **2013**, *118*, 698–712. [CrossRef]
13. Stoliker, D.L.; Repert, D.A.; Smith, R.L.; Song, B.; LeBlanc, D.R.; McCobb, T.D.; Conaway, C.H.; Hyun, S.P.; Koh, D.-C.; Moon, H.S. Hydrologic controls on nitrogen cycling processes and functional gene abundance in sediments of a groundwater flow-through lake. *Environ. Sci. Technol.* **2016**, *50*, 3649–3657. [CrossRef] [PubMed]
14. Barthel, R.; Banzhaf, S. Groundwater and surface water interaction at the regional-scale–a review with focus on regional integrated models. *Water Resour. Manag.* **2016**, *30*, 1–32. [CrossRef]
15. Kollet, S.J.; Maxwell, R.M. Integrated surface–groundwater flow modeling: A free-surface overland flow boundary condition in a parallel groundwater flow model. *Adv. Water Resour.* **2006**, *29*, 945–958. [CrossRef]

16. Kolditz, O.; Bauer, S.; Bilke, L.; Böttcher, N.; Delfs, J.-O.; Fischer, T.; Görke, U.J.; Kalbacher, T.; Kosakowski, G.; McDermott, C. OpenGeoSys: An open-source initiative for numerical simulation of thermo-hydro-mechanical/chemical (THM/C) processes in porous media. *Environ. Earth Sci.* **2012**, *67*, 589–599. [CrossRef]
17. Brunner, P.; Simmons, C.T. HydroGeoSphere: A Fully Integrated, Physically Based Hydrological Model. *Groundwater* **2012**, *50*, 170–176. [CrossRef]
18. Xia, J.; Wang, Q.; Zhang, X.; Wang, R.; She, D. Assessing the influence of climate change and inter-basin water diversion on Haihe River basin, eastern China: A coupled model approach. *Hydrogeol. J.* **2018**, *26*, 1455–1473. [CrossRef]
19. Wu, B.; Zheng, Y.; Tian, Y.; Wu, X.; Yao, Y.; Han, F.; Liu, J.; Zheng, C. Systematic assessment of the uncertainty in integrated surface water-groundwater modeling based on the probabilistic collocation method. *Water Resour. Res.* **2015**, *50*, 5848–5865. [CrossRef]
20. Feng, D.; Yi, Z.; Mao, Y.; Zhang, A.; Xin, W. An integrated hydrological modeling approach for detection and attribution of climatic and human impacts on coastal water resources. *J. Hydrol.* **2017**, *557*, 305–320. [CrossRef]
21. Xue, L.; Ye, S.Y.; Wei, A.H.; Zhou, P.P.; Wang, L.H. Modelling the response of shallow groundwater level to combined climate and water-diversion scenarios in Beijing-Tianjin-Hebei Plain, China. *Hydrogeol. J.* **2017**, *25*, 1733–1744.
22. Kitanidis, P.K. Persistent questions of heterogeneity, uncertainty, and scale in subsurface flow and transport. *Water Resour. Res.* **2015**, *51*, 5888–5904. [CrossRef]
23. Lu, D.; Zhang, G.; Webster, C.; Barbier, C. An improved multilevel Monte Carlo method for estimating probability distribution functions in stochastic oil reservoir simulations. *Water Resour. Res.* **2016**, *52*, 9642–9660. [CrossRef]
24. Liao, Q.; Zhang, D.; Tchelepi, H. A two-stage adaptive stochastic collocation method on nested sparse grids for multiphase flow in randomly heterogeneous porous media. *J. Comput. Phys.* **2017**, *330*, 828–845. [CrossRef]
25. Razavi, S.; Tolson, B.A.; Burn, D.H. Review of surrogate modeling in water resources. *Water Resour. Res.* **2012**, *48*, W07401. [CrossRef]
26. Asher, M.J.; Croke, B.F.; Jakeman, A.J.; Peeters, L.J. A review of surrogate models and their application to groundwater modeling. *Water Resour. Res.* **2015**, *51*, 5957–5973. [CrossRef]
27. Zhang, J.; Lin, G.; Li, W.; Wu, L.; Zeng, L. An iterative local updating ensemble smoother for estimation and uncertainty assessment of hydrologic model parameters with multimodal distributions. *Water Resour. Res.* **2018**, *54*, 1716–1733. [CrossRef]
28. Wang, X.; Liu, T.; Zheng, X.; Peng, H.; Xin, J.; Zhang, B. Short-term prediction of groundwater level using improved random forest regression with a combination of random features. *Appl. Water Sci.* **2018**, *8*, 125. [CrossRef]
29. Chakraborty, M.; Sarkar, S.; Mukherjee, A.; Shamsudduha, M.; Ahmed, K.M.; Bhattacharya, A.; Mitra, A. Modeling regional-scale groundwater arsenic hazard in the transboundary Ganges River Delta, India and Bangladesh: Infusing physically-based model with machine learning. *Sci. Total Environ.* **2020**, *748*, 141107. [CrossRef]
30. Fu, Y.; Cao, W.; Pan, D.; Ren, Y. Changes of groundwater arsenic risk in different seasons in Hetao Basin based on machine learning model. *Sci. Total Environ.* **2022**, *817*, 153058. [CrossRef]
31. Peng, J.; Li, S.; Qi, L. Study on river regulation measures of dried-up rivers of Haihe River basin, China. *Water Sci. Technol.* **2013**, *67*, 1224–1229. [CrossRef]
32. Jiang, B.; Wong, C.P.; Lu, F.; Ouyang, Z.; Wang, Y. Drivers of drying on the Yongding River in Beijing. *J. Hydrol.* **2014**, *519*, 69–79. [CrossRef]
33. Hu, H.; Mao, X.; Yang, Q. Impacts of Yongding River ecological restoration on the groundwater environment: Scenario prediction. *Vadose Zone J.* **2018**, *17*, 1–15. [CrossRef]
34. Luo, Z.; Zhao, S.; Wu, J.; Zhang, Y.; Liu, P.; Jia, R. The influence of ecological restoration projects on groundwater in Yongding River Basin in Beijing, China. *Water Supply* **2019**, *19*, 2391–2399. [CrossRef]
35. Sun, K.; Hu, L.; Guo, J.; Yang, Z.; Zhai, Y.; Zhang, S. Enhancing the understanding of hydrological responses induced by ecological water replenishment using improved machine learning models: A case study in Yongding River. *Sci. Total Environ.* **2021**, *768*, 145489. [CrossRef] [PubMed]
36. Han, Y.; Zhai, Y.; Guo, M.; Cao, X.; Lu, H.; Li, J.; Wang, S.; Yue, W. Hydrochemical and Isotopic Characterization of the Impact of Water Diversion on Water in Drainage Channels, Groundwater, and Lake Ulansuhai in China. *Water* **2021**, *13*, 3033. [CrossRef]
37. Liu, C.; Pan, C.; Chang, Y.; Luo, M. An integrated autoregressive model for predicting water quality dynamics and its application in Yongding River. *Ecol. Indic.* **2021**, *133*, 108354. [CrossRef]
38. Xie, Y.; Jinjin, G.; Guo, Y.; Peng, W.; Wang, L. Temporal and Spatial Variation of water quality in the Yongding River Basin. *Proc. IOP Conf. Ser. Earth Environ. Sci.* **2021**, *831*, 012052. [CrossRef]
39. Hou, L.; Peng, W.; Qu, X.; Chen, Q.; Fu, Y.; Dong, F.; Zhang, H. Runoff changes based on dual factors in the upstream area of Yongding river basin. *Pol. J. Environ. Stud.* **2019**, *28*, 143–152. [CrossRef] [PubMed]
40. Litang, H.; Jianli, G.; Shouquan, Z.; Kangning, S.; Zhengqiu, Y. Response of groundwater regime to ecological water replenishment of the Yongding River. *Hydrogeol. Eng. Geol.* **2020**, *47*, 5–11.
41. Mingzhu, L.; Seyf-Laye, A.-S.M.; Ibrahim, T.; Gbandi, D.-B.; Honghan, C. Tracking sources of groundwater nitrate contamination using nitrogen and oxygen stable isotopes at Beijing area, China. *Environ. Earth Sci.* **2014**, *72*, 707–715. [CrossRef]
42. Yu, Y.; Ma, M.; Zheng, F.; Liu, L.; Zhao, N.; Li, X.; Yang, Y.; Guo, J. Spatio-temporal variation and controlling factors of water quality in Yongding River replenished by reclaimed water in Beijing, North China. *Water* **2017**, *9*, 453. [CrossRef]
43. Wang, L.; Wang, Z.; Koike, T.; Yin, H.; Yang, D.; He, S. The assessment of surface water resources for the semi-arid Yongding River Basin from 1956 to 2000 and the impact of land use change. *Hydrol. Process. Int. J.* **2010**, *24*, 1123–1132. [CrossRef]

44. Qiang, Y.; Yu, Z.; Cuiping, Y.; Dong, H. Review and suggestions on river ecological management in urban section of Yongding River in Beijing. *China Flood Drought Manag.* **2018**, *28*, 28–32.
45. Guo, H.; Chen, X.; Liu, G. Study on the practices of water rights trading in South-to-North Water Diversion Middle Route Project. *South North Water Transf. Water Sci. Technol.* **2018**, *16*, 175–182.
46. Huan, H.; Wang, J.-S.; Zhai, Y.-Z.; Zheng, J.-Q. Chemical Characteristics and Evolution of Groundwater in the Yongding River Alluvial Fan of Beijing Plain. *Acta Geosci. Sin.* **2011**, *32*, 357–366.
47. Li, D.; Xiao, W.; Zhao, M.; Zhang, J.; Liu, S.; Liu, J. Flood dispatching numerical simulation for detention basins of yongding river. *J. Tianjin Univ.* **2015**, *48*, 76–86.
48. Hu, H.; Mao, X.; Yang, Q. Development of a groundwater flow and reactive solute transport model in the Yongding River alluvial fan, China. *Front. Earth Sci.* **2019**, *13*, 371–384. [CrossRef]
49. Qi-chen, H.; Jing-li, S.; Yu, L.; Zhenhua, X.; Yali, C. Optimization of artificial recharge of groundwater system based on parallel genetic algorithm—A case study in the alluvial fan of Yongding River in Beijing. *South North Water Transf. Water Sci. Technol.* **2015**, *13*, 67–71.
50. Xiaofang, Y.; Mingyu, W.; Liya, W.; Jianhui, Z. Investigation of key controlling factors and numerical simulation uncertainty of the groundwater level companying with Yongding river ecological restoration. *J. Univ. Chin. Acad. Sci.* **2015**, *32*, 192–199.
51. Zhang, K.; Gann, D.; Ross, M.; Robertson, Q.; Sarmiento, J.; Santana, S.; Rhome, J.; Fritz, C. Accuracy assessment of aster, srtm, alos, and tdx dems for hispaniola and implications for mapping vulnerability to coastal flooding. *Remote Sens. Environ.* **2019**, *225*, 290–306. [CrossRef]
52. Beijing Water Authority. *Beijing Water Resources Bulletin*; Beijing Water Authority: Beijing, China, 2018. (In Chinese)
53. Beijing Water Authority. *Beijing Water Resources Bulletin*; Beijing Water Authority: Beijing, China, 2019. (In Chinese)
54. Tianjin Water Authority. *Tianjin Water Resources Bulletin*; Beijing Water Authority: Beijing, China, 2018. (In Chinese)
55. Tianjin Water Authority. *Tianjin Water Resources Bulletin*; Beijing Water Authority: Beijing, China, 2019. (In Chinese)
56. Hebei Water Authority. *Hebei Water Resources Bulletin*; Beijing Water Authority: Beijing, China, 2018. (In Chinese)
57. Hebei Water Authority. *Hebei Water Resources Bulletin*; Beijing Water Authority: Beijing, China, 2019. (In Chinese)
58. Haihe River Water Conservancy Commission. *Haihe River Basin Water Resources Bulletin*; Haihe River Water Conservancy Commission: Tianjin, China, 2018. (In Chinese)
59. Haihe River Water Conservancy Commission. *Haihe River Basin Water Resources Bulletin*; Haihe River Water Conservancy Commission: Tianjin, China, 2019. (In Chinese)
60. Institute of Hydrogeology and Environmental Geology, Chinese Academy of Geological Sciences. *1:50000 Environmental Geological Survey in Non-Capital Functional Relief Area*; Institute of Hydrogeology and Environmental Geology, Chinese Academy of Geological Sciences: Shijiazhuang, China, 2019. (In Chinese)
61. Institute of Hydrogeology and Environmental Geology, Chinese Academy of Geological Sciences. *Investigation Report on Water Resources in Haihe River Basin*; Institute of Hydrogeology and Environmental Geology, Chinese Academy of Geological Sciences: Shijiazhuang, China, 2022. (In Chinese)
62. Zadeh, L.A. Fuzzy sets. *Inf. Control* **1965**, *8*, 338–353. [CrossRef]
63. Coronel, E.G.; Alesso, C.A.; Bollero, G.A.; Armstrong, K.L.; Martin, N.F. Field-specific yield response to variable seeding depth of corn in theMidwestField-water supply response to variable seeding depth of corn in the Midwest. *Agrosystems Geosci. Environ.* **2020**, *3*, e20034. [CrossRef]
64. Mcdonald, M.G. A Modular Three-Dimensional Finite-Difference Ground-water Flow Model. *Tech. Water Resour. Investig.* **1988**, *83-875*, 83–875. [CrossRef]
65. Gong, C.; Zhang, Z.; Wang, W.; Duan, L.; Wang, Z. An assessment of different methods to determine water supply for estimating groundwater recharge using lysimeters. *Sci. Total Environ.* **2021**, *788*, 147799. [CrossRef] [PubMed]
66. Jingli, S.; Ling, L.; Yali, C.; Zhaoji, Z. Groundwater flow simulation and its application in groundwater resource evaluation in the North China Plain, China. *Acta Geol. Sin. Engl. Ed.* **2013**, *87*, 243–253. [CrossRef]
67. Verikas, A.; Gelzinis, A.; Bacauskiene, M. Mining data with random forests: A survey and results of new tests. *Pattern Recognit. J. Pattern Recognit. Soc.* **2011**, *44*, 330–349. [CrossRef]
68. Xi, C.; Ishwaran, H. Random forests for genomic data analysis. *Genomics* **2012**, *99*, 323–329.
69. Molnar, C. *Interpretable Machine Learning*; Lulu Press: Morrisville, NC, USA, 2020.
70. Zijian, J.; Yali, C.; Shouquan, Z.; Wan, C.; Jingli, S. Evaluation of the Impact of EcologicalWater Supplement on Groundwater Restoration Based on Numerical Simulation: A Case Study in the Section of Yongding River, Beijing Plain. *Water* **2021**, *13*, 3059. [CrossRef]
71. Karian, W.; Huawei, C.; Shidong, F.; Fulin, L.; Zhen, W.; Dandan, X. Analysis of exploitation control in typical groundwater over-exploited area in North China Plain. *Hydrol. Sci. J.* **2021**, *66*, 851–861.
72. Yixuan, W.; Yanjun, S.; Ya, G.; Hang, L. Research progress on the changes of environmental and water resources in the upper Yongding River Basin. *South North Water Transf. Water Sci. Technol.* **2021**, *4*, 656–668.
73. Kai, L.; Xueke, L.; Xin, L. Trends in groundwater changes driven by precipitation and anthropogenic activities on the southeast side of the Hu Line. *Environ. Res. Lett.* **2021**, *16*, 094032. [CrossRef]

74. Alemu, Y.; Fenta, N.; Ashebir, B.; Minyahl, D.; Marc, C.; Kristine, W. Groundwater recharge and water table response to changing conditions for aquifers at different physiography: The case of a semi-humid river catchment, northwestern highlands of Ethiopia. *Sci. Total Environ.* **2020**, *748*, 142243. [CrossRef]
75. Maréchal, J.C.; Dewandel, B.; Ahmed, S.; Galeazzi, L.; Zaidi, F.K. Combined estimation of specific yield and natural recharge in a semi-arid groundwater basin with irrigated agriculture. *J. Hydrol.* **2006**, *329*, 281–293. [CrossRef]

water

MDPI

Article

Mining Scheme for Small Rivers near Water Sources—A Case Study of Liuan River in Linquan County, China

Zhenyu He [1], Bo Kang [2,*], Yuezan Tao [1] and Li Qin [3]

1 School of Civil Engineering, Hefei University of Technology, Hefei 230009, China; hezhenyu707@163.com (Z.H.); 2019030008@mail.hfut.edu.cn (Y.T.)
2 School of Resource and Environmental Engineering, Hefei University of Technology, Hefei 230009, China
3 School of Resource and Environmental Engineering, Anhui Technical College of Water Resources and Hydroelectric Power, Hefei 231603, China; 13696789253@163.com
* Correspondence: kangbo@hfut.edu.cn; Tel.: +86-18963794495

Abstract: During dry seasons or years, the runoff processes from small rivers influence the safety of riverside groundwater source fields. Water source exploitation has a considerable effect on river runoff. In this study, the riverside source field of the Liuan River in the Linquan County, Anhui Province, was analyzed. The effect of mining on the ecological flow of the river under river runoff conditions in different typical dry years was quantified. This was undertaken using numerical simulations of the groundwater flow to provide guidance for the establishment of mining schemes for riverside source fields. In 95% of typical dry years, the water supply of small rivers is insufficient. The improved 7Q10 method used to calculate the ecological flow in different dry years revealed that mining water had little effect on the ecological flow. However, during the pumping process, the groundwater level of the water source area decreased greatly. The establishment of riverside source fields can aid in reducing excessive development and use of deep groundwater. The planning, construction, and implementation of the "Divert water from the Yangtze River to the Huaihe River" project can effectively reduce the economic losses that have occurred due to severe drought in the local area.

Keywords: small rivers; riverside source field; drought; ecological flow; Visual MODFLOW; water balance

check for updates

Citation: He, Z.; Kang, B.; Tao, Y.; Qin, L. Mining Scheme for Small Rivers near Water Sources—A Case Study of Liuan River in Linquan County, China. *Water* **2022**, *14*, 1921. https://doi.org/10.3390/w14121921

Academic Editors: Dimitrios E. Alexakis and Richard C. Smardon

Received: 28 April 2022
Accepted: 8 June 2022
Published: 15 June 2022

1. Introduction

In many arid and semi-arid regions, the long-term exploitation of groundwater has caused a range of environmental problems [1]. Researchers worldwide have studied groundwater overexploitation and highlighted that it can lead to a considerable drop in the water level, the formation of groundwater falling funnels, ground subsidence due to ground cracks, groundwater pollution, and deterioration of the groundwater quality [2]. In 2003, the Ministry of Water Resources of the People's Republic of China formulated the "Opinions on Strengthening the Management of Water Resources in Groundwater Overexploited Areas" with the aim of reducing groundwater exploitation [3]. Based on the reduction in groundwater exploitation, some areas that use groundwater as a source of municipal water may need to adjust their strategies for water supply [4]. Many researchers have suggested the use of riverside source fields to secure urban water supplies [5]. Riverside source fields refer to water sources in which the lateral river runoff is the main water source for pumping wells that are distributed along rivers. Compared with other groundwater and surface water sources, the exploitation of riverside source fields is more complex. It depends not only on the hydrogeological conditions near the rivers but also on the hydraulic connection with aquifers and additional hydrogeological parameters [6]. Sufficient water in the riverside source field is recharged from the river [7], resolving surface water quality problems through the water–rock reaction in the flow path, which meets the safe drinking water standard in some areas [8]. The advantage in terms of water supply from the riverside source is that it can be used for regulation and improving the storage capacity of the aquifer, as well as mutual adjustment with the river.

Many researchers have studied the joint regulation of surface water and groundwater in large river basins [9]. However, small rivers are often an important water resource in arid and semi-arid areas. Rivers with protected areas of less than 6.6 km^2 in size are generally classified as small rivers. There are few studies that have been undertaken on the joint regulation and impact mechanisms of groundwater and surface water in small river basins. Small rivers in arid and semi-arid areas have low levels of flow and low water levels. There is also a strong seasonal influence and ecological effect on river flow in arid and semi-arid areas. Small rivers can play a key role in meeting the local demand for water supplies as the riverside source field. However, small rivers may suffer from periodic water shortages during the dry season, and different exploitation strategies need to be formulated for implementation during dry years [10]. During these dry years, exploitation of water sources near the river will lead to declining water levels in small rivers, as well as changes in the hydraulic gradient [11]. If the water level of the small rivers is lower than the ecological water level, it will cause irreversible ecological and environmental problems in these river basins. With improvement in groundwater management systems, groundwater management is not limited to the pumping capacity. There are also strict control standards for changes in the groundwater level and the associated ecological impact. With the proposal of the ministry of water resources of China that the protection of ecological flow of rivers and lakes should be put in the first place, the protection of ecological flow should be paid more attention in the water resource management of the Huaihe River Basin [12]. Due to the uneven distribution of rainfall in the Huaihe River Basin, the ecological flow of the river should be paid more attention in the dry season. Liuan River, as a small tributary of the Huaihe River Basin, will also have an important impact on the ecological flow due to the decline of water level in the process of taking water near the river. In arid and semi-arid areas, there are frequent exchanges between the surface water and the groundwater. The dynamic changes in groundwater levels and the river levels should also form a key focus while ensuring maintenance of the water supply [13]. Therefore, the exploitation strategy of small shallow rivers should be adjusted under climate change [14]. In calculating the water supply capacity of riverside water sources in small rivers, the influence from changes in the boundary conditions on the water supply should be a key focus [15].

The hydraulic changes and ecological impacts from abstraction of water near the river in small river basins are not currently clear. This study focuses on a typical small river in the Huaibei Plain of China and uses it as a case study to examine the characteristics of small rivers. The numerical simulation method is used to study the hydraulic changes and their ecological effects from the water intake area near the Liuan River.

In this study, the riverside source field of the Liuan River in the Huaihe River Basin (Figures 1 and 2) was used as the case study and the hydrogeological conditions were systematically analyzed. A numerical model was established to analyze the effects of the water intake in the riverside source area of the Liuan River in 95%, 75%, and 50% typical dry years. The model was used to simulate and predict the effects of the water supply on the riverside source area and the ecological flow. Based on the results of this study, this paper proposes a mining scheme that is suitable for small rivers near water sources [16]. Compared with the water supply of large rivers, when studying the water supply of small rivers, the impact of pumping on the river runoff and ecological flow, and changes in the groundwater level should be a key focus [17], as well as the planning, construction, and implementation of the "Divert water from the Yangtze River to the Huaihe River" project. The "Divert water from the Yangtze River to the Huaihe River" project is a large-scale trans-basin water diversion project, which mainly focuses on urban and rural water supplies. The project also encompasses the development of navigation in the Yangtze River and the Huaihe River, combined with the replenishment of irrigation water and the improvement of the water ecological environment of the Chaohu Lake and the Huaihe River. The study aims to provide suggestions for solving the problem of water supplies in arid and semi-arid areas, and to prevent economic losses caused by the imbalance between the supply and demand of water resources.

Figure 1. Location map showing the Huaihe River Basin.

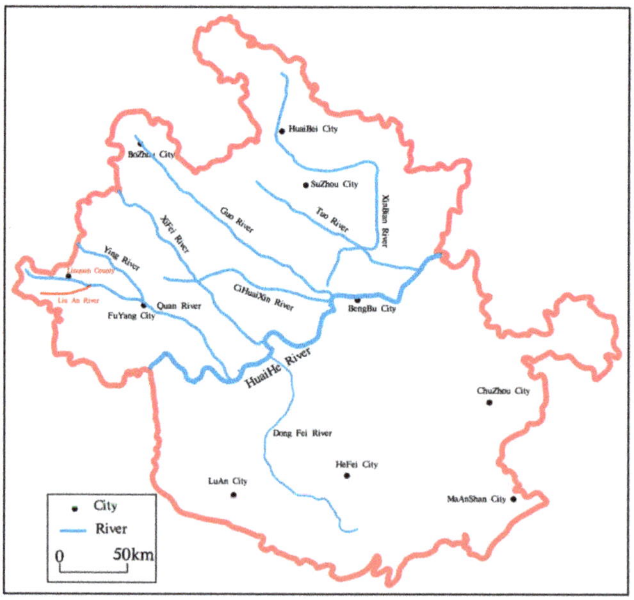

Figure 2. Map of the Huaihe River drainage system.

2. Study Area

2.1. Geographical Location

The research area is in Linquan County, Fuyang City, Anhui Province. The terrain is relatively low in the northeast and high in the southwest, with an elevation ranging from 34 to 40 m asl. The research area is approximately 4 km^2 (Figures 3 and 4) with an accumulation plain formed by the flooding of the Yellow and the Huaihe rivers.

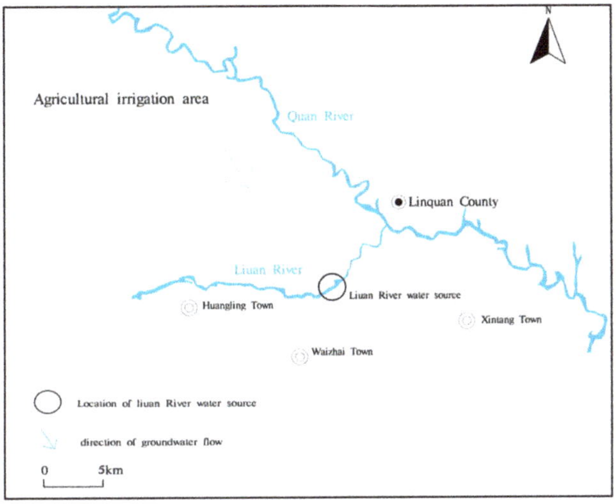

Figure 3. Location map of the main Liuan River.

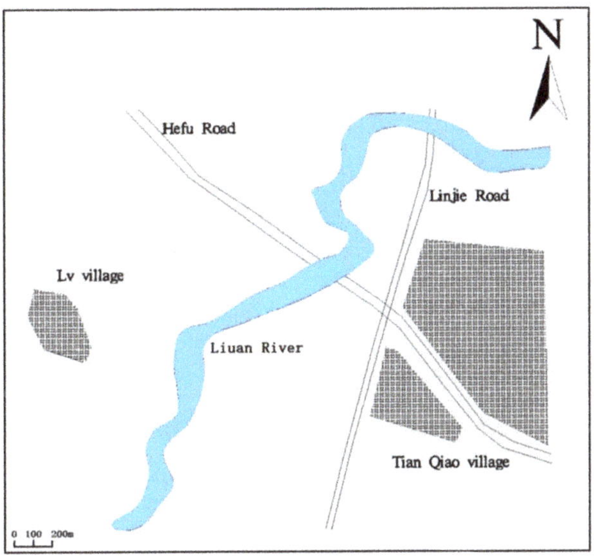

Figure 4. Location map of the main Liuan River water source.

2.2. Meteorological and Hydrological Conditions

Linquan County is adjacent to the Quanhe River in the north and Honghe River in the south. Linquan County is in the transition zone between the warm temperate and subtropical zones and forms part of the semi-humid continental climate zone of the warm temperate zone. The average annual precipitation in the county is 906.2 mm.

According to the Liuan River System governance planning report in Linquan County, the Liuan River is one of the main tributaries of the Quanhe River. It covers an area of 406.40 km^2 and has a total length of 41.60 km. The Liuan River length of the study area is approximately 3500 m, the width of the riverbed is approximately 100 m, and the length of the riverbed can reach 150 m locally. The average annual water level of the Liuan River is 34 m asl, the average annual runoff is 14.3 m^3/s, and the minimum daily flow is 2.5 m^3/s. The maximum annual river flow generally occurs in August with the minimum flow occurring in January.

2.3. Hydrogeological Conditions

Based on the relative age of the formations, the genetic groundwater type, the hydraulic characteristics, and the burial conditions, the study area can be classified into the Quaternary Holocene microconfined fissure and the pore phreatic aquifer group (Figure 5). The water-bearing sections were divided into the Middle Holocene (Q_4^2) fissure and the pore phreatic water-bearing section. The groundwater runoff in the study area is controlled by the regional topography, from northwest to southeast, which is basically consistent with the river flow direction.

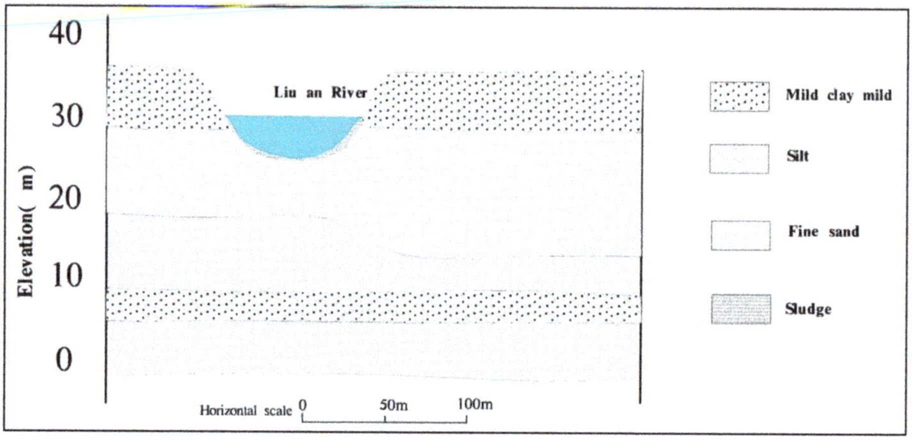

Figure 5. Hydrogeological section of the Liuan River water source.

The upper section of the formation is a sand-bearing sub-clay layer with a bluish yellow color and a thickness of 7–10 m, which is the exploitable layer with fissures in the strata. The water level is at approximately 2–4 m, which is relatively deep in the bank of the modern valley, and relatively shallow in the inter-river area at a depth of approximately 1–2 m.

This layer is extensively exposed along the river, with Q_4^2 being a recharge channel for the aquifer section.

The Q_4^2 aquifer comprises two to three layers of silt and fine sand aquifers. In some areas, four to five or more layers have been recorded. The burial depth of the top and bottom layers ranges from 7 to 10 m and from 20 to 25 m, respectively. The thickness of a single layer is 1–5 m and the average accumulated thickness is 5–15 m. The northern and eastern areas of the county mainly have fine sand aquifers with relatively few layers, but the thickness of a single layer is approximately 5–10 m, and can reach more than 15 m. Aquifers alternate with sandy loam soil and clay (claypan), and generally have low water pressure or no water, except for in the sandy loam soil layer, in the vertical section part of the opposing waterproof layer. However, because of variable development in the ancient riverbed, the lateral variation of the claypan is so high that the aquifer creates a water-bearing system with hydraulic connections at different depths. Based on pumping data obtained from several wells, the water inflow can reach 40 m^3/h when the depth has been reduced by 3–6 m.

The study area is mainly a broad and flat accumulation plain, and its main movement mode is vertical, continuously alternating the circulation of seepage and evaporation, leading to low mineralization of the groundwater [18]. The hydrochemistry of the water in the study area is relatively simple. The anionic content (mainly HCO_3) is stable, whereas the cation (main cations are Ca and Na) content differs.

Groundwater is mainly abstracted for artificial mining. There is a weak permeable layer on the surface and the consumption of water through phreatic evaporation is low. The bed of the Liuan River is within the phreatic aquifer, which is a typical groundwater source adjacent to the river.

3. Methods

3.1. Model Setup and Validation

The strata found in the study area dates from the Quaternary and Tertiary periods. The hydrogeological conditions of the Cenozoic strata in the study area are consistent with the regional hydrogeological conditions. In this study, the porewater aquifer group was analyzed. From top to bottom, this group can be divided into one weakly permeable layer group and two aquifer groups.

The groundwater in the study area is in an unstable state, the aquifer is heterogeneous and anisotropic, and the water flow obeys Darcy's law. The flow direction of groundwater is basically parallel to the flow direction of the river, and the boundary on both sides of the river is generalized as the general head boundary. The water level of Liuan River is stable, so it is generalized into the boundary of fixed water level. A diving model to determine the leakage recharge was established according to the hydrogeological conceptual model [19]

$$\frac{\partial}{\partial x}\left(k_x h\frac{\partial H}{\partial x}\right)+\frac{\partial}{\partial y}\left(k_y h\frac{\partial H}{\partial y}\right)+k\frac{h-h_1}{m}+W=S\frac{\partial H}{\partial t} \quad (1)$$

$$H(x,y,t)|_{t=0}=H_0(x,y) \quad (x,y)\in D \quad (2)$$

$$H(x,y,t)|_{\Gamma 1}=H_1(x,y,t) \quad (x,y)\in \Gamma_1 \quad (3)$$

where k_x is the horizontal permeability coefficient (m/d); k_y is the vertical permeability coefficient (m/d); h and h_1 respresent the shallow groundwater level and the confined aquifer level (m), respectively; m is the thickness of the weakly permeable layer (m); W is the unit volume flow, which is used to represent the amount of water flowing into the source or out of the sink (m^3); S is the gravity feed water; H_0 is the initial groundwater level (m); H_1 is the groundwater level at the boundary of the simulated period (m); T is the time (d); D is the simulated range; and Γ_1 is the Dirichlet function.

The research area is relatively small, so the water level in the groundwater observation hole on 1 July 2020 is set as the initial groundwater level. In the model, the Liuan River is defined as the river boundary. According to the topography, the groundwater depth, and the rainfall characteristics of Linquan County, the rainfall infiltration coefficient in this simulation is calculated using the following formula:

$$Q=10^{-1}\times P\times\alpha\times F \quad (4)$$

where Q is the rainfall infiltration amount (m^3/a), P is the rainfall (mm), and α is the rainfall infiltration recharge coefficient. In the study area, the value is 0.22, with F being the study area (km^2). The rainfall infiltration data are collected monthly. See Table 1 for rainfall.

Table 1. Monthly rainfall in the study area.

Month	7	8	9	10	11	12	1	2	3	4	5	6
Rainfall (mm)	229.8	134.3	68.9	40.5	42.2	17.1	22.6	28.5	56.5	64.1	71.9	135.8

The groundwater depth in the study area is approximately 3.5 m, and the evaporation is high. This can be calculated as follows:

$$E = E_0 (1 - \frac{h}{h_0})^n \tag{5}$$

where E is the evaporation intensity of the shallow groundwater (mm/d); E_0 is the river surface evaporation intensity (mm/d); h is the shallow groundwater level buried deep (m); h_0 is the shallow groundwater evaporation limit according to depth (m).

The evaporation in the study area is affected considerably according to the season, so the evaporation is assessed quarterly. See Table 2 for the evaporation.

Table 2. Monthly evaporation in the study area.

Month	7–9	10–12	1–3	4–6
Evaporation capacity (mm)	187	97	87	220

Visual MODFLOW software was used to divide the study area in Linquan County into 10,000 cells (100 rows × 100 columns). Each cell was 480 m long and 270 m wide and covered an area of 0.13 km^2 based on observations of the groundwater level in Linquan County. The water level of the Zhangying monitoring well at the Linquan County National Monitoring Station recorded from 1 July 2019 to 30 June 2020 was used for model verification. The initial operation date of the model is 1 July 2019, and the operation ended on 30 June 2020.

Based on research data in the study area, the trial estimation correction method was used to adjust the existing parameters, which is one of the indirect methods used to reverse the parameters [20]. The water level error was more than 70% within a range of 0.5 m (Figure 6). The simulated water level matched the water level at the monitoring point. This indicates that the mathematical model established is correct, can accurately represent the hydrogeological conditions of the study area, and can be used to simulate the effect of pumping on the river flow.

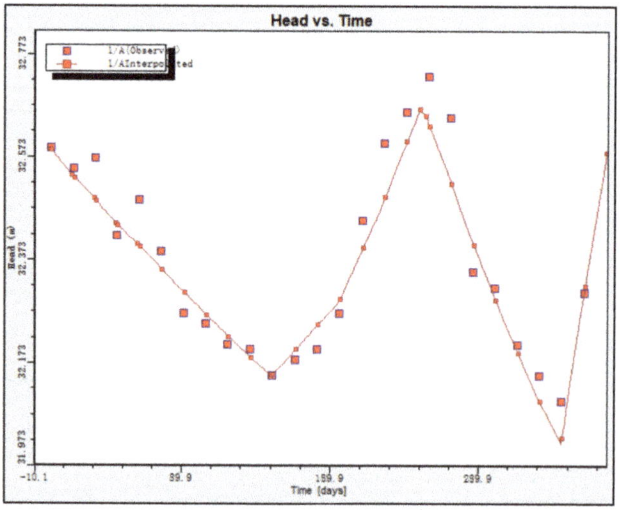

Figure 6. Diagram showing the water level fit for the Zhangying 50671120 monitoring well in Linquan County.

3.2. Parameter Sensitivity Analysis

Parameter sensitivity analysis is used to increase or decrease the variable x according to a specific proportion observing the degree of change of function f(x). The calculation formula is as follows:

$$S_{ij} = \frac{\partial F_i(x)}{\partial x_j}\Big| x = x^j \; (i = 1, 2, \ldots, m; \; j = 1, 2, \ldots, n) \tag{6}$$

where x is the design variable, S_{ij} is the sensitivity of function $F_i(x)$ to variable x. The larger the S_{ij} value, the more sensitive the variable x is. Sensitivity analysis was carried out on the four parameters of the permeability coefficient, gravity feed water, rainfall infiltration amount, and the evaporation intensity in the numerical model. The parameters selected are increased by 5% and 10%, and the range of variation in the water level after operation of the observation model is shown in Figure 7.

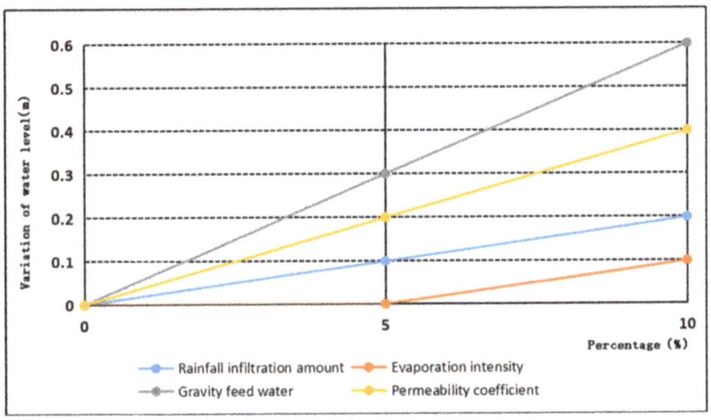

Figure 7. Results of the parameter sensitivity analysis.

The gravity feed water and permeability coefficient of the aquifer have a pronounced impact on the groundwater level, while the effects of evaporation and rainfall infiltration are relatively small (Figure 7). This is consistent with the groundwater in the study area mainly being recharged by the lateral runoff of the river.

After establishing and verifying a numerical model for Linquan County, the hydrogeological parameters were obtained. The model was then refined to establish a numerical model for the riverside source field of the Liuan River.

Results show that the current water supply demand near the riverside source field of the Liuan River is approximately 25,000 m^3/d. Based on the calculation of the exploitable groundwater reserves in Linquan County and the results of the model operation [21], the riverside source field of the Liuan River met the local water demand for Linquan County under normal conditions.

3.3. Adjustment Method and Control Conditions

The balance of water resource supply and demand must be adjusted every ten days during typical irrigation years. Based on the current situation and the water inflow and water consumption projected for 2020, water regulation calculations were synchronously carried out for the Xiaqiao to Chengxi sluices. The water level of the Liuan River was analyzed by calculating the water level of the Chengxi sluices [22].

Based on the water balance principle, the following calculation was used:

$$V_i = V_{i-1} + W_{\text{Interval}} - W_{\text{Irrigation}} - W_{\text{Drainage}} \tag{7}$$

where $V_i < V_{Dead}$, $V_i = V_{Dead}$, and V_i and V_{i-1} represent the water volume stored in the regulation reservoir at the end of the ith and $i-1^{th}$ months, respectively; $W_{Interval}$ is the interval inflow in the current month (without evaporation loss); $W_{Irrigation}$ is the irrigation water capacity in the current month; $W_{Drainage}$ is the discharge volume; V_0 is the reservoir capacity corresponding to the initial water level; and V_{Dead} is the dead storage.

The Liuan River is the main tributary of the Quanhe River. A rubber dam was built upstream of the Liuan River Bridge to increase the storage level of the main river channel of the Liuan River and to hold the surface water in the Liuan River from the Huangling to the Chengxi sluices. Here, the normal storage level was 34.50 m, and the elevation of the riverbed was 28.50 m.

When the water storage capacity at the west gate is less than the corresponding dead storage capacity of 3.5 million m^3, there is no water in the river channel.

3.4. Analysis of the Incoming Flow

The incoming flow is computed as the sum of water flow from the upper section and the interval inflow. In the upper section, the water flow is from the agricultural irrigation water in the upper section of the Huangling sluice. Given that there is little water during drought years and the irrigation is exhausted, the amount of water coming from the upper section is zero in the absence of measured data from the hydrology station.

The interval inflow can be calculated by subtracting the water storage and water consumption from the precipitation yield

$$Q_{Interval} = P \times \alpha \times F \tag{8}$$

where P is the monthly precipitation, α is the monthly runoff coefficient, and F is the interval water production area. The monthly runoff coefficients of the intervals are shown in Table 3.

Table 3. Values of the monthly runoff coefficients in the study area.

Monthly Total Precipitation (mm)	$P \geq 100$	$100 > P \geq 50$	$50 > P \geq 30$	$30 > P \geq 20$	$P < 20$
Monthly runoff coefficient	0.26	0.22	0.12	0.07	0

3.5. Calculation Method of Minimum Ecological Water Demand

At present, the most sensitive problem in arid and semi-arid areas is the ecological water use of rivers. In addition, the contradiction between industrial and agricultural production water and ecological environment water is the largest. Due to the lag of ecological environment problems, it is necessary to pay attention to the impact of pumping on ecological environment water use in the process of taking water near the river.

The minimum ecological flow is the flow that maintains the basic morphology of the river bed, prevents the flow interruption, maintains the self-purification ability of the water body and avoids irreversible damage to the water organisms in the river [23]. The calculation of ecological water demand of Liuan River mainly considers evaporation and leakage in the river channel [24]. At present, there are several methods to calculate the minimum ecological flow: Tennant method, aquatic biological base flow method, RVA method, wetted perimeter method, 7Q10 method. At present, there are some disputes about these calculation methods. The Tennant method and the 7Q10 method are widely used in the Huaihe River Basin. Therefore, the 7Q10 method is used to calculate the ecological flow in the calculation process of this section [25].

In this paper, the statistical data of Yangqiao hydrological station in Liuan River were selected to collect the flow data of Liuanhe River from 1970 to 2021, 90% typical dry years were selected, and the ecological flow was calculated by using the 7Q10 method [26]. This method is based on hydrological parameters. The average water volume in the driest month for seven consecutive days with a 90% guarantee rate is used as the minimum ecological flow of the river, and then the minimum ecological water demand in a single day is calculated.See Table 4 for the minimum ecological water demand of Liuan River.

Table 4. The minimum ecological water demand of Liuan River.

Q_1/(m^3/s)	Q_2 (m^3)
2.66	229,824

Q_1: Ecological flow; Q_2: Minimum Ecological Environment Water Demand.

4. Results

The current situation for the urban and rural water supply in Linquan County depends on the groundwater at medium and low depths. A large area of overexploitation has formed in the confined aquifer at a depth of 50 m. In 2015, the area of restricted groundwater exploitation centered on the county was 415.6 km^2. Based on current data, the shallow groundwater exploitation volume in Fuyang City is 571 million m^3/a, which accounts for 34.2% of the shallow groundwater resources. The average shallow groundwater exploitation volume in Fuyang City is 990 million m^3/a, and the recoverable modulus is 102,000 $m^3/(km^2 \times a)$.

The groundwater exploitation volume in Linquan County is approximately 78,000 m^3/d. Because of the condition of the water resources in the region and the influence of policies, the exploitation of pore-confined water is currently restricted and further development and use should not be considered. Based on the water supply plans of Linquan County, the water shortage will be 127,400 m^3/a in 2025. Based on the urban water supply plans of the county, the current water supply scale is 100,000 m^3/a and water shortages will reach 25,000 m^3/d in 2025 before the completion of the "Divert water from the Yangtze River to the Huaihe River" project. The construction of the riverside source field of the Liuan River will solve problems with respect to land subsidence in the Fuyang area as well as insufficient water supplies in the city.

Therefore, future water supply demand from the city can be met by the construction of the riverside source fields along the Liuan River.

4.1. *Water Supply and Water Balance in Different Years*

(1) Water supply analysis in 95% typical dry years

Based on the flood regulation calculation, the water quantity in the Liuan River is insufficient for 4 months in 95% typical dry years given that the river becomes dry. The Liuan River riverside source field was simulated numerically.

Based on the local hydrogeological conditions and the degree of water richness in the aquifer, the water yield per well was set in the model to be 1000 m^3/d. The results of the model operation show that the wells begin to dry up on the tenth day of model operation and that the water level stabilizes after 32 days in 95% typical dry years. Eight wells were dry and the remaining wells can maintain a stable water supply of approximately 16,000 m^3/d. The change in the groundwater level is shown in Figures 8 and 9.

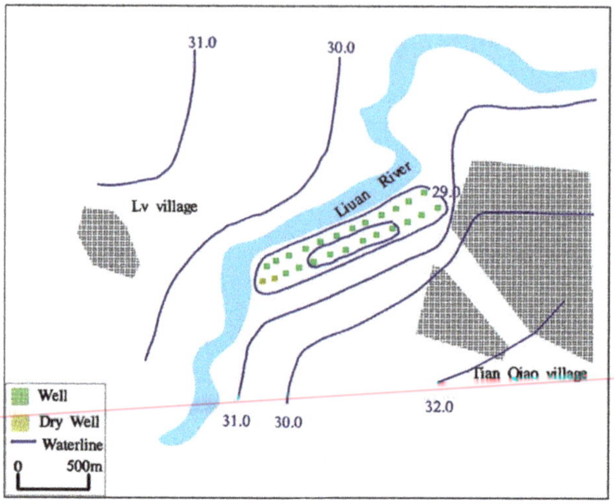

Figure 8. Water level flow field 10 days after pumping.

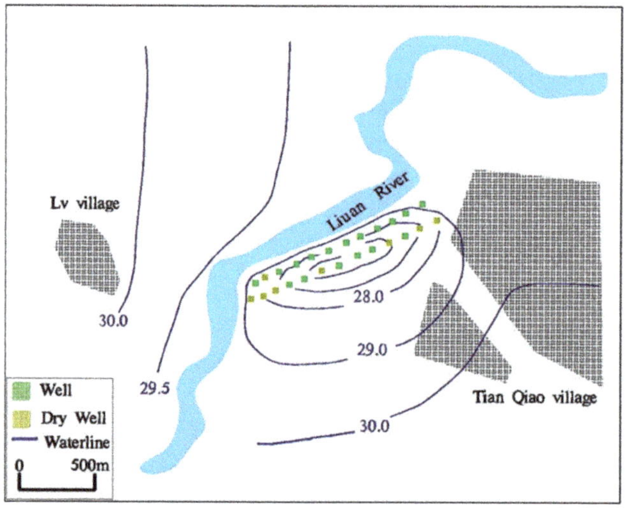

Figure 9. Water level flow field 32 days after pumping.

To determine the flow direction of the groundwater under the pumping conditions of a 95% typical dry year, the particle-tracking module of the Visual MODFLOW software was used to set the particles in the Liuan River. The model run was performed to observe the particle movement. The movement track section is illustrated in Figure 10.

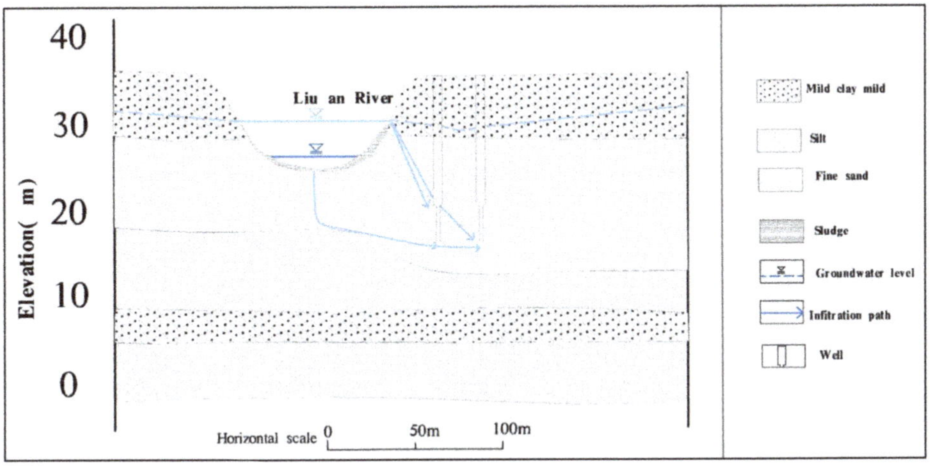

Figure 10. Flow direction of the groundwater under pumping conditions in a 95% typical dry year.

Results indicate that the groundwater level was stable for 32 days. The profile shows that the river continuously supplies water to the pumping wells, whereas the water supply from the side (other than the riverside source field) was insufficient. This led to the pumping wells drying up because of the river's lateral runoff supply to ensure the normal water supply of some of the pumping wells.

To determine changes in the groundwater quantity during the model operation, the ZBud module of the Visual MODFLOW software was used. The water balance data [27] are shown in Table 5.

Table 5. Water balance data from a 95% typical dry year.

Date	Pump Output (m^3)	River in Flow (m^3)			Proportion of Pumping Water (m^3)	Groundwater Exchange Rate (m^3)
		The River Drains into the Pumping Wells (m^3)	The River Seeps into the Groundwater (m^3)	Total		
1	25,000	1049.1	2560	3609.1	14%	41,641.42
12	19,000	3025.8	13,746	16,771.8	88%	489.91
22	17,000	3030.3	12,987.6	12,150.9	94%	722.68
32	16,000	3066.5	12,924.2	15,990.7	99%	719.8
42	16,000	3086.9	12,901.14	15,988.04	99%	747.86
52	16,000	3099.5	12,891.14	15,990.64	99%	769.86
62	16,000	3107.9	12,885.1	15,993	99%	784.9
72	16,000	3113.8	12,881.46	15,995.26	99%	795.54

The data in Table 5 show that the pumping well can supply 16,000 m^3 of water when the water level stabilizes on day 32 of the model operation. Based on the analysis of the amount of water (Table 5), the water supplied by the pumping well primarily originates from the infiltration of the river, which is a typical riverside source field. The river water directly infiltrates into the pumping well. Most of the water flows underground and then through the underground pumping wells. This process is predominantly related to the hydrogeological conditions of the study area. The Liuan River is between the sub-clay layer and the aquifer on the surface. Therefore, the river water mainly replenishes the groundwater through infiltration to ensure a normal water supply in the wells. Water exchange occurs between the groundwater and the river water on the first day under drought conditions. During the pumping process, the groundwater storage gradually decreases. After the water volume stabilizes, the water supply is maintained by the infiltration of the river.

(2) Water supply analysis of 75% typical dry years

Based on the flood regulation calculation, the Liuanhe River can guarantee sufficient water throughout the year, with a 75% guarantee rate. A simulation was carried out for the riverside source field of the Liuanhe River for years with a 75% guarantee rate. The results are shown in Figure 11.

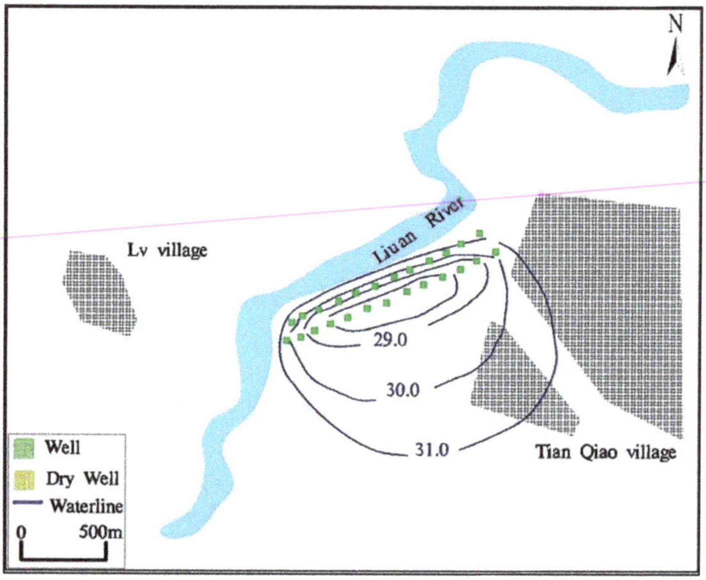

Figure 11. Water level flow field 60 days after pumping.

Based on the precipitation analysis for years with a 75% guarantee rate, the month with the least rainfall was selected for the model prediction. After 60 days, the water level stabilized and the pumping well did not dry up; therefore, it met the water demand.

The model was analyzed using the ZBud module. The results were compared with the river leakage in years with a 95% guarantee rate, as shown in Table 6.

Table 6. Water balance data from a 75% typical dry year.

Date	Pump Output (m^3)	River in Flow (m^3)				Groundwater Exchange Rate (m^3)
		The River Drains into the Pumping Wells (m^3)	The River Seeps into the Groundwater (m^3)	Total	Proportion of Pumping Water (m^3)	
1	25,000	2143	2281	4424	17%	47,607
12	25,000	3289.7	16,645.7	19,935.4	80%	1533.3
22	25,000	3461.6	17,495.8	20,957.4	83%	2042.2
32	25,000	3539.6	18,441.5	21,981.1	87%	2297.5
42	25,000	3573.8	19,417.2	22,991	91%	2416.8
52	25,000	3589.4	19,406.3	22,995.7	91%	2473.7
62	25,000	3596.9	20,401.6	23,998.5	95%	2500.4
72	25,000	3600.4	20,398.5	23,998.9	95%	2512.5

The water balance data for 75% typical dry years show that the water supplied to the pumping wells originates from river infiltration.

Based on comparison with typical dry years, 75% typical dry years met the local water supply requirements. In 95% typical dry years, the water level stabilizes within 32 days and 65% of the water supply scale could be met. In extremely dry years, reduction in the irrigation water based on the actual situation can ensure a basic water supply for residents.

4.2. Groundwater Level Change

According to the results from different typical dry year models, the central water level of the pumping wells decreases by the same amplitude, except the water level changes considerably during the early days of exploitation. After 10 days, the water level decreases slowly, and after 40 days, the water level remains unchanged, as shown in Table 7. Therefore, the stability of the regional groundwater level can be ensured by using the capacity of aquifer regulation and replenishing the shortage with abundance.

Table 7. Groundwater levels for different typical dry years.

		Date							
		1	12	22	32	42	52	62	72
95% typical dry year	groundwater level (m)	31	27.8	27.5	27.3	27.2	27.2	27.2	27.2
75% typical dry year		31	28	27.8	27.5	27.4	27.4	27.4	27.4
50% typical dry year		34	30.3	29.9	29.7	29.6	29.6	29.6	29.6

4.3. Effects of Ecological Water Demand

Given that the tributary being studied is part of the Huaihe River system, people pay attention to the ecological problems and the effect of ecological flow has become an important part of water resource management in the region. The time series runoff data of the river is limited.

Therefore, the ecological flow is calculated by calculating a different guaranteed arid reliability year exchange capacity for the groundwater accounts for the proportion of ecological flow, mining the impact on the ecological flow analysis alongside the river source. The effects of the water intake on the ecological flow were analyzed by calculating the proportion of the groundwater exchange capacity to the ecological flow in years with different drought guarantee rates. The proportion of the daily maximum water exchange to the minimum ecological water consumption in years with different guaranteed rates was small. The 95%, 75%, and 50% typical dry years accounted for 6.73%, 11.20%, and 11.45% of the minimum ecological water consumption, respectively, as shown in Figure 12.

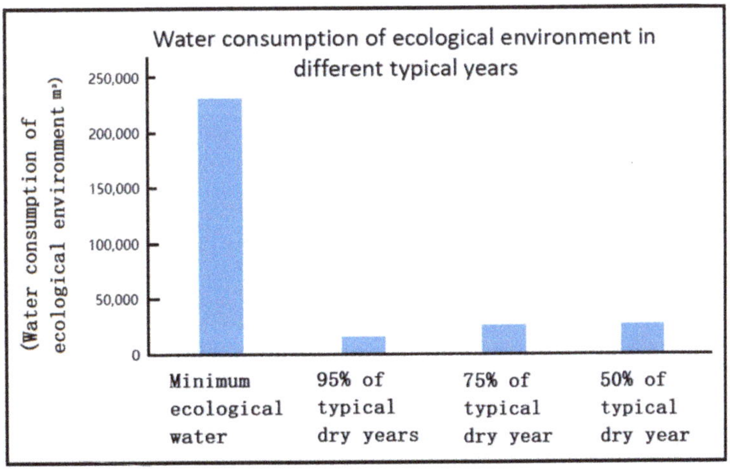

Figure 12. Maximum eco-environmental water consumption in different typical years.

4.4. Summary of Results

In this study, the water supply of the water source area near the Liuan River in different typical dry years was analyzed under different drought conditions. A numerical model was used to simulate and predict the groundwater under various drought conditions. In 95% typical dry years, the water supply can reach 16,000 m^3/d. In 75% and 50% typical dry years, the water supply can reach 25,000 m^3/d. The results simulated indicate that local water supply plans can be met, except for in particularly dry years. The groundwater level changes and the effect on the ecological flow in the riverside source field in extreme drought years were simulated. Based on the results, the following conclusions can be drawn:

(1) Based on the hydrogeological conditions in the study area, the water source mainly comes from underground river infiltration, which can attain more than 80% of the water supply, and the other water sources originate from lateral infiltration of the river. The water seeps into the aquifer and then into the pumping wells.

(2) The water supply from small rivers is insufficient during 95% typical dry years. In other years, it meets the mining plans. Therefore, the water supply guarantee must be considered when small rivers are used to establish riverside source fields, especially in dry years.

(3) Because of the lack of long-term hydrological data for small rivers, the improved 7Q10 method was used to calculate the ecological flow during different dry years. The findings of the study reveal that mining water has little effect on the ecological flow.

5. Discussion

5.1. Baseline Analysis

Based on analysis of the conditions in Linquan county, deep groundwater overexploitation significantly affects the local ecological environment. Therefore, based on the establishment of riverside source fields, sustainable development of industrial policies in the Linquan county can be implemented and excessive development and use of deep level groundwater can be effectively reduced.

5.2. Small River Riverside Water Sources

Based on research and analysis of the riverside source fields of the Liuan River, this paper puts forward the following suggestions, namely to establish a riverside source area for the Liuan River, replace the task of water supply with small rural drinking water plants in the region, and gradually close the rural self-provided water source wells. Unified water supply and distribution can effectively guarantee the safety of drinking water for rural residents and the scale of irrigation, which is not only conducive to the management of water conservancy departments, but also can increase economic benefits. Compared with large rivers, it is more important to pay attention to the influence of pumping on river runoff and whether the change in the groundwater level is within the red line of water resource management in the process of mining the riverside water sources of small rivers.

6. Conclusions

Liuan River, as a tributary of the Huaihe River, has obvious hydrological and ecological significance. Under the condition of the basic completion of the water diversion project from the Yangtze River to the Huaihe River, the supply of water resources in Linquan county cannot reach the balance between supply and demand, which is a blank point under the condition of inter-basin water transfer. Its research is representative and typical, and it is reasonable to take Liuan River Basin as a research representative.

Small rivers used to establish riverside source fields can not only alleviate the problem of regional water resource shortages, but also optimize the regional water supply through the allocation of multiple water sources. Through the study on the exploitation scheme of the riverside source field of the Liuan River, it can be found that the water source mainly comes from river infiltration underground, which can attain more than 80% of the water supply. In 95% typical dry years, the Liuan River cannot meet the water supply task.

Therefore, it is necessary to adjust the water supply strategy in the regional water supply to ensure the assurance rate of water supply under special drought conditions.

The impact on the ecological flow is small in the process of pumping water from the riverside source field of the Liuan River. However, unlike large rivers, water abstraction from small rivers requires a strong focus on the impact of pumping on the river water level. If the river water level is lower than the ecological water level, it will have an irreversible impact on the small river basin. Therefore, in the process of mining small rivers, we need to ensure that they can not only meet the needs of production and domestic water, but also protect the ecological environment.

With the promulgation and implementation of groundwater management regulations, the sustainable use of water resources in small river basins is vital to maintaining normal supplies of regional water resources, and the use and protection of water resources in small river basins require more strategic development strategies.

Author Contributions: Writing—original draft preparation, Z.H., B.K. and L.Q.; writing—review and editing, Z.H., B.K.; supervision, Y.T. All authors have read and agreed to the published version of the manuscript.

Funding: This research is financially supported by the National Natural Science Foundation of China (Grant No. 42107162), the Natural Science Foundation of Anhui Province (Grant No. 1908085QD168), The Fundamental Research Funds for the Central Universities of China (Grant Number: PA2021 KCPY0055)and Key R&D Plan of Anhui Province (No. 202104g01020010).

Data Availability Statement: The data presented in this study are available on request from the corresponding author.

Conflicts of Interest: The authors declare no conflict of interest. The funders had no role in the design of the study; in the collection, analyses, or interpretation of data; in the writing of the manuscript, or in the decision to publish the results.

References

1. Sathya, N.; Muralidhara, P. Groundwater development in fragile water resource ecosystems in arid or semi-arid areas in northwest Indian Himalayas: Implication for sustainable development. *J. Geosci. Res. Northeast. Asia* **1998**, *1*, 50–51. [CrossRef]
2. Han, Z. Groundwater for urban water supplies in northern China—An overview. *Hydrogeol. J.* **1998**, *6*, 416–420.
3. Ministry of Water Resources of the People's Republic of China. *Opinions on Strengthening the Management of Water Resources in Groundwater Overexploited Areas*; Ministry of Water Resources of the People's Republic of China: Beijing, China, 2003.
4. Uddameri, V.; Kakarlapudi, C.; Hernandez, E.A. A GIS enabled nested simulation-optimization model for routing groundwater to overcome spatio-temporal water supply and demand disconnects in South Texas. *Environ. Earth Sci.* **2014**, *71*, 2573–2587. [CrossRef]
5. Yi, S.; Chi, B.; Wu, F.; Zhou, Y. Research on Numerical Simulation and Evaluation of Riverside Source Field—A Case Study of Liguanpu Riverside Source Field in Shenyang City. *J. Nat. Resour.* **2006**, *21*, 154–160. (In Chinese)
6. Haque, M.A.M.; Jahan, C.S.; Mazumder, Q.H.; Nawaz, S.M.S.; Mirdha, G.C.; Mamud, P.; Adham, M.I. Hydrogeological condition and assessment of groundwater resource using visual modflow modeling, Rajshahi city aquifer, Bangladesh. *J. Geol. Soc. India* **2012**, *79*, 77–84. [CrossRef]
7. Hancock, P.J. Human Impacts on the Stream-Groundwater Exchange Zone. *Environ. Manag.* **2002**, *29*, 763–781. [CrossRef] [PubMed]
8. Zhu, Y.; Zhai, Y.; Du, Q.; Teng, Y.; Wang, J.; Yang, G.; Jinsheng, W. The impact of well drawdowns on the mixing process of river water and groundwater and water quality in a riverside well field, Northeast China. *Hydrol. Process.* **2018**, *33*, 945–961. [CrossRef]
9. Wu, X.; Zheng, Y.; Wu, B.; Tian, Y.; Han, F.; Zheng, C. Optimizing conjunctive use of surface water and groundwater for irrigation to address human-nature water conflicts: A surrogate modeling approach. *Agric. Water Manag.* **2016**, *163*, 380–392. [CrossRef]
10. Molina-Navarro, E.; Hallack-Alegría, M.; Martínez-Pérez, S.; Ramírez-Hernández, J.; Mungaray-Moctezuma, A.; Sastre-Merlín, A. Hydrological modeling and climate change impacts in an agricultural semiarid region. Case study: Guadalupe River basin, Mexico. *Agric. Water Manag.* **2016**, *175*, 29–42. [CrossRef]
11. Lorenzo-Lacruz, J.; Vicente-Serrano, S.M.; López-Moreno, J.I.; Beguería, S. The impact of droughts and water management on various hydrological systems in the headwaters of the Tagus River (central Spain). *J. Hydrol.* **2010**, *386*, 13–26. [CrossRef]
12. Ministry of Water Resources of the People's Republic of China. The Letter of the Ministry of Water Resources on the Issuance of the First Batch of Key River and Lake Ecological Flow Guarantee Targets. *China Water Resour.* **2020**, *15*, 5–7.

13. Lewandowski, J.; Lischeid, G.; Nuetzmann, G. Drivers of water level fluctuations and hydrological exchange between groundwater and surface water at the lowland River Spree (Germany): Field study and statistical analyses. *Hydrol. Process* **2010**, *23*, 2117–2128. [CrossRef]
14. Alifujiang, Y.; Abuduwaili, J.; Ge, Y. Trend Analysis of Annual and Seasonal River Runoff by Using Innovative Trend Analysis with Significant Test. *Water* **2021**, *13*, 95. [CrossRef]
15. Saenger, N.; Kitanidis, P.K.; Street, R.L. A numerical study of surface-subsurface exchange processes at a riffle-pool pair in the Lahn River, Germany. *Water Resour. Res.* **2005**, *41*. [CrossRef]
16. Brunner, P.; Therrien, R.; Renard, P.; Simmons, C.T.; Franssen, H.J. Advances in understanding river-groundwater interactions. *Rev. Geophys.* **2017**, *55*, 818–854. [CrossRef]
17. Liu, C.; Zhao, C.; Xia, J.; Sun, C.; Wang, R.; Liu, T. An instream ecological flow method for data-scarce regulated rivers. *J. Hydrol.* **2011**, *398*, 17–25. [CrossRef]
18. Zhu, G.; Su, Y.; Feng, Q. The hydrochemical characteristics and evolution of groundwater and surface water in the Heihe River Basin, northwest China. *Hydrogeol. J.* **2008**, *16*, 167–182. [CrossRef]
19. Zume, J.; Tarhule, A. Simulating the impacts of groundwater pumping on stream–Aquifer dynamics in semiarid northwestern Oklahoma, USA. *Hydrogeol. J.* **2008**, *16*, 797–810. [CrossRef]
20. Karan, S.; Engesgaard, P.; Looms, M.C.; Laier, T.; Kazmierczak, J. Groundwater flow and mixing in a wetland—Stream system: Field study and numerical modeling. *J. Hydrol.* **2013**, *488*, 73–83. [CrossRef]
21. Chung, I.M.; Lee, J.W.; Kim, N.W. Estimating Exploitable Groundwater Amount in Musimcheon Watershed by Using an Integrated Surface Water-Groundwater Model. *Econom. Environ. Geology* **2011**, *44*, 433–442. [CrossRef]
22. Gao, Q.; Du, H.; Zu, R. The balance between supply and demand of water resources and the water-saving potential for agriculture in the hexi corridor. *Chin. Geogr. Sci.* **2002**, *12*, 23–29. [CrossRef]
23. Yan, D.; Wang, G.; Wang, H.; Qin, T. Assessing ecological land use and water demand of river systems: A case study in Luanhe River, North China. *Hydrol. Earth Syst.* **2012**, *16*, 2469–2483. [CrossRef]
24. Wan, F.; Zhang, F.; Zheng, X.; Xiao, L. Study on Ecological Water Demand and Ecological Water Supplement in Wuliangsuhai Lake. *Water* **2022**, *14*, 1262. [CrossRef]
25. Gu, H. Analysis on key and difficult points of determination and guarantee of river lake ecological flow in Huaihe River Basin. *China Water Resour.* **2020**, *15*, 47–49.
26. Li, H.; Qu, X.; Tao, J.; Hu, C.; Zuo, Q. The optimal allocation of water resources oriented to prioritizing ecological needs using multiple schemes in the Shaying River Basin (Henan Section), China. *Water Supply* **2022**, *22*, 1593–1610. [CrossRef]
27. Istanbulluoglu, E.; Wang, T.; Wright, O.M.; Lenters, J.D. Interpretation of hydrologic trends from a water balance perspective: The role of groundwater storage in the Budyko hypothesis. *Water Resour. Res.* **2012**, *48*, 273–279. [CrossRef]

water

MDPI

Article

Hydro-Geochemical Characteristics of the Shallow Alluvial Aquifer and Its Potential Artificial Recharge to Sustain the Low Flow of the Garonne River

Nazeer Asmael [1,*], Alain Dupuy [1], Paul McLachlan [2] and Michel Franceschi [1]

[1] ENSEGID-Bordeaux INP, EPOC UMR 5805, University of Bordeaux, CNRS, 1 allée Fernand Daguin, 33600 Pessac, France; alain.dupuy@ensegid.fr (A.D.); michel.franceschi@ensegid.fr (M.F.)

[2] Department of Geoscience, Aarhus University, Nordre Ringgade 1, Aarhus C, 8000 Aarhus, Denmark; pm@geo.au.dk

* Correspondence: nazeer.asmael@ensegid.fr

Abstract: The complex and interconnected water challenges linked to global climate change and natural and anthropogenic water resources pressure have become major challenges in the 21st century. The Garonne River and its accompanying alluvial aquifers are considered the most important source for agricultural activities in the Garonne Valley, Nouvelle-Aquitaine Region, southwest France. The water is used for irrigation in summer and to reduce frost damage in spring. The alluvial shallow aquifer is recharged by rainfall, lateral inflow from the hillside, and seepage from the riverbed during the flood periods. The aquifer maintains the flow of the river during dry periods. Moreover, the potential recharge of this aquifer is particularly sensitive to annual climatic fluctuations and consequently affects surrounding ecosystems and related socio-economic activities. The increasing impacts of climate change have increased the concern about the availability of these resources. Various adaptation strategies have been considered to mitigate and adapt to the new situation in southwest France. The artificial recharge of the alluvial aquifer is one such regional adaptation strategy to adapt to climate change. The study has two main objectives: to assess the natural and anthropogenic influence on the groundwater chemistry, and to model water infiltration, and understand the aquifer response and, consequently, the effects on river baseflow. The TAG (Technopole Agen-Garonne) project aims to increase the economic wealth of the region while respecting the region's agricultural traditions. Runoff water from the TAG zone is collected in retention basins and is a potential source to recharge the shallow alluvial aquifer. Sampling campaigns were carried out during the summer of 2019 to collect groundwater samples from several observation wells. Groundwater levels were measured in 132 wells/boreholes to determine the groundwater level fluctuations and create piezometric maps. Piper, spatial distribution, and ionic ratio plots were used to determine the dominant hydrochemical processes and to delineate the hydrochemical facies in the study area. The groundwater chemistry is controlled by silicate weathering and anthropogenic influence. Groundwater quality appears to be affected by the river water in the wells located in the low plain area. The measurements showed that the groundwater levels in the wells located near the river increase more than 2 m after a flood event. The artificial recharge has increased the groundwater level by more than 1 m close to the infiltration basin after a rainstorm. Similarly, a three-dimensional (3D) groundwater model shows a similar magnitude aquifer response to the induced infiltration. The modeling-obtained result shows that the infiltrated water would take about 4 months to reach the Garonne River, which is an appropriate time to maintain the river's low-flow and thermal buffering capacity, and thus the functioning of its ecosystems during dry periods.

Keywords: Garonne River; alluvial aquifer; climate change; water resources management; MAR (managed aquifer recharge); groundwater modeling; France

Citation: Asmael, N.; Dupuy, A.; McLachlan, P.; Franceschi, M. Hydro-Geochemical Characteristics of the Shallow Alluvial Aquifer and Its Potential Artificial Recharge to Sustain the Low Flow of the Garonne River. *Water* **2023**, *15*, 2972. https://doi.org/10.3390/w15162972

Academic Editors: Yuanzheng Zhai and Jin Wu

Received: 28 June 2023
Revised: 24 July 2023
Accepted: 15 August 2023
Published: 18 August 2023

1. Introduction

The French alluvial aquifers, where most fertile agricultural lands and big cities are located, play an important role in groundwater use in France. Current and future research in hydrogeology and water resources cannot avoid the effects of climate change, especially on these shallow aquifers.

Located in the southwest of France, the Nouvelle-Aquitaine Region is characterized by rich history, landscapes, and economy, which results from mixing climate and geography. This vast sedimentary area at the foothills of the Pyrenees and the Massif Central Mountains is bordered by a vast sandy coastline, ending in the extreme south with a rocky coastline in the Basque area. It comes into contact with the sea only through the Gironde Estuary in the Arcachon Basin, the Adour Estuary, and the ports between the cities of Capbreton and Hendaye. This region is also considered the largest cultivated forest in Europe, a wine-growing area, with rich and varied agriculture and a huge variety of natural ecosystems [1]. In the Garonne Valley, located in the central part of the Nouvelle-Aquitaine Region where agricultural activities are well developed, irrigation allows crop production where water would otherwise be a limiting factor [2]. The Garonne alluvial aquifer is an important reservoir and is extensively used for agricultural purposes due to its wide extension, high productivity, and the ease and low cost of extraction. This aquifer generally maintains a direct hydraulic connection with the river, which drains or recharges it according to the season. It was formerly used as a drinking water source, but it has now been abandoned for this purpose due to its high vulnerability to pollution, particularly from agricultural activities. Several studies have outlined the possible negative impacts of climate change on the Garonne River discharge and the recharge (infiltration ratio) of its surrounding alluvial aquifers [1,3–8].

The interactions of groundwater with surface waters take place in a range of different settings [9,10]: (1) Rivers gain water from groundwater inflow through the streambed (gaining stream); (2) they lose water to groundwater by outflow through the streambed (losing stream); or (3) they do both, gaining water in some reaches and losing water in others, or gaining and losing water in the same reach at different times depending on groundwater and surface water levels as well as the permeability of riverbed and aquifers. The dynamic exchange scale between the Garonne River and its alluvial aquifer is identified as between 3 and 4 km wide and limited mainly to the low plain [11,12]. The groundwater naturally feeds the springs and watercourses in the Garonne Valley, particularly during low river level periods. This groundwater discharge depends mainly on winter and spring rainfall. Anthropogenic activities such as groundwater and surface water extraction also affect the hydraulic gradients between the aquifer and the river system. Since the 1960s, anthropogenic impacts on the river's flow regime have significantly increased, particularly irrigation practices that coincide with a low-flow period [13]. The decrease in the river discharges is generally related to low precipitation and flood magnitude, consequently reducing the water storage in the aquifers, and a decrease in snow and ice storage in the high Pyrenees Mountains [13,14]. The decline in river flows and groundwater levels will become greater as the pressures on water resources continue to grow due to human population growth and climate change impacts. These could impact water availability and temperature, impacting the related ecosystems [15,16]. According to [17], global warming in the southwest of France is likely to increase the frequency of heatwaves, the duration of summer droughts, and the number of violent storm-type events. It is also likely to decrease the duration, extent, and quantity of snow cover, summer precipitation, average summer discharges of rivers, and groundwater levels. Water resources can be considered a crucial issue to fight against climate change and adaptation measures that can be taken.

The objective is to slow the rainfall trajectory towards the sea, prevent runoff, and enhance infiltration. Taking part of the existing surface water to be recharged into the aquifers for later different use can be a pragmatic strategy for adapting to climate change and sustaining the low flow in the Garonne River. This economically interesting solution can improve the quantity and quality of the alluvial aquifer, which is significantly exposed to pollution. Furthermore, this solution can also assist local water management and allocation problems. The weathering processes of the alluvium deposits play an important role in the groundwater quality of the unconfined alluvial aquifer. Hydrochemical data can be used to characterize this aquifer and specify different processes in the system and the effects of anthropogenic activities on the groundwater. The nitrate concentration of the alluvial aquifer is highly related to the hydrogeological context and surface land use. In the Garonne Valley, the alluvial aquifer is characterized by high nitrate concentration compared with the surface water [18]. The geological features of this aquifer and its shallow depth present a favorable condition that strongly affects its mineralization and vulnerability. In the Garonne Valley, almost 70% of water abstraction during low-water periods is used for irrigation [7]. Hence, the alluvial aquifer sustains extensive agricultural activities using fertilizers and a comprehensive irrigation system. It is degraded by nitrate pollution due to leaching from fertilizers in the return flow and consequently pollution is diffused into the groundwater system.

A large socio-economic project (TAG) of 240 ha in the study area has developed close to Agen, in southwest France. This project involves using large retention basins to collect runoff water for recharging the alluvial aquifer. This will allow reconstitution from the end of winter to the end of spring by using the alluvial aquifer for groundwater storage. Thus, it will increase the aquifer potential flow toward the Garonne River to sustain its low flow. Based on our measurements in August 2019, the river temperature was measured locally as 29 °C while the temperature of the drained groundwater into the river was 15 °C in the same period. Subsequently, the potential increase in drainage from the aquifer into the river during the dry period will ensure the functioning of its ecosystems. The shallow depth of the unconfined alluvial aquifer can be an advantage, allowing rapid and economical storage.

Hydrochemical data can be used to identify hydrogeological processes and understand the flow system behavior. Local-scale groundwater modeling can also be used to identify the response of the alluvial aquifer to local artificial recharge and to delineate the flow patterns. Hence, the main objectives of this paper are: (i) to assess the natural and anthropogenic influence on the groundwater chemistry and determine the main hydrochemical process controlling the groundwater chemistry; (ii) to model water infiltration, and understand the quantitative aquifer response to the artificial recharge and, consequently, the role of this methodology in maintaining river base flow; and (iii) to enhance the relationship between the hydro-system and the aquatic ecosystem along the river against climate change. This work will help to assess the potential of this experimental approach as a water management tool across the Garonne Valley as a nature-based, viable, and sustainable solution for extreme hydro-climatic events.

2. Description of the Study Area

The study site is located in the southwest of France, in the Lot-et-Garonne Department on the left side of the Garonne River near the city of Agen in the Nouvelle-Aquitaine Region (Figure 1). It covers an area of approximately 20 km^2. The Garonne Valley forms a vast plain, 5 to 8 km wide, bordered by asymmetrical hillsides that form an integral part of the valley. The hillsides on both sides are regularly cut by tributaries of the Garonne. The flat valley floor includes the river's major bed, structured by the alluvial terraces that form its limits. The average altitude of these terraces ranges between 10 and 15 m, allowing the buildings to be relatively protected against floods. The TAG area is a large development zone that is listed in strategic documents of the Lot-et-Garonne Department and meets an economic need in the Agen Agglomeration by preserving agricultural activities in this area according to the planned economic development and the allocated infrastructure. Small stream networks, ditches, marshes, old gravel pits, and small wetlands all contribute to the biodiversity of this area. The proximity of the Garonne River to this area constitutes its principal ecological connection. This economic project considers sustainable development by preserving considerable green spaces, collecting runoff water to be recharged into the shallow alluvial aquifer, using completely solar-powered lighting, and giving particular attention to the existing flora and fauna.

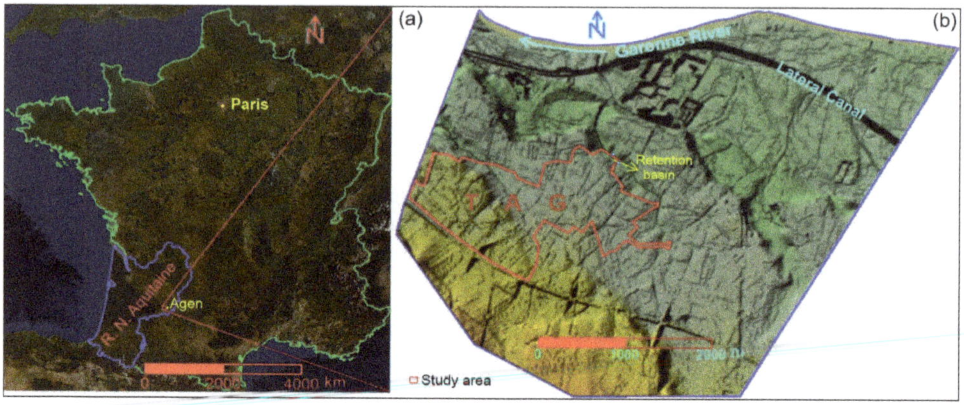

Figure 1. (**a**) Map of France, prepared by QGIS 3.16.12, free and open-source software, showing the location of the study area in Nouvelle-Aquitaine Region; (**b**) the topographic units of the study area according to IGN (French national geographic institute, https://geoservices.ign.fr/, accessed on 20 September 2021) spatial satellite data with a 5 m resolution (**b**) [19].

Three large retention basins have been created in the TAG area, with a total area of 0.8 ha. The catchment area of these basins varies between about 0.7 and 2 km^2, taking into consideration that this catchment area extends outside of project area. Run-off water from public spaces and existing houses in the project area is collected by lateral ditches along the roads and directed towards the retention basins. Runoff water from parcels greater than 1 hectare is collected by a small retention basin located on the parcel while the runoff water from parcels less than or equal to 1 ha is collected by the lateral ditches and drained to the large retention basins.

The study area is bordered to the south by hills, where the altitude is around 83 m, and to the north by the Garonne River, where the altitude is around 32 m. Three geographical units can be identified in the area (Figure 2). The first one occupies the southern and southwestern part with a steep slope, the second is located in the middle of the study area and characterized by moderate slopes, and the third extends to the north and northeast towards the Garonne River, where the slope remains gentle.

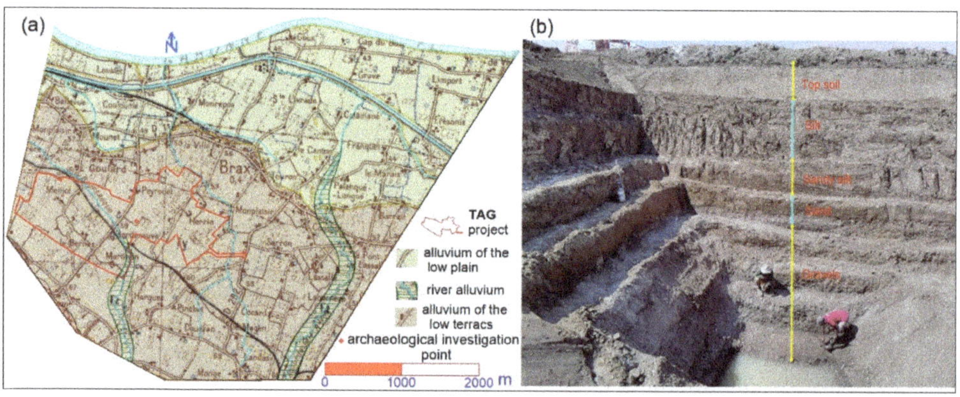

Figure 2. The different geological formation outcrop in the study area (**a**), based on [19,20]; a stratigraphic section of the alluvium of low terrace showing an archaeological investigation structure (**b**), based on [21].

From a geological point of view, at the end of the Mesozoic Era, tectonic movements began increasing until the middle of the Cenozoic Era. They have caused a gradual uplift of the Pyrenees Mountains, and deformations and faults in the ground. A succession of marine transgressions and regressions corresponding to climate changes has left an alternation of sedimentary rocks in the Nouvelle-Aquitaine Region. Less permeable clay and marl layers form a barrier to groundwater circulation.

The Garonne Valley is considered one of the largest alluvial plains in France [22]. The Garonne River sank into the Mesozoic deposits during the Quaternary and intensively deposited alluvium. The Quaternary alluvium is found along the Garonne River and its main tributaries. We can distinguish between current, recent, and ancient alluvial deposits, which form the major bed, the low plains, and the terraces, respectively [22]. The terraces were formed as a result of the successive excavation and infilling process in the valley. They are laid directly on the impermeable molassic bedrock. The ancient alluvial deposits are made up of silts, clays, and sandy gravel, and are represented by the low terrace in the study area. In contrast, the recent alluvium deposits, represented by the low plain in this area, consist of gravel and large pebbles intercalated with a sandy matrix. According to the geological map of France [20], the boundaries between different geological units are well observed in the study area (Figure 2). Different lithological formations have been distinguished during an archaeological study [21] conducted on the TAG area (Figure 2). The bottom of the structure was reached at a depth of 10 m, where the groundwater level was found to be 9.20 m on July 2018.

Figure 3 shows a geological cross section in the Garonne Valley between Aiguillon and Saint-Pierre-de-Buzet about 25 km downstream of the city of Agen, representing the vertical and horizontal extent of the alluvial deposits in the Garonne Valley.

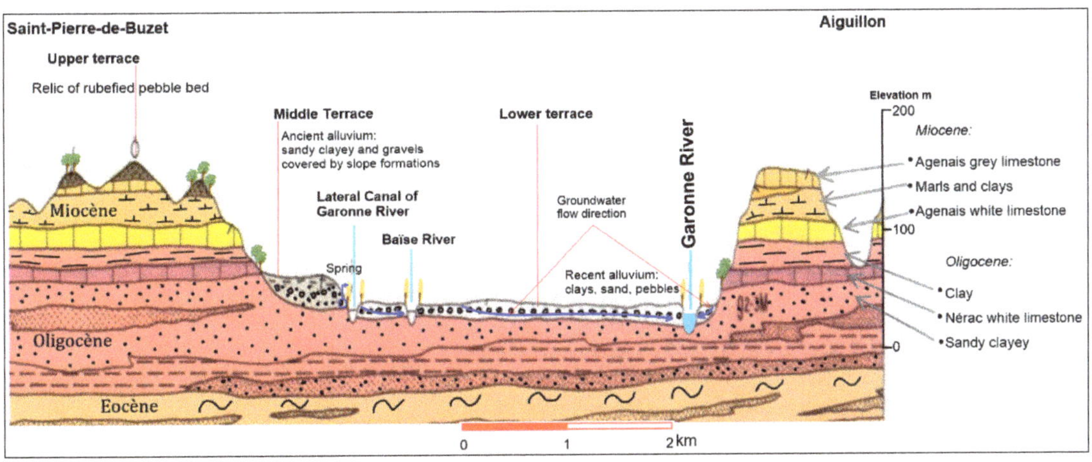

Figure 3. Geological cross-section in the Garonne Valley; from BRGM.

The Garonne River, with a drainage basin of 57,000 km^2 [23], is the third largest French Atlantic river in terms of discharge. The French and Spanish Pyrenees Mountains, the foothills of the Massif Central Mountains, and the plain between them feed the river. The hydrological behavior of the river is quite variable. It reflects the influence of Mediterranean and oceanic weather conditions and the effects of different landscapes and associated altitudes and slopes [24,25]. Hence, its flow is always influenced by the melting of the Aneto glacier, located in the Spanish Garonne, snow, rain, and drainage of the accompanying shallow alluvial aquifer. The Tonneins gauging station is about 40 km downstream from the city of Agen, and it is the lowest station not influenced by the tidal action of the ocean. The minimum objective discharge depends on station location; for Tonneins station, which is about 8 km after the junction point between the Garonne River and its tributary, the Lot River, this discharge is 110 m^3/s. According to the data measured in Tonneins station between 1989 and 2022, the mean annual discharge rate of the river is 517 m^3s^{-1}. The lowest value measured in this period was 34.1 m^3s^{-1} in August 1989, and the highest was 6190 m^3s^{-1} during the floods of February 2021 (https://www.hydro.eaufrance.fr, accessed on 17 October 2022).

The river is usually flooded in winter or spring after the snow melts. This flood is sometimes followed by a significant low discharge in summer. According to [3], the possible effect of climate change on the river is as follows: an average 11% river discharge decrease during the low-flow period (July–October) and an increase in winter discharge followed by a decrease during the spring. A program to supply water to sustain its low flows and to satisfy the users' needs and the aquatic environment was initiated by the SMEAG (Syndicat Mixte d'Études et d'Aménagement de la Garonne) organization about 30 years ago using 12 reservoirs located in the mountain area. Around 70 Mm^3/year, mainly from hydroelectric reservoirs in the Pyrenees Mountains, are mobilized for this purpose. This supply can represent up to 50% of the river's discharge from mid-August to mid-September. In 2022, the Garonne River had the longest low-flow period in its recent history, which lasted almost four months continuously. Figure 4 shows the dates on which the target low flow was exceeded between 1960 and 2022, according to the data measured at the Tonneins station. This figure shows that the river appears to have an extremely low flow quite early since 1984 (the earliest one took place on 6 June 2006), and thus, the number of years in which the objective low flow has not been exceeded has decreased.

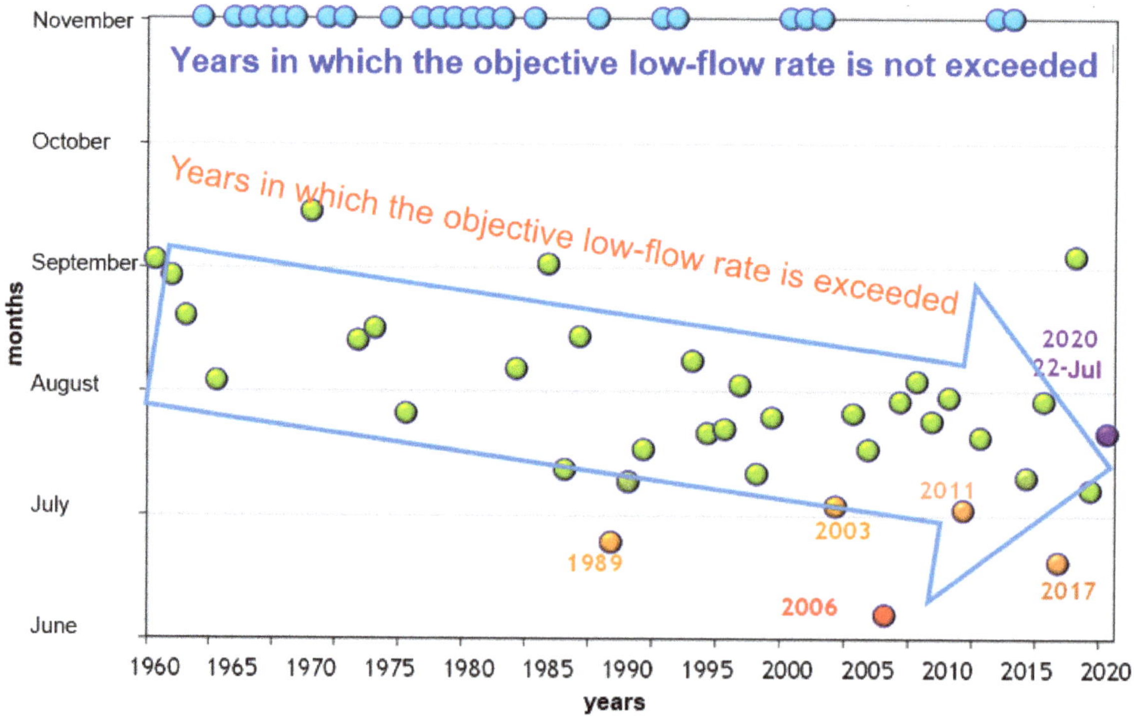

Figure 4. Low-flow index of the Garonne River at the Tonneins station; data source: SMEAG (https://www.smeag.fr, accessed on 20 September 2021), red and orange colours indicate that the river begins to experience low flow conditions early in the corresponding years.

The alluvial aquifer in the Garonne Valley is integrated into the river system. This aquifer plays an important role in water storage and hydraulic connection as a transmission zone between itself and the adjacent hillsides on one side and the Garonne River on the other side. The thickness of the aquifer varies from 4 to 10 m covered by a sandy-clayey and silty layer of 2 to 7 m. This aquifer tends to be locally semi-confined depending on the thickness of the clay and silt cover. The ancient alluvial deposits have undergone much more weathering than the recent formations. Thus, they are characterized by more clay deposits than the recent ones, which reduces the permeability of the aquifer. Hence, the low plain of recent alluvial deposits constitutes an abundant unconfined aquifer in direct hydraulic connection with the river. It feeds the river and sustains its base flow, particularly in the summer period, and it is recharged by the river during flood periods, reducing the river's flood risk. Subsequently, the plains alluviums constitute a buffer zone controlling the exchanges between the alluvial aquifer and the river. The aquifer–river interface comprises a complex mosaic of surface and subsurface flows of varying lengths, depths, directions, and residence times [26,27].

Consequently, the flow direction between the river and the surrounding alluvial aquifer can be reversed according to groundwater and surface water gradient. Since the Garonne River flows directly on the substratum molassic, the seepage rate will be limited by the hydraulic conductivities of the river banks. The underground part of the alluvial aquifer is located at a depth that varies from 10 to 15 m. This portion of the aquifer corresponds to an impermeable molassic bedrock from the Oligocene which constitutes the bottom of the aquifer. Groundwater seepages into the riverbed have been observed at several sites in the study area during summer. The aquifer contribution to the river base flow differs between the wet and dry years. The exchange between the alluvial aquifer and the river is illustrated in Figure 5.

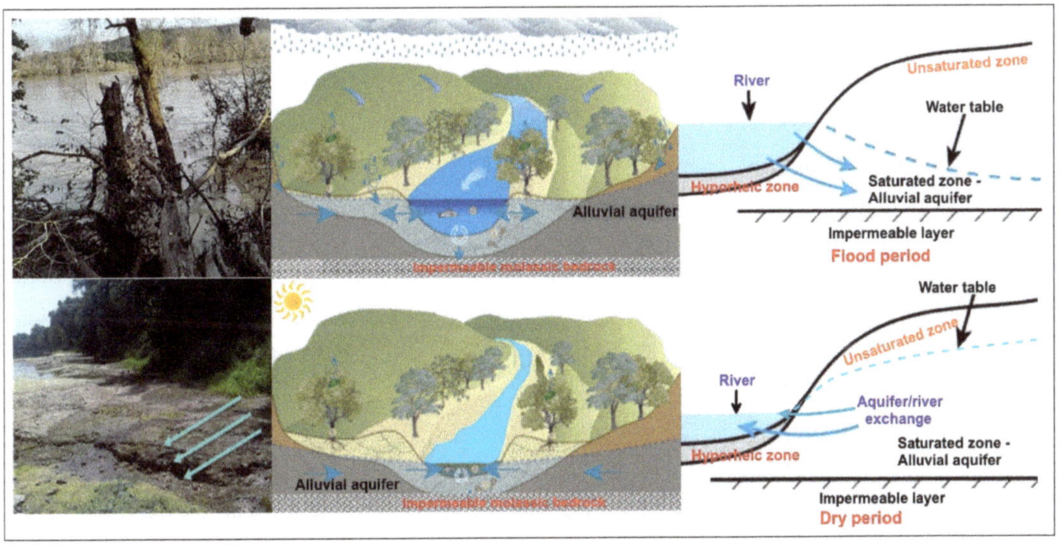

Figure 5. Alluvial aquifer/Garonne River interface and exchange patterns during flood and dry periods, based on [4,19,28].

3. The Concept of Artificial Groundwater Recharge in the TAG Area

The groundwater volume of an aquifer should be preserved, so the exploitation should only concern the yearly renewable volume of this aquifer. In the case of the shallow alluvial aquifer of the Garonne Valley, the quantitative imbalance could be caused by insufficient recharge of the aquifer by autumn or winter rainfall and is accentuated by significant withdrawals in the summer period to meet the agricultural needs. Artificial groundwater recharge of this aquifer could be integrated into water resources management strategy to maintain long-term water sustainability and related ecosystems. Based on the observed hydraulic relationship between the Garonne River and the adjacent alluvial aquifer, the artificial recharge of this aquifer has been initiated to sustain the low flow of the river.

4. Materials and Methods

A total of 132 wells and boreholes, distributed across the study area (see Figures 6 and 7), were used to measure the groundwater levels. According to the owners of the wells, most of them reach the top of the molassic bedrock. The groundwater levels were measured monthly over 20 months (July 2018–February 2020) using a manual piezometer sensor. The piezometric map of March 2019 is shown in Figure 7. A total of 45 groundwater samples were collected during the fieldwork campaign in September 2018. Prior to sampling, each well was pre-pumped for 10 to 15 minutes to renew the groundwater. All samples were analyzed in the ENSEGID (Graduate School of Environmental, Geological, and Sustainable Development Engineering) Institute laboratory in Bordeaux, France. Nitrate measurement was carried out in the field using a HORIBA B-742 LAQUAtwin Nitrate meter. The pH and temperature were also measured in situ using a WTW MultiLine Multi 3630 IDS Multi-Parameter handheld box. The major ions of 45 samples were analyzed by ion chromatography using a Dionex ICS-900. HCO_3^- was analyzed with a Metrohm 716 DM Titrimeter by titration with 0.1 N hydrochloric acid. The ionic mass balance of the samples shows that the majority of errors (75%) were lower than +/−5%. The geochemical software PHREEQC (version 3.6.2.15100) [29] was used to calculate the saturation indices (S.I.) of the minerals present in the groundwater system.

Runoff water is collected in lateral ditches in the TAG area and directed to retention basins. This water is considered a potential source to recharge the shallow alluvial aquifer and its later use for preserved green areas in this industrial zone. The monitoring of groundwater levels in the study area over 20 months provided essential information on the hydrodynamic regime of the aquifer, the recharge/discharge zones, and the effects of abstraction and floods on the groundwater level.

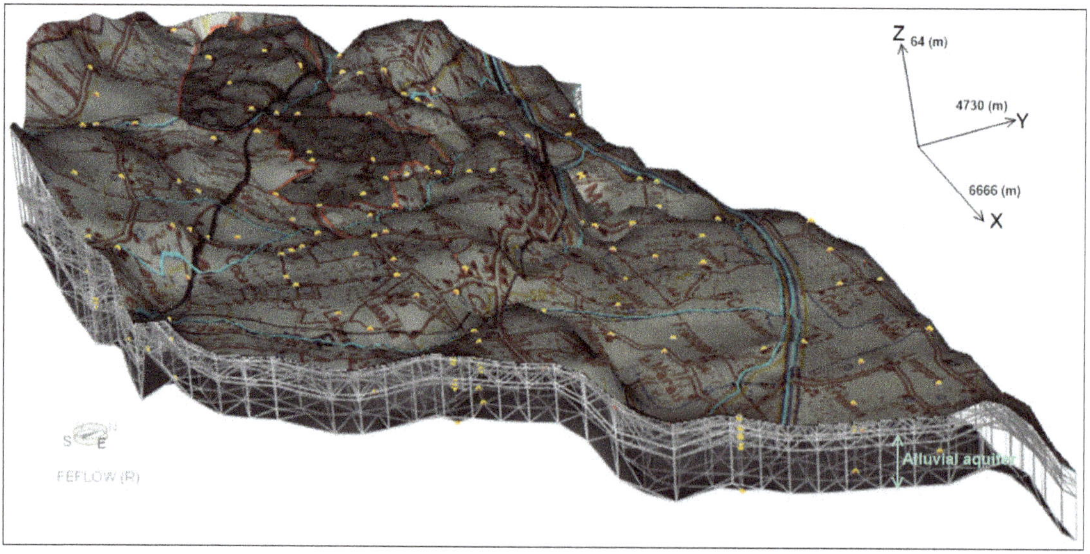

Figure 6. The three-dimensional (3D) representation model of the study area showing four layers and five slices, constituting the model domain [18].

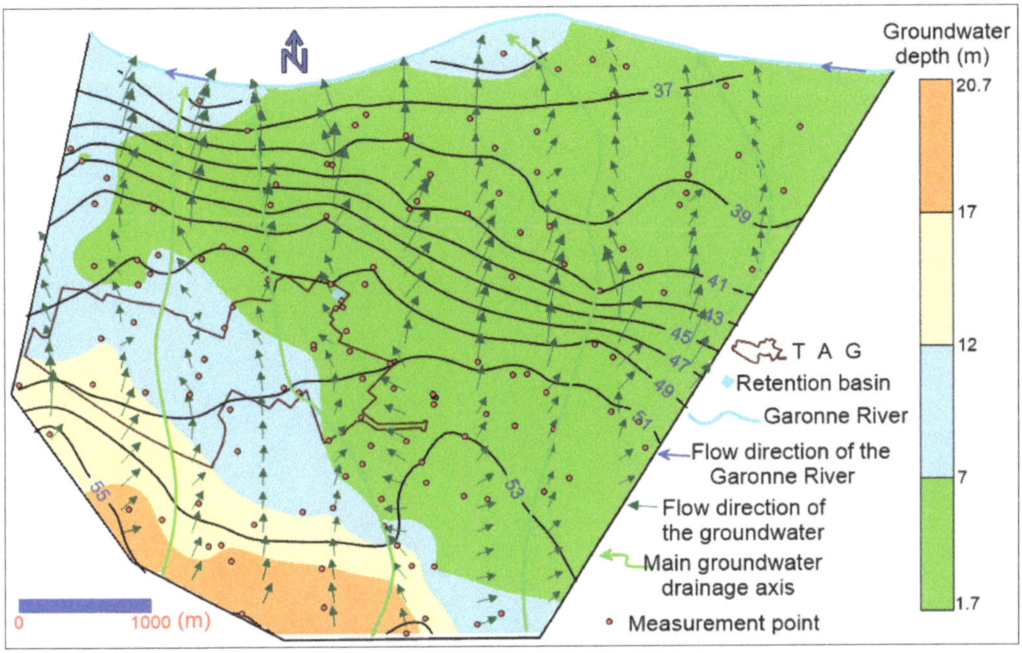

Figure 7. Location map of observation wells and the piezometric map showing groundwater flow direction and water table depth from the surface for March 2019.

The hydrodynamic model was developed to study the flow regime of the alluvial aquifer after implementing the artificial recharge system using an infiltration basin. Using FEFLOW 7.1 by DHI-WASY [30], a finite element subsurface flow and transport simulation system, the framework of super elements was defined, and then the 2D horizontal mesh of finite-element was generated using the automatic triangle option. The model top layer topography was determined based on IGN spatial satellite data with a 5 m resolution. Manual corrections concerning the auto-route bridges as well as the bridges of the lateral canal of the Garonne River located in the study area were made, respecting the topography around these bridges. After that, and based on the available information, three-dimensional (3D) slice elevation, layer properties, and boundary conditions were defined. Then, the advanced subsurface flow system was investigated under the steady state condition. The 3D model grid consists of four layers (Figure 6), and the thickness of each layer varies according to the topography and the geological formation. The first layer represents the topsoil, which is 1–3 m thick. The second layer consists of silt and sandy silt with a thickness varying between 1 and 7 m, increasing towards the hillslopes. The third layer represents a thin sandy layer of 1–2 m. The fourth layer consists of gravel, pebbles, and sandy clay housed in the molassic formations with a thickness varying between 6 and 10 m. The alluvial aquifer is represented by the third and fourth layers. QGIS (Geographic Information System) was used to visualize and pre-process the geographic data required for the model.

5. Result and Discussion

5.1. Piezometric Map Delineation and Groundwater Level Fluctuation

The piezometric map provides important information about the groundwater flow direction and the variation of the hydraulic gradient. The piezometric map for March 2019 is shown in Figure 7. This map shows that the piezometric levels vary between 55 m.a.s.l in the low terrace in the south and southwest of the study area, near the hillsides, and 35 m.a.s.l in the northern part in the low plain near the Garonne River. The piezometric surface of the study area shows that there is hydraulic continuity between the two hydrogeological units of the low plain and low terrace. The hydraulic gradient is slightly marked in the southwest towards the hillsides, while it is more important between the low terrace and the low plain.

The significant variations in thickness and lithology of the alluvium lead to significant variations in the permeability of the alluvial aquifer in the study area. According to the groundwater levels measurement over 20 months, the northern part close to the river has shown more groundwater level fluctuation in both dry and flood periods according to the river level changes, and hence the exchange between the river and the aquifer. The groundwater flow direction is mainly oriented south–north towards the main drainage axis to the Garonne River. The hydraulic gradient is very variable from one sector to another of the study area, and it is 2.5 $m{\cdot}km^{-1}$ in the south and 8.5 $m{\cdot}km^{-1}$ in the southwest, where the isopiezes are parallel to the boundary, indicating a groundwater supply from the hillside. In the TAG area, the lowest value of hydraulic gradient (1.8 $m{\cdot}km^{-1}$) was measured, indicating a high permeability of the aquifer. In the sector between the low terrace and the low plain, oriented southeast–northwest, where the isopiezes are generally equidistant, the highest hydraulic gradient value of about 11 $m.km^{-1}$ was measured. This value was measured at 3 $m{\cdot}km^{-1}$ in the northeast of the study area, while it was measured to be 8 $m{\cdot}km^{-1}$ to the northwest of this area. Groundwater depth changed between 1.7 and 20.7 m from the surface in the study area during March 2019.

Figure 8 shows groundwater level fluctuations at five measurement points in the study area. The groundwater levels measured in two boreholes located close to the retention basin show that the groundwater level increased between 1 and 0.5 m in the boreholes PZ1 and PZ6, respectively, after the rainfall event of November 2019 (Figure 8). The groundwater level measured in PZ10, located about 500 m upstream of the retention basin, showed a response due to natural input (Figure 8). Following the storm rainfall in the upstream part of the river's basin in November 2019, the Garonne River was flooded. The effect of this flood on the piezometric levels was measured in the two wells (FP84, FP92), located at 70 and 150 m from the river bank, respectively (Figure 8). These wells show a response important to this flood event as the groundwater levels increased by about 2 m in these wells. This observation shows a clear exchange between the Garonne River and its accompanying alluvial aquifer at the end of the dry period, where the river has well recharged this aquifer.

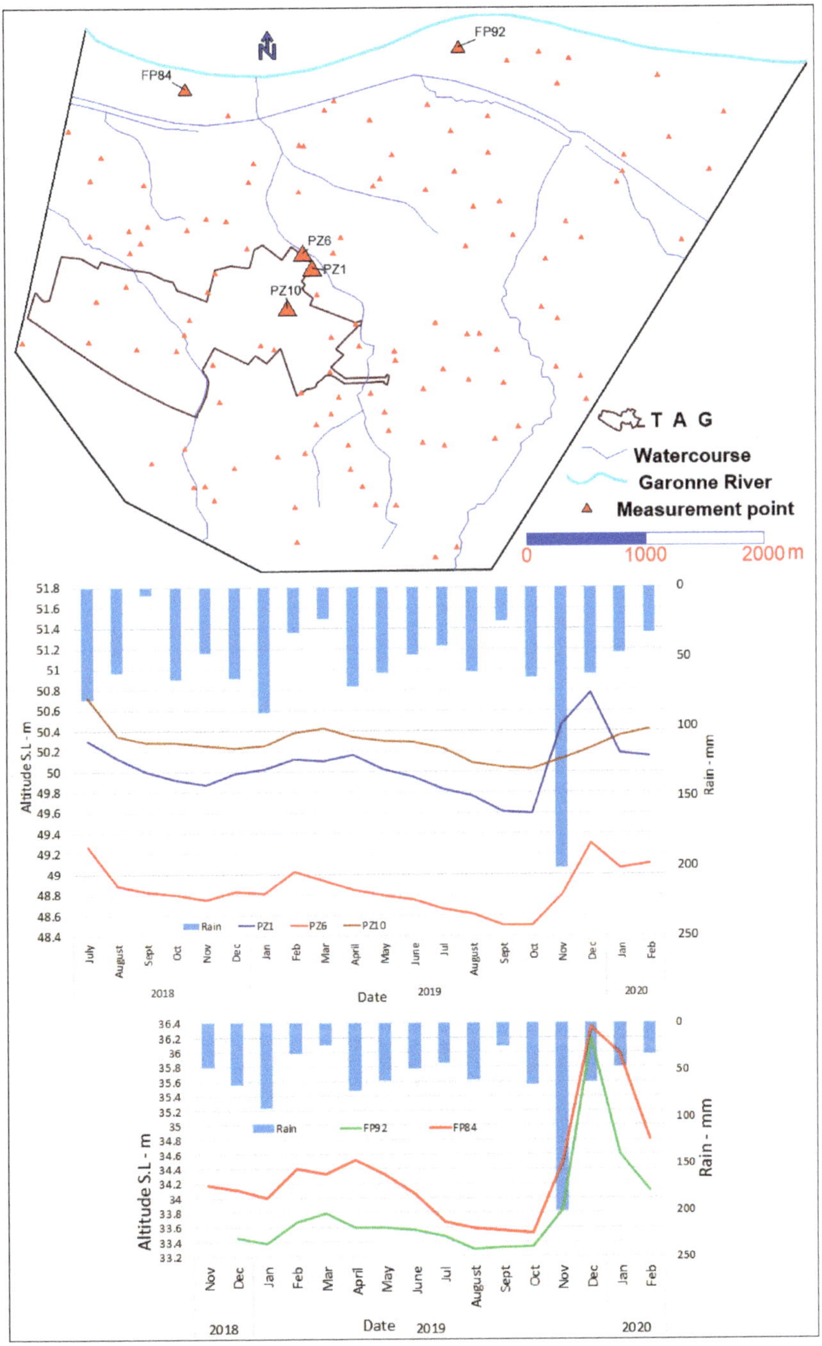

Figure 8. Monthly groundwater level fluctuations in two boreholes located close to the retention basin, one well about 500 m upstream of the infiltration basin, two wells close to the river, and monthly rainfall variation measured at Agen station close to the TAG area.

5.2. Groundwater Hydrochemistry

Different processes, such as groundwater flow, recharge and discharge, and water-rock reactions, control groundwater hydrochemistry. Along the groundwater flow direction, mineral weathering affects hydrochemistry during residence time [31,32]. Knowledge about groundwater quality is an important part of studying the hydrodynamics functioning of the aquifer.

5.2.1. Nitrates (NO_3^-)

The nitrate concentration was measured in situ in September 2018 in 50 wells (Figure 9). The measured values ranged between 12 and 117 mgL^{-1}. Additionally, more than two-thirds of the samples exceeded the drinking water threshold of 50 mgL^{-1}. Relatively high concentrations were measured in the central part of the study area. Nevertheless, nitrate concentrations were measured to be relatively low in the low plain of the Garonne River, where there is an exchange between the river and the aquifer. These low nitrate concentrations might have resulted from groundwater denitrification due to the input of dissolved organic carbon from the surface water or groundwater dilution by the Garonne River's water.

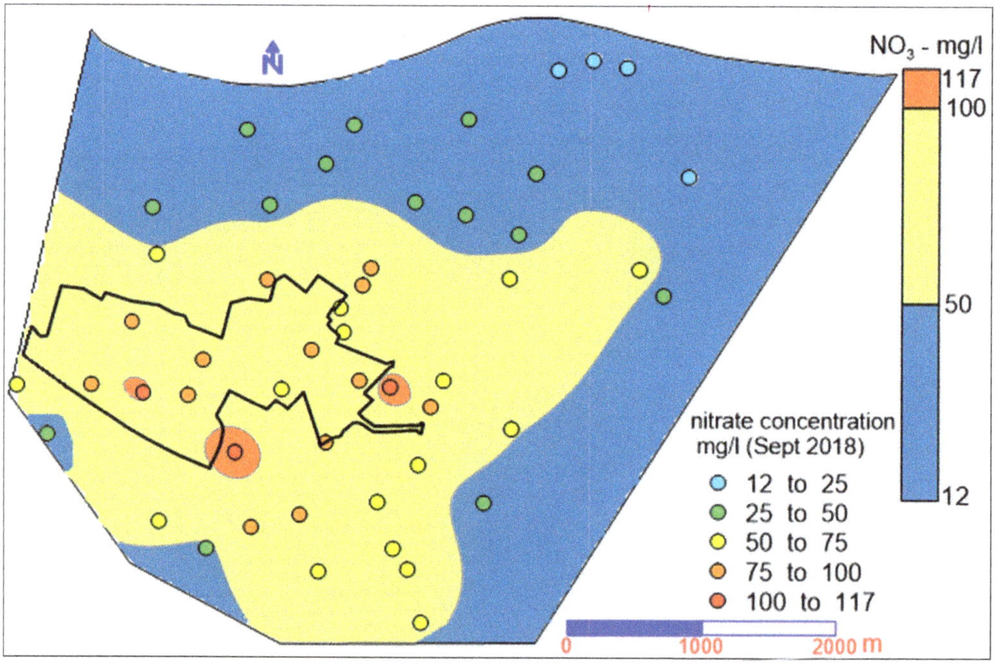

Figure 9. Spatial distribution of nitrate concentrations (zonation and points) measured in the alluvial aquifer of the study area, September 2018.

5.2.2. Dominated Hydrochemical Processes in the Alluvial Aquifer

To better understand the hydrogeological processes involved in the hydrochemical evolution of the aquifer system, groundwater mineral equilibrium calculations can be useful to predict the presence of reactive minerals and estimate the mineral reactivity of the groundwater system [33,34]. The saturation indices of certain minerals have been calculated and are presented in Table 1.

Table 1. Saturation index of minerals (anhydrite, carbonate, gypsum, halite, and quartz) in the alluvial aquifer calculated with PHREEQC.

Mineral Saturation Index	Min	Max	Average	Standard Deviation
Anhydrite	−2.9	−1.9	−2.3	0.2
Calcite	−1.3	0.1	−0.6	0.3
Dolomite	−2.8	−0.5	−1.7	0.5
Gypsum	−2.5	−1.5	−1.9	0.2
Halite	−8.5	−6.6	−7.5	0.4
Quartz	−0.1	0.3	0.1	0.1

Calcium and magnesium concentrations have been plotted as a function of calculated pCO_2 by PHREEQC, and the pure calcite and dolomite dissolution determined by PHREEQC (Figure 10).

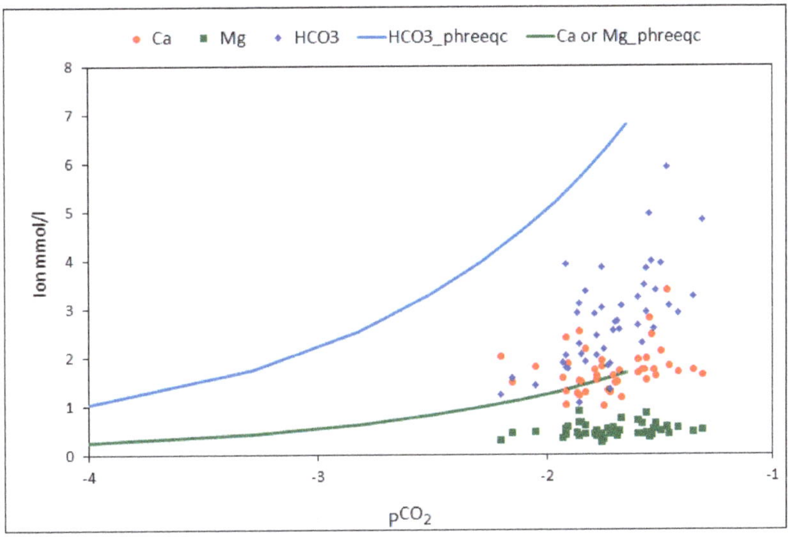

Figure 10. Calcium, magnesium, and alkalinity concentration in $mmolL^{-1}$ as a function of pCO_2 (atm) calculated values; solid lines are the alkalinity, calcium, and magnesium determined with PHREEQC for pure calcite and dolomite dissolution.

Most water samples are near or below the PHREEQC line concerning the calcium, while all samples are below the PHREEQC lines for bicarbonate and magnesium (Figure 10). This means that the sources of calcium supply in groundwater, such as the dissolution of calcite, dolomite, and gypsum, are not the main process controlling groundwater chemistry in this case.

5.2.3. Hydrochemical Facies of the Groundwater

To better understand the hydrogeological processes involved in the aquifer system, the water type, according to Stuyfzand classification [35], is depicted together with a Piper diagram [36], as shown in Figure 11. The main water type and subtype have been determined for each water sample based on chloride, alkalinity, and the most important ions values. Concerning the chloride concentration, the sample can take the code H, S, b, B, f, F, g, and G with respect to increasing chloride concentration. In our case, groundwater is classified as oligohaline and soft (Cl changes between 10 and 88 mgL^{-1}), so the code varies between g and F. The g factor (relatively low chloride concentration) is mainly found in the low plain, where the aquifer is directly connected with the river, and the exchange between them is well observed by the drainage during the dry periods and increasing the groundwater level during the flood periods. Regarding the alkalinity, the sample can take the code * or values from 1 to 9 corresponding to the increasing of alkalinity concentration. The result in our case shows that 7% of the samples were classified as slightly high, 71% as moderate, and 22% as low (alkalinity varying between 1 and 6 meqL^{-1}, the code changes between 1, 2, and 3, and the groundwater classifies as soft and very soft).

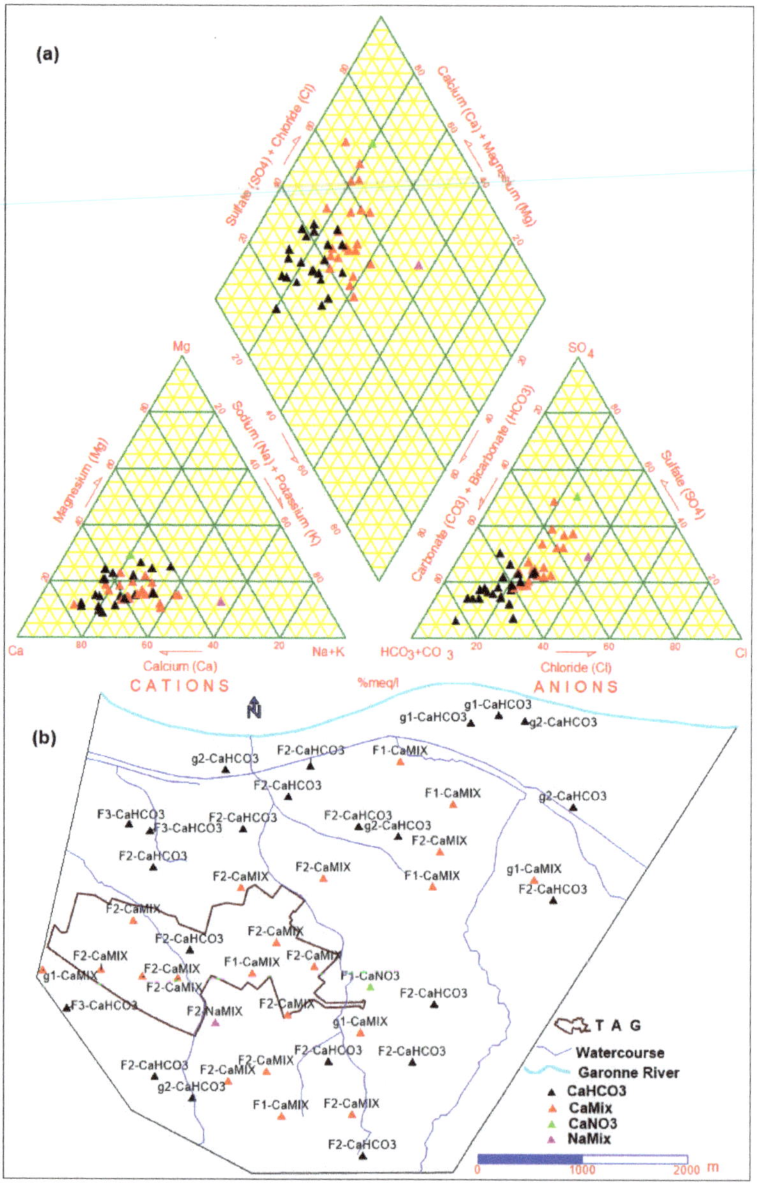

Figure 11. (**a**) Modified Piper diagram depicting hydrochemical facies of groundwater in the study area; (**b**) spatial distribution and classification of these facies based on Stuyfzand classification.

The water type changes between Ca-HCO_3 and Ca-Mix. The Ca-HCO_3 water type is mainly found in the low plain near the Garonne River and in the southern part of the study area, where lateral groundwater inflow can occur along the hillsides. The "Mix" anions water type refers to the water in which no anion family makes up more than 50% of the sum of anions. This water type indicates that more salt dissolution and reactions such as the oxidation of organic matter that increase the HCO_3^- might occur [37]. The local appearance of the water type Ca-NO_3, indicates that NO_3 is more dominant than the other anions, and strong influence of anthropogenic activities on the water quality is presented.

The Piper trilinear diagram can help to visualize the main hydrochemical facies of groundwater and compare their evolution. The spatial distribution of groundwater samples on the Piper trilinear (Figure 11) confirms the water type classification based on Stuyfzand (Ca-HCO_3 and Ca-mixed types). The dominance of Ca water type in the aquifer system might indicate more liberation of Ca^{2+} ions from silicate minerals [38,39]. This water type reveals the influence of the recharge mechanism by the Garonne River or the direct infiltration from the surface and the effects of ion exchange or reversed ion exchange on water quality. The source of HCO_3^- in the groundwater might be due to intense chemical weathering processes in the aquifers or from natural processes such as incongruent dissolution of silicates in the groundwater reacting with carbon dioxide that might release bicarbonates [39,40].

5.2.4. Silicate Hydrolysis

The dissolved silica derives primarily from the weathering and subsequent dissolution of silicates and aluminosilicates in bedrocks and soils [41,42]. This process is irreversible, so the silica is retained in an aquatic solution, influenced by thermodynamic factors in the dissolution process, pH, and adsorption of silicate minerals [39,43]. The spatial distribution of silicate values measured in the groundwater collected in September 2018 is presented in Figure 12. The relatively low SIO_2 values were mainly measured in the wells near the hillsides and the Garonne River. These values can be explained by a shorter path of the water in the aquifer and by mixing of groundwater with the surface water, respectively. The relatively high values were measured in the central part of the study area, indicating the dissolution processes of aluminosilicate minerals in this area. They might have a longer residence time of the groundwater.

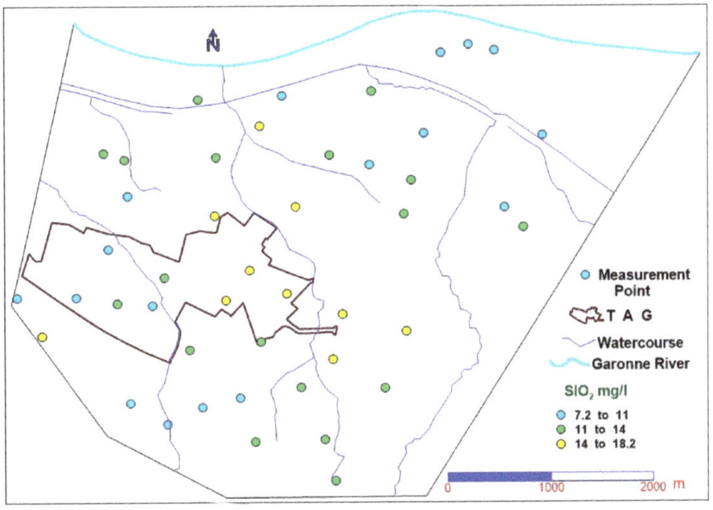

Figure 12. Classification and spatial distribution of SiO_2 concentration (mgL^{-1}) measured in groundwater, September 2018.

The mineral stability diagram is another approach to test the proposed hydrochemical evolution of silicate minerals and their stability [44,45]. These diagrams are usually used to assess the degree of equilibrium and water–rock interaction [46,47]. The relationships between ($[Ca^{2+}]/[H^+]^2$), ($[Na^+]/[H^+]$) and log $[H_4SiO_4]$ is shown in Figure 13. The stability fields for Ca-Plagioclase (anorthite) and its possible weathering products, gibbsite, kaolinite, and Ca-montmorillonite, as a function of log ($[Ca^{2+}]/[H^+]^2$) and log $[H_4SiO_4]$ show that the kaolinite is more likely to be stable than, for example, gibbsite, as a result of anorthite weathering products. However, the stability fields for Na-Plagioclase (albite) and its potential weathering products, paragonite, gibbsite, kaolinite, and pyrophyllite, as a function of log ($[Na^+]/[H^+]$) and log $[H_4SiO_4]$ show that the main reaction is the conversion of albite to kaolinite. The primary silicate minerals, plagioclase and clinopyroxene, can be dissolved and weathered to kaolinite, which is more likely to be in equilibrium in the system.

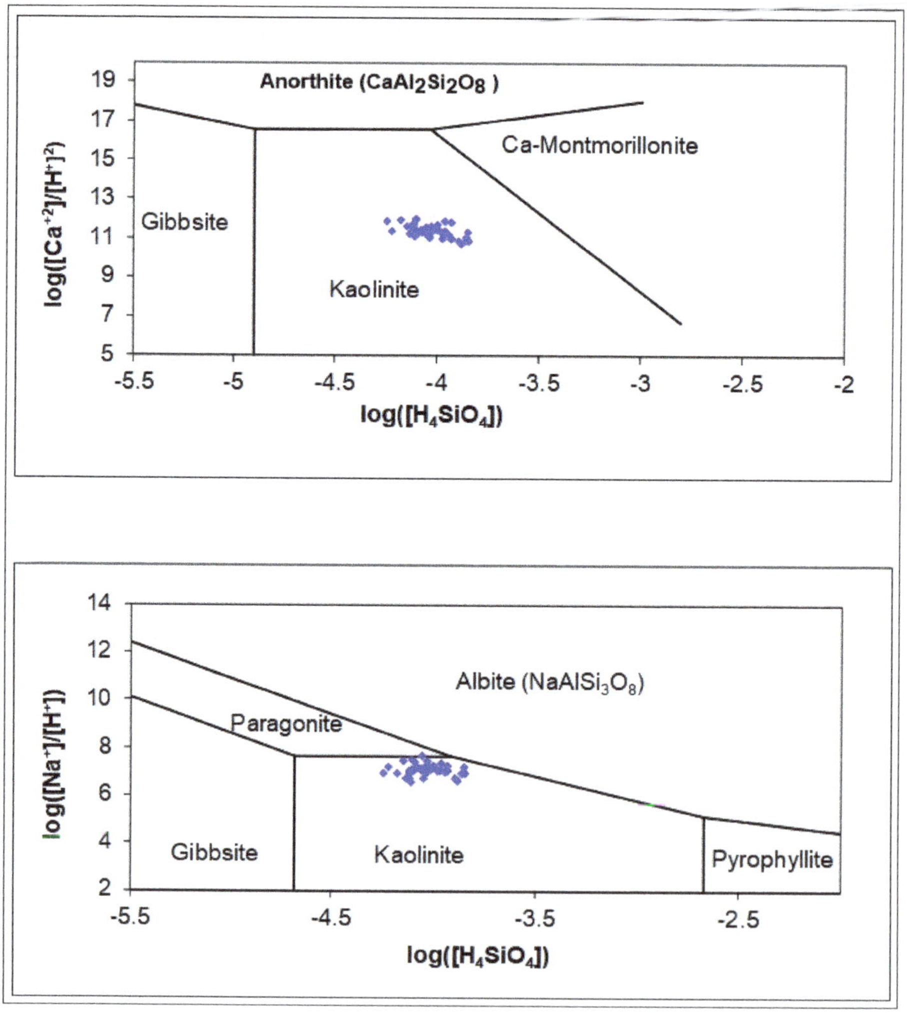

Figure 13. Stability diagram of Ca-Plagioclase (anorthite) and Na-Plagioclase (albite) and their possible weathering products.

Nevertheless, the equilibrium between clay and primary silicate minerals is probably the main process controlling groundwater chemistry in the study area. The two diagrams (Figure 13) show that the primary silicate minerals, anorthite and albite, tend to dissolve and transform into secondary minerals in the groundwater system. However, the groundwater is not in equilibrium with either anorthite or albite, and both minerals will decompose into kaolinite, as shown in the following equations:

$$CaAl_2Si_2O_8 + 2H_2CO_3 + H_2O \rightarrow Al_2Si_2O_5(OH)_4 + Ca^{2+} + 2CO_3{}^2 + 2H^+$$

$$2NaAlSi_3O_8 + 2H_2CO_3 + 9H_2O \rightarrow Al_2Si_2O_5(OH)_4 + 2Na^+ + 4H_4SiO_4 + 2HCO_{3-}$$

Ref. [48] have explained the effect of rock dissolution on the groundwater composition through the scattered plots of Ca^{2+} + mg^{2+}) versus ($HCO_3{}^-$ + $SO_4{}^{2-}$), as shown in Figure 14.

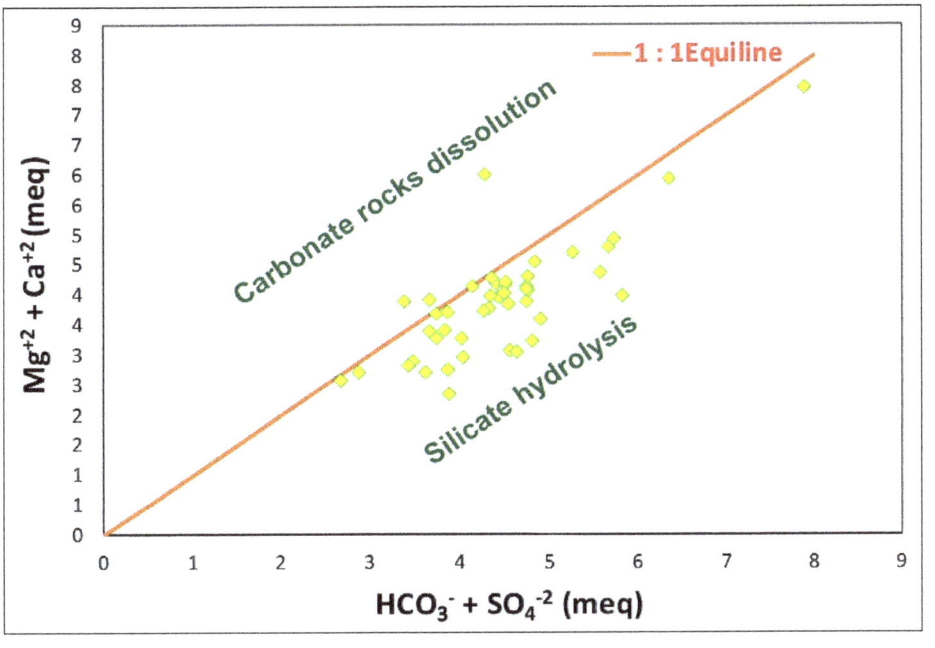

Figure 14. Scattered plots of (Ca^{2+} + Mg^{2+}) vs. ($HCO_3{}^-$ + $SO_4{}^{2-}$) showing the effect of carbonate rock dissolution and silicate hydrolysis on groundwater composition.

Figure 14 shows that the majority of the water samples are located near or below the equiline on the $HCO_3{}^-$ + $SO_4{}^{2-}$ side, indicating that, on the one hand, silicate hydrolysis is the main hydrochemical process controlling the chemistry of these samples and, on the other hand, the excess $HCO_3{}^-$ + $SO_4{}^{2-}$ can be derived from other processes such as ion exchange.

The effect of the evaporation process on groundwater chemistry was investigated using the molar relation of Na^+ and Cl^- (Figure 15). When this ratio is close to one, this indicates that halite dissolution contributes mainly to Na^+ concentration in the groundwater. In contrast, most samples have a ratio greater than 1, which signifies silicate weathering or ion exchange [49,50]. In other words, an excess of Na^+ in the aquifer system likely results from silicate weathering and ion exchange.

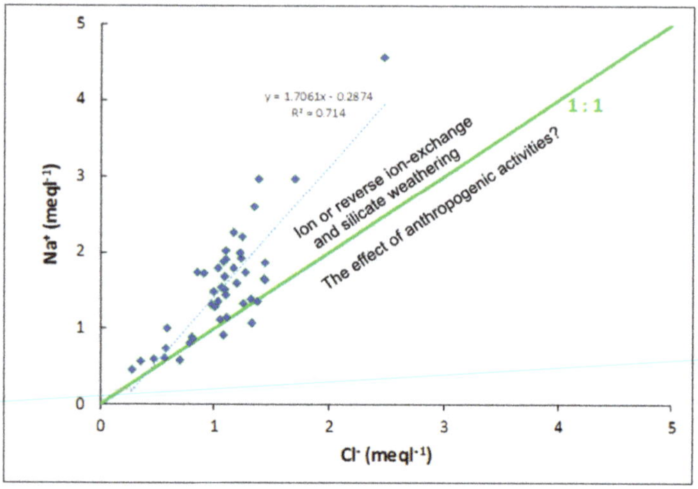

Figure 15. Scatter plot of Na^+ and Cl^- molar ratio.

5.3. *Modeling of Groundwater Flow*

The model mesh and structure have been generated based on all basic geometric information. The finite element method is characterized by its flexibility in generating the mesh and refining the nodal grid. There are no inactive nodes in the generated mesh because the elements match the model limit. The model domain consists of 17,936 mesh elements and 11,545 mesh nodes (4484 elements per layer and 2309 nodes per slice); the triangulation method was used for mesh generation. A reasonable refinement has been applied to the TAG area to achieve better precision on groundwater flow in this area for potential future infiltration from other retention basins.

5.3.1. Boundary Conditions

Defining suitable model boundary conditions is an essential step in building numerical groundwater models; it is largely responsible for the flow mechanism and the calculation in the system [51,52]. The model boundary conditions were determined according to hydrogeological conditions. The model southern boundary forms the geological limits between the alluvium of the low terrace and the molassic formations of the Aquitanian and Stampian ages, whereas in the north towards the Garonne River, the boundary is naturally established based on the topography and water level of the river in this period. The eastern and western boundaries were assigned as no-flow boundaries since the equipotential contours intersect with these boundaries almost in perpendicular form (Figure 7). The potentiometric map was created based on groundwater levels measurements of March 2019 using the ordinary kriging algorithm method [53]. This was the map for the steady-state model simulation. The south and southwest boundaries upstream of the model area are prescribed constant hydraulic heads (B.C.s) that correspond to manual groundwater level measurements. At the northern boundaries near the river, and since this one flows directly on the substratum molassic, the initial hydraulic heads were taken according to the linear interpolation of the Garonne Riverbed altitude, taking into account the water level of the river during this period. Since the alluvial aquifer overlies an impermeable molassic layer, there is no upward leakage from the model bottom. The direct recharge of the alluvial aquifer by infiltration depends strongly on the rainfall and the permeability of the unsaturated zone. The topsoil cover is globally homogeneous over the model area (vegetal soil, silt, and sandy silt), and a homogeneous annual recharge (70 mmy^{-1}) was applied to the first model layer. The lower and upper horizontal model boundaries are limited by the surface of the substratum molassic and the topography of the model top layer, respectively. According to the water level fluctuations in the wells located near the watercourses in the study area and the piezometric curve shapes, there is no hydrodynamic relationship between the alluvial aquifer and the hydrographic network. The exception is the hydrodynamic relationship between this aquifer and the Garonne River, where the river recharges the aquifer during the flood periods.

5.3.2. Hydraulic Conductivity

Hydraulic conductivity is an essential parameter to evaluate the aquifer's potential productivity, response to recharge, and groundwater circulation. The meandering system strongly influences the heterogeneity of the alluvial deposits. The meandering crossing and channel abandonment processes, which allow the conservation of dense clay plugs, are the main cause of sediment heterogeneity in the alluvial plain. As a result, there is significant local variation in the hydraulic conductivity values of the alluvial aquifer. Thus, these values can be changed within a few meters. This variation has strongly affected the productivity of the aquifer from one well to another. This can be explained by the alteration of the alluvial deposits and the presence of paleochannels or clay lenses characterized by low permeability. The hydraulic conductivity values were assigned to the model based on the steady-state trial and error calibration method to simulate the measured hydraulic head and subsequently archive a reasonable agreement between the measured and simulated groundwater levels. These values change between 1.4×10^{-2} and 6×10^{-5} m.s^{-1}, and reflect a large hydraulic heterogeneity even at a small scale in the model area (Figure 16). For the other layers, uniform hydraulic conductivity values were assigned for each layer because of their small effect on the groundwater flow system of the alluvial aquifer.

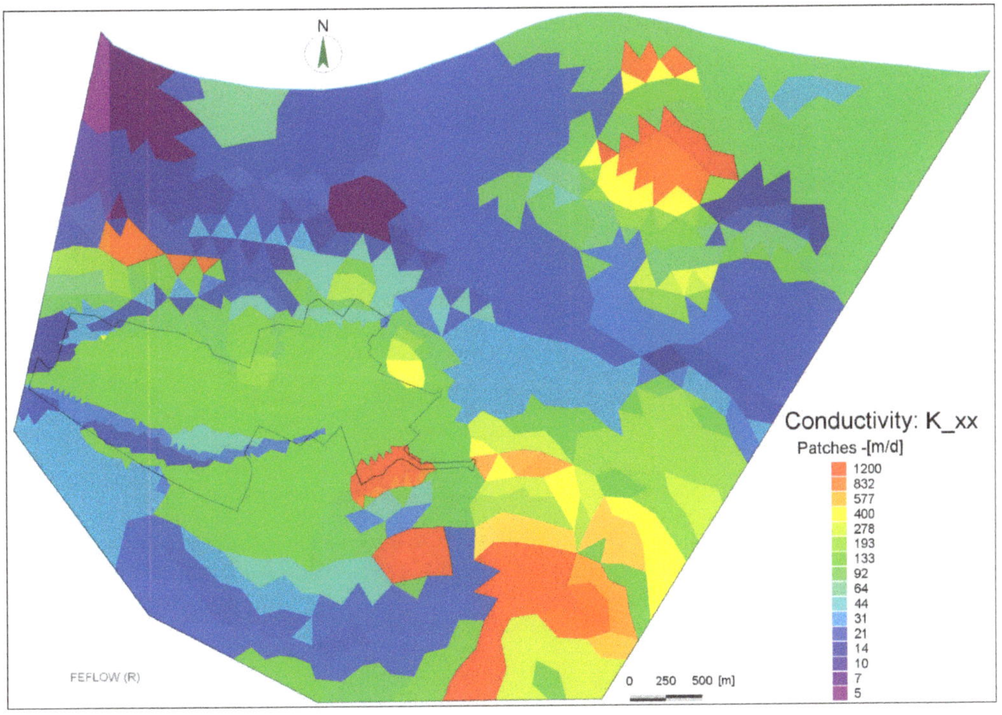

Figure 16. A spatial distribution map of the hydraulic conductivity of the alluvial aquifer resulted from the model calibration.

5.3.3. Steady-State Calibration

The calibrated model allows us to check the consistency of the data governing the hydraulic performance of the aquifer system, mainly the recharge, permeability, and piezometric levels. The first step in model calibration was to adjust the hydraulic conductivity values using the inverse method in steady-state to simulate the hydraulic heads measured in the field.

The superposition of the measured and calculated groundwater level contours (Figure 17) indicates a good agreement. The calibration is mainly performed by adjusting the hydraulic conductivities to obtain piezometric contours as close as possible to the actual piezometer. The calibrated model defines the initial groundwater level, which is used to run it in the transient regime considering the effect of artificial groundwater recharge from the retention basin.

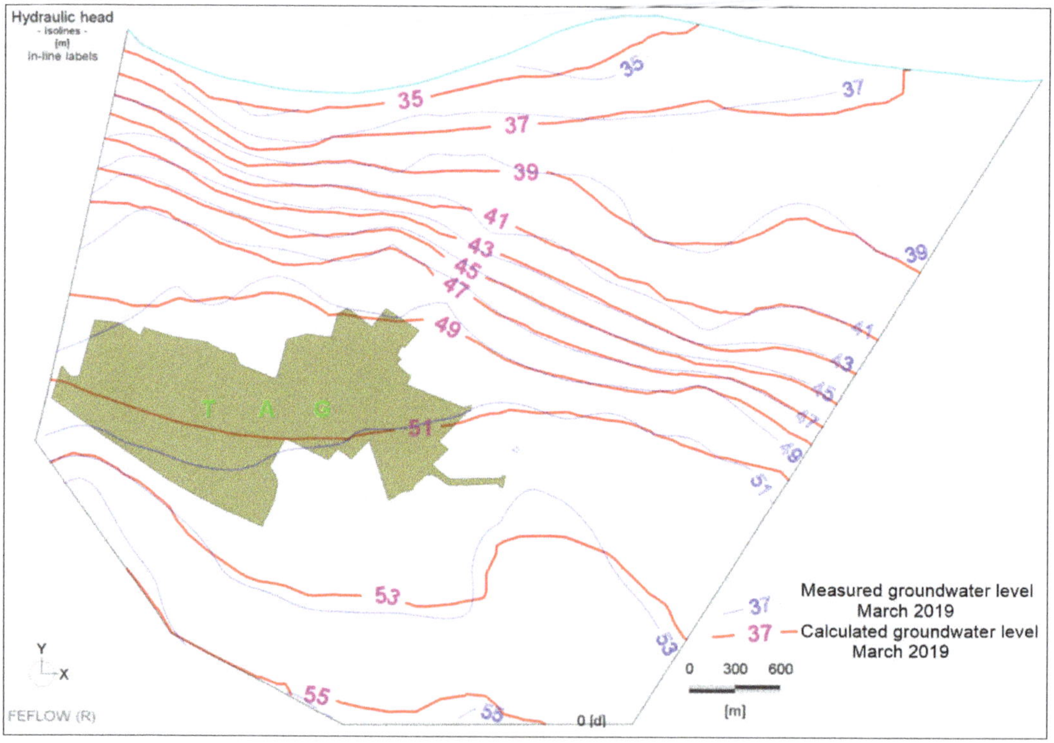

Figure 17. Measured and calculated piezometric maps, March 2019.

The scatter plot and regression analysis of the calculated and measured hydraulic heads are shown in Figure 18. The result indicates good agreement between the simulated and measured hydraulic heads.

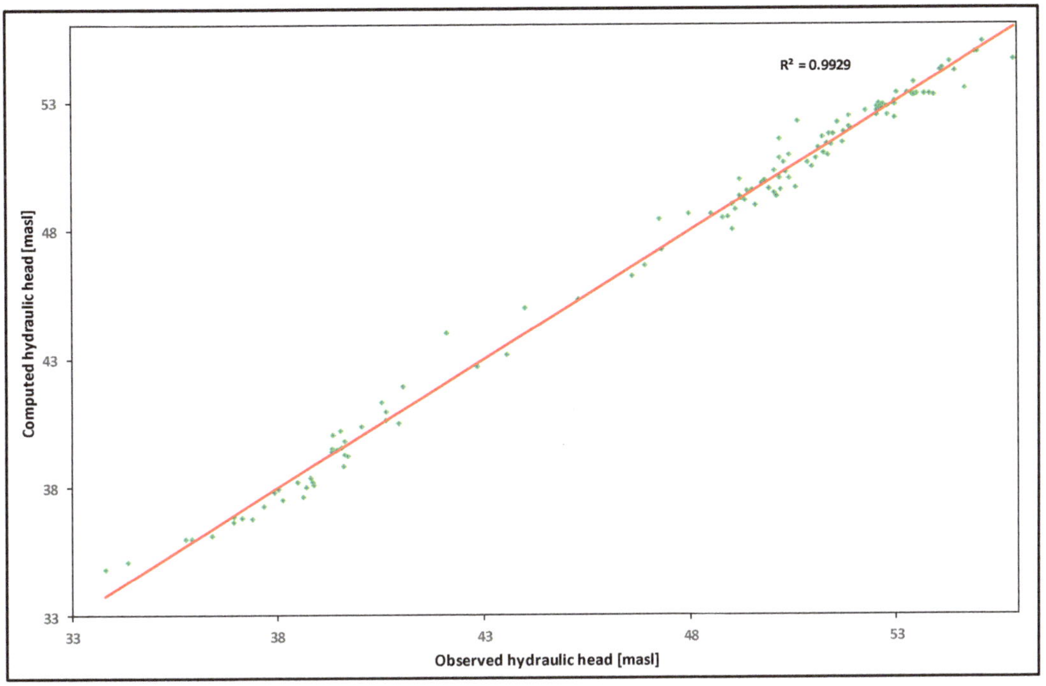

Figure 18. Scatter plot of measured and simulated initial hydraulic heads for March 2019.

5.3.4. Results of Transient Modeling

After model calibration, the model was turned into daily time steps to simulate the groundwater flow after the artificial recharge has integrated into the hydrodynamic system of the alluvial aquifer in the TAG area. After sufficient eventual precipitation storm events, the retention basin filled up. Subsequently, the infiltrated water increased the water level in the adjacent wells, as we have measured in two wells near this basin. The calculated equipotential lines show that a piezometric dome was created close to the retention basin after runoff water infiltration.

To simulate groundwater flow patterns after applying groundwater artificial recharge, we selected the nodes corresponding to the infiltration basin in the TAG area in the model mesh. After that, we applied a forward process in a transient state, associated with time series corresponding to groundwater levels measured close to the retention basin, to delineate groundwater path lines in flow direction and to estimate the travel times. From Figures 17 and 19, we can see that the initial equipotential contour located downstream of the retention basin (contour 49 m.a.s.l), moved about 500 m toward a groundwater flow direction after the model was run in the transient state. It was estimated that the infiltrated water would take about four months to reach the Garonne River, which is an appropriate time to sustain its low flow if the recharge occurs between March and April.

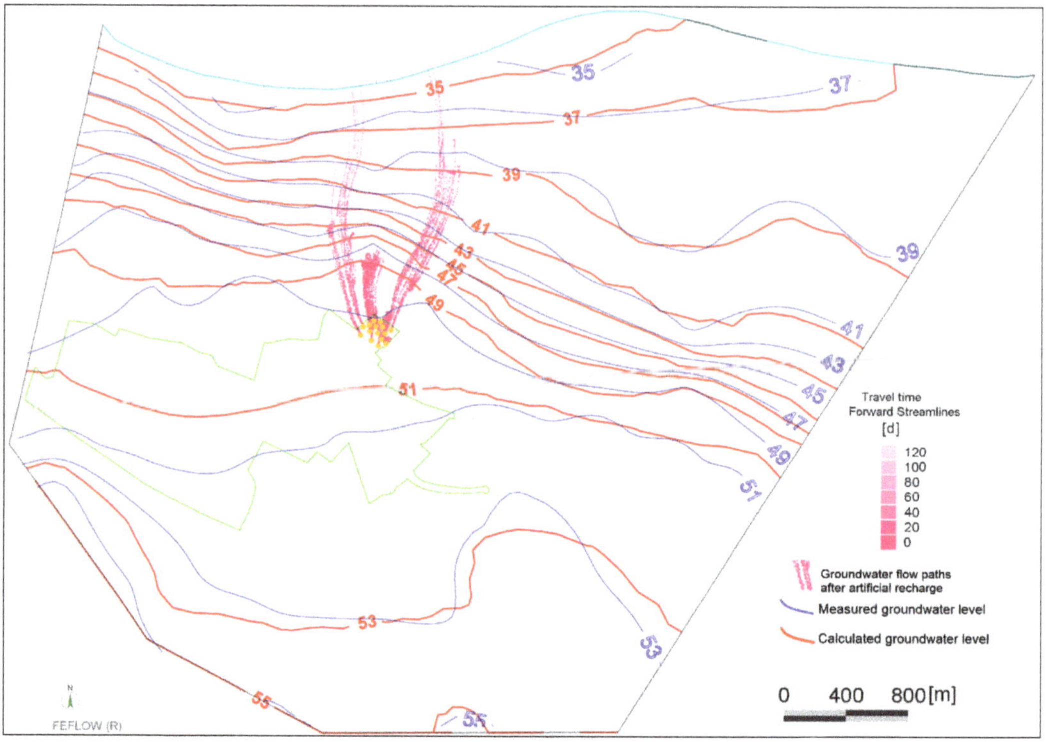

Figure 19. Groundwater flow patterns towards the Garonne River after artificial groundwater recharge applied in the TAG area, based on [19].

6. Conclusions

Despite the very complex and partially known geological and hydrogeological features of the alluvial aquifer in the Garonne Valley, coupling of hydrochemical characteristics and hydrodynamic modeling supports a better understanding of the mechanisms controlling the water quality and the aquifer dynamic as well as the relationship between this aquifer and the Garonne River. The dynamic exchange between the river and the alluvial aquifer can be considered a major long-term management tool, considering the effects of climate change.

Groundwater level changes show that the variation is greater in the low plain in the northern part near the Garonne River, where the alluvial aquifer drains into the river during the dry periods and is recharged by the river during flood periods. The water levels measured in two wells located near the river indicate the effect of the river on the groundwater level during the flood period. The major drainage axis of the groundwater is generally oriented from south to north toward the Garonne River, which is the main outlet for the aquifer. The piezometric curves show no hydraulic connection between the aquifer and the watercourses in the study area except for the Garonne River.

The nitrate concentrations measured in the alluvial aquifer have confirmed the contamination of this aquifer by nitrates. This contamination is related directly to the strong anthropic pressure, particularly the use of nitrogenous fertilizers in agricultural activities, considering the excess irrigation water and the high intrinsic vulnerability of the alluvial aquifer.

According to the saturation indices values calculated with PHREEQC, the groundwater has negative saturation indices (SI < 0) concerning carbonate minerals (calcite $CaCO_3$ and dolomite $CaMg(CO_3)_2$), as well as evaporitic minerals (gypsum $CaSO_4.2H_2O$, anhydrite $CaSO_4$, and halite NaCl), indicating that the system is either depleted of such minerals, or the duration of the water–rock interaction (the residence time) is relatively short. However, the dissolution of carbonate rocks is not the major process controlling groundwater chemistry in the study area, and these minerals will continue to dissolve in the groundwater if they are present in the system. In contrast, the quartz saturation index shows a positive value for most of the sample, indicating that the dissolution of aluminosilicate minerals significantly affects the alluvial aquifer's water quality. Hence, silicate hydrolysis is the main hydrochemical process controlling the chemistry of the alluvial aquifer in the study area.

The groundwater samples fall within the Ca-HCO_3 and Ca-mixed types, and just two samples have shown Ca-NO_3 and Na-Mix types according to Stuyfzand classification. Silicate weathering and ion exchange are the main processes controlling groundwater chemistry in the study area.

The groundwater level changes measured in two boreholes near the retention basin show groundwater levels increasing between 0.5 and 1 m after a storm and runoff water infiltration. The modeling results show that the infiltrated water would take about four months to reach the Garonne River, which is an appropriate time to sustain its low flow if the recharge occurs in spring. This result shows that the alluvial aquifer in the study area is potentially interesting for the groundwater artificial recharge to maintain the groundwater level and sustain the low flow of the Garonne River. This strategy can contribute to ensuring the proper function of the river environment and creating humid zones in the alluvial Garonne Valley. This pragmatic and potentially eco-responsible management tool can be considered in the context of sustainable development as an adaptation strategy to climate change in the region. Excess water generated by the flood or snowmelt in the Garonne basin, which is characterized by good quality, could be used for artificial recharge of the shallow alluvial aquifer, well connected with the river, once suitable sites have been identified.

Future work should focus on obtaining more data about the aquifer's hydraulic conductivity, flood event, groundwater temperature measurement, groundwater extraction for agricultural use, and other water consumption in the area. This information can be useful for groundwater hydrodynamic and thermal modeling and to calculate the study area's water budget, as well as to show the effect of artificial recharge on the river discharge. More groundwater sampling before and after the recharge must be considered to specify the qualitative effect. Thus, this work will involve measuring the thermal plume induced by recharged water within the aquifer. These measurements will also be modelled in a coupled hydrodynamic and thermal model in order to simulate the aquifer's buffering effect on temperature. The coupled model will also serve as a prospective tool for quantifying the effect of recharge under warmer atmospheric conditions.

Author Contributions: N.A. performed field activities and wrote the original draft of the paper. N.A., A.D., P.M. and M.F. designed the methodological framework. N.A. and A.D. performed the simulation and analyzed the results. All co-authors collaborated on the redaction, review, and editing of the manuscript. All authors have read and agreed to the published version of the manuscript.

Funding: This research has been supported by the Nouvelle Aquitaine Region (File no 2017-1R20104) and by SMEAG, Syndicat mixte d'études et d'aménagement de la Garonne (File no CT_2019-068).

Institutional Review Board Statement: The study was conducted in accordance with the Declaration of Helsinki, and approved by the Institutional Review Board.

Informed Consent Statement: Not applicable.

Data Availability Statement: Data is unavailable due to privacy restrictions.

Conflicts of Interest: The authors declare no conflict of interest.

References

1. Le Treut, H. *Les Impacts du Changement Climatique en Aquitaine. Nouvelle Édition*; Presses Universitaires de Bordeaux: Pessac, France, 2013; 368p, ISBN 9791030003741.
2. EEA (European Environment Agency). *Water Resources across Europe—Confronting Water Scarcity and Drought*; European Environment Agency: Luxembourg, 2009.
3. Caballero, Y.; Voirin-Morel, S.; Habets, F.; Noilhan, J.; LeMoigne, P.; Lehenaff, A.; Boone, A. Hydrological sensitivity of the Adour-Garonne river basin to climate change. *Water Resour. Res.* **2007**, *43*, 19P. [CrossRef]
4. Hendrickx, F.; Sauquet, E. Impact of warming climate on water management for the Ariège River basin (France). *Hydrol. Sci. J.* **2013**, *58*, 976–993. [CrossRef]
5. Agence de l'eau Adour-Garonne, Eau et Changements Climatiques en Adour-Garonne, Les Enjeux Pour la Ressource, les Usages et les Milieux, 31078 Toulouse CEDEX 4, France. 2014. Available online: http://oai.eau-adour-garonne.fr/oai-documents/60721/GED_00000000.pdf (accessed on 21 February 2022).
6. Guivarch, M.; Dany, A.; Toulzac, N.; Guiheux, A.; Le Gall, G. *État Initial du SAGE Vallée de la Garonne*; Schéma d'Aménagement et de Gestion des Eaux, Vallée de la Garonne: Montigny le Bretonneux, France, 2015; 430p, Available online: https://www.gesteau.fr/sites/default/files/2-sage_garonne_etat_initial_fev15.pdf (accessed on 22 February 2022).
7. Rousset, A.; Le Treut, H. *Anticiper les Changements Climatiques en Nouvelle-Aquitaine*; Rapport AcclimaTerra; Editions Région Nouvelle-Aquitaine: Bordeaux, France, 2018; Available online: http://www.acclimaterra.fr/uploads/2018/05/Rapport-AcclimaTerra.pdf (accessed on 22 February 2022).
8. Valette, P. Changements Environnementaux et Adaptation des Sociétés Dans la Moyenne Vallée de la Garonne (XVIIIe siècle à aujourd'hui). Sud-Ouest Européen, Presses Universitaires du Mirail-CNRS, Aléas, Vulnérabilités et Adaptations des Sociétés du Passé. 2011, Volume 32, pp. 35–51. Available online: http://journals.openedition.org/soe/668 (accessed on 22 February 2022).
9. Winter, T.C.; Harvey, J.W.; Franke, O.L.; Alley, W.M. *Ground Water and Surface Water—A Single Resource*; U.S. Geological Survey Circular 1139; U.S. Geological Survey: Reston, VA, USA, 1998; 79p. Available online: https://pubs.usgs.gov/circ/1998/1139/report.pdf (accessed on 21 February 2022).
10. Ivkovic, K.M. A top-down approach to characterise aquifer-river interaction processes. *J. Hydrol.* **2009**, *365*, 145–155. [CrossRef]
11. Gandolfi, J.M.; Danneville, L.; Petit, V.; Tilloloy, F. *Connaissance, Évaluation et Protection des Aquifères Alluviaux de Tarn-et-Garonne*; 82, BRGM Report/R-39543; Bureau de Recherches Géologiques et Minières: Toulouse, France, 1997.
12. Le Cointe, P.; Nuttinck, V.; Rinaudo, J.D. A Tool to Determine Annual Ground-Water Allocations in the Tarn-et-Garonne Alluvial Aquifer (France). In *Sustainable Groundwater Management*; Rinaudo, J.D., Holley, C., Barnett, S., Montginoul, M., Eds.; Global Issues in Water Policy; Springer: Cham, Switzerland, 2020; Volume 24, pp. 253–274.
13. Steiger, J.; James, M.; Gazelle, F. Channelization and consequences on floodplain system functioning on the Garonne River, UK. *Regul. Rivers Res. Mgmt.* **1998**, *14*, 13–23. [CrossRef]
14. Lambert, R.; Lami, J.M.; Senges, F. *Etude Hydraulique de la Garonne. La Sécheresse de 1989 dans le Bassin de la Garonne (à L'amont du Mas d' A gênais)*; Université de Toulouse-le-Mirail, Institute de Géographie, et du Ministère de l'Environnement, Délégation de bassin Adour-Garonne: Toulouse, France, 1990; Volume 2, 75p.
15. Poff, N.L.; Allan, J.D.; Palmer, M.A.; Hart, D.D.; Richter, B.D.; Arthington, A.H.; Rogers, K.H.; Meyers, J.L.; Stanford, J.A. River flows and water wars: Emerging science for environmental decision making. *Front. Ecol. Environ.* **2003**, *1*, 298–306. [CrossRef]
16. Sophocleous, M. The science and practice of environmental flows and the role of hydrogeologists. *Ground Water* **2007**, *45*, 393–401. [CrossRef]
17. Pörtner, H.-O.; Roberts, D.C.; Tignor, M.; Poloczanska, E.S.; Mintenbeck, K.; Alegría, A.; Craig, M.; Langsdorf, S.; Löschke, S.; Möller, V.; et al. (Eds.) *Climate Change 2022: Impacts, Adaptation, and Vulnerability. Contribution of Working Group II to the Sixth Assessment Report of the Intergovernmental Panel on Climate Change*; Cambridge University Press: Cambridge, UK, 2022.
18. Weng, P.; Sánchez-Pérez, J.; Sauvage, S.; Vervier, P.; Giraud, F. Assessment of the quantitative and qualitative buffer function of an alluvial wetland: Hydrological modelling of a large floodplain (Garonne River, France). *Hydrol. Process.* **2003**, *17*, 2375–2392. [CrossRef]
19. Asmael, N.; Dupuy, A.; Leroy, B. Artificial Recharge of the Shallow Alluvial Aquifer as an Adaptation Strategy in the Garonne Valley, France. *IAHS* **2023**, *100*, 1–6.
20. BRGM Infoterre. Available online: https://infoterre.brgm.fr (accessed on 22 February 2022).
21. Prodéo, F.; Hanry, A.; Sainte-Colombe-en-Bruilhois—Faurat, Montplaisir, notice archéologique, ADLFI. Archéologie de la France. Available online: http://journals.openedition.org/adlfi/100204 (accessed on 17 August 2023).
22. Danneville, L. Eau-Aménagement-Environnement Contribution des Eaux Souterraines Aux Débits et à la Qualité des Eaux de Surface: Exemple de la Garonne, de ses Sources à la Confluence du Tarn. Documents du BRGM 262 Editions BRGM. Ingénierie de l'environnement. Université Paris XI—Orsay, France. PhD Report. 1995. Available online: https://theses.hal.science/tel-01420437/file/2012_These_Daneville_Garonne.pdf (accessed on 22 February 2022).
23. Lancaster, R.R. Fluvial Evolution of the Garonne River, France: Integrating Field Data with Numerical Simulations. Master's Thesis, Louisiana State University, Baton Rouge, LA, USA, 2005.
24. Probst, J.L. Hydrologie du Bassin de la Garonne: Modèles de Mélange, Bilan de l'Erosion, Exportation des Nitrates et des Phosphates. Ph.D. Thesis, University Toulouse, Toulouse, France, 1983.

25. Grusson, Y.; Anctil, F.; Sauvage, S.; Sánchez Pérez, J.M. Coevolution of Hydrological Cycle Components under Climate Change: The Case of the Garonne River in France. *Water* **2018**, *10*, 1870. [CrossRef]
26. Flipo, N.; Mouhri, A.; Labarthe, B.; Biancamaria, S.; Rivière, A.; Weill, P. Continental hydrosystem modelling: The concept of nested stream—Aquifer interfaces. *Hydrol. Earth Syst. Sci.* **2014**, *18*, 3121–3149. [CrossRef]
27. Weill, P.; Labarthe, B.; Jost, A.; Amestoy, J.; Cojan, I.; Deleplancque, B.; Flipo, N. *Complexité Hydrosédimentaire de la Plaine Alluviale de la Bassée. [Rapport de Recherche] Programme PIREN-Seine, Rapport de Synthèse 2011–2015, Vol. 2 "Interfaces Nappe-Rivière"*; MINES ParisTech: Paris, France, 2016; pp. 49–95.
28. Queensland Government. *Groundwater Dependent Ecosystem Pictorial Conceptual Model 'Alluvia': Version 1.5*; Queensland Government: Brisbane, Australia, 2017. Available online: https://wetlandinfo.des.qld.gov.au/resources/static/pdf/ecology/gde/170303-gde-pcm-alluvia.pdf (accessed on 22 February 2022).
29. Parkhurst, D.L.; Appelo, C.A.J. PHREEQC for windows A Computer Program for Speciation, Batch-Reaction, One-Dimensional Transport, and Inverse Geochemical Calculations. *U.S. Geol. Surv. Tech. Methods* **2013**, *6*, 497.
30. Diersch, H.J.G. *FEFLOW. Finite Element Simulation System for Subsurface Flow and Transport Processes*; Springer: Berlin/Heidelberg, Germany, 2018.
31. Wen, X.H.; Wu, Y.Q.; Wu, J. Hydrochemical characteristics of groundwater in the Zhangye Basin, Northwestern China. *Environ. Geol.* **2008**, *55*, 1713–1724. [CrossRef]
32. Zhang, B.; Zhao, D.; Zhou, P.; Qu, S.; Liao, F.; Wang, G. Hydrochemical Characteristics of Groundwater and Dominant Water–Rock Interactions in the Delingha Area, Qaidam Basin, Northwest China. *Water* **2020**, *12*, 836. [CrossRef]
33. Deutsch, W.J. *Groundwater Geochemistry: Fundamentals and Application to Contamination*; CRC Press: Boca Raton, FL, USA, 1997.
34. Si, J.; Feng, Q.; Wen, X.; Su, Y.; Xi, H.; Chang, Z. Major ion chemistry of groundwater in the extreme arid region northwest China. *Environ. Geol.* **2009**, *57*, 1079–1087.
35. Stuyfzand, P.J. A new Hydrochemical Classification of Water Type. In *Regional Characterization of Water Quality*; IAHS Publication: Oxford, UK, 1989; Volume 182, pp. 89–98.
36. Piper, A.M. A graphic procedure in the geochemical interpretation of water analyses. *Trans. Am. Geophys. Union* **1944**, *25*, 914–928. [CrossRef]
37. Coetsiers, M.; Walraevens, K. Chemical characterization of the Neogene Aquifer, Belgium. *Hydrogeol. J.* **2006**, *14*, 1556–1568. [CrossRef]
38. Nakagawa, K.; Amano, H.; Asakura, H.; Berndtsson, R. Spatial trends of nitrate pollution and groundwater chemistry in Shimabara, Nagasaki, Japan. *Environ. Earth Sci.* **2016**, *75*, 234. [CrossRef]
39. Vinnarasi, F.; Srinivasamoorthy, K.; Saravanan, K.; Gopinath, S.; Prakash, R.; Ponnumani, G.; Babu, C. Chemical weathering and atmospheric carbon dioxide (CO_2) consumption in Shanmuganadhi, South India: Evidences from groundwater geochemistry. *Environ. Geochem. Health* **2020**, *43*, 771–790. [CrossRef]
40. Srinivasamoorthy, K.; Vasanthavigar, M.; Chidambaram, S.; Anandhan, P.; Manivannan, R.; Rajivgandhi, R. Hydrochemistry of groundwater from Sarabanga minor basin, Tamilnadu, India. *Int. Acad. Ecol. Environ. Sci.* **2012**, *2*, 193–203.
41. Drever, J.I. *The Geochemistry of Natural Waters: Surface Water and Groundwater Environments*, 3rd ed.; Prentice-Hall: Upper Saddle River, NJ, USA, 1998; 436p.
42. Brielmann, H. Recharge and Discharge Mechanism and Dynamics in the Mountainous Northern Upper Jordan River Catchment, Israel. Ph.D. Thesis, Ludwig-Maximilians-University, Munich, Germany, 2008.
43. Huang, S.b.; Han, Z.t.; Zhao, L.; Kong, X.K. Risk assessment and prediction of heavy metal pollution in groundwater and river sediment: A case study of a typical agricultural irrigation area in Northeast China. *Int. J. Anal. Chem.* **2015**, *2015*, 921539. [CrossRef]
44. Drever, J.I. *The Geochemistry of Natural Waters*; Prentice-Hall: Englewood Cliffs, NJ, USA, 1988; Volume 437.
45. Güler, C.; Thyne, G.D. Hydrologic and geologic factors controlling surface and groundwater chemistry in Indian Wells-Owens Valley area, southeastern California, USA. *J. Hydrol.* **2004**, *285*, 177–198. [CrossRef]
46. Giggenbach, W.F. Geothermal Solute Equilibria. Derivation of Na-k-Mg-Ca Geoindicators. *Geochim. Cosmochim. Acta* **1988**, *52*, 2749–2765. [CrossRef]
47. Xing, L.; Guo, H.; Zhan, Y. Groundwater hydrochemical characteristics and processes along flow paths in the North China Plain. *J. Asian Earth Sci.* **2013**, *70–71*, 250–264. [CrossRef]
48. Datta, P.S.; Tyagi, S.K. Major Ion Chemistry of Groundwater in Delhi Area: Chemical Weathering Processes and Groundwater Flow Regime. *J. Geol. Soc. India* **1996**, *47*, 179–188.
49. Mayback, M. Global chemical weathering of surficial rocks estimated from river dissolved loads. *Am. J. Sci.* **1987**, *287*, 401–428. [CrossRef]
50. Rajmohan, N.; Elango, L. Identification and Evolution of Hydrogeochemical Processes in the Groundwater Environment in an Area of the Palar and Cheyyar River Basins, Southern India. *Environ. Geol.* **2004**, *46*, 47–61. [CrossRef]
51. Elango, L. *Numerical Simulation of Groundwater Flow and Solute Transport*, 1st ed.; Allied Publishers Pvt. Ltd.: New Delhi, India, 2005.

52. Asmael, N.M.; Dupuy, A.; Huneau, F.; Hamid, S.; Coustumer, P.L. Groundwater Modeling as an Alternative Approach to Limited Data in the Northeastern Part of Mt. Hermon (Syria), to Develop a Preliminary Water Budget. *Water* **2015**, *7*, 3978–3996. [CrossRef]
53. Krige, D.G. A statistical approach to some basic mine valuation problems on the Witwatersrand. *J. Chem. Metall. Min. Soc. S. Afr.* **1951**, *52*, 119–139.

water

MDPI

Article

Coupling Simulation and Prediction of Sustainable Utilization of Water Resources in an Arid Inland River Basin under Climate Change

Xiaofan Qi [1], Wenpeng Li [1,*], Yuejun Zheng [1], Huqun Cui [2], Weidong Kang [3], Zhenying Liu [4] and Xinmin Shao [5]

1 China Institute of Geo-Environment Monitoring, Beijing 100081, China
2 Center for Hydrogeology and Environmental Geology, CGS, Baoding 071051, China
3 Department of Geology, Northwest University, Xi'an 710069, China
4 School of Water Resources and Environment, Hebei GEO University, Shijiazhuang 050031, China
5 Zhejiang Monitoring Center of Geology Environment, Hangzhou 310007, China
* Correspondence: liwenpeng@mail.cgs.gov.cn

Abstract: The arid endorheic basin of northwest China is characterized by rich land resources, water shortage, and a fragile ecological environment. The establishment of a credible coupling model of groundwater and surface water based on multi-source observation data is an effective means to study the change in basin water cycles and the sustainable utilization of water resources in the past and future. Based on the latest understanding of hydrogeological conditions, hydrology and water resource utilization data in the middle reaches of the Heihe River Basin (HRB), this paper constructs an up-to-date coupling model of surface water and groundwater to study the water balance change of the basin. The water resources data series under historical replay and CMIP5 climate model prediction are constructed to predict future changes in water resources. The study shows that, under the joint influence of natural conditions and human activities, the average annual recharge of groundwater in the study area from 1990 to 2020 is 17.98×10^8 m^3/a, the average annual discharge is 18.62×10^8 m^3/a, and the difference between recharge and discharge is -0.64×10^8 m^3/a. The total groundwater storage is -19.99×10^8 m^3, of which the groundwater storage from 1990 to 2001 was -17.52×10^8 m^3 and from 2002 to 2020 was -2.47×10^8 m^3. Abundant water from 2002 to 2020 in the basin significantly improved the loss of groundwater storage. Under the prediction of historical reappearance and the CMIP5 CNRM-CM5 model RCP4.5 and RCP8.5 pathways, the groundwater level of the Heihe River–Liyuanhe River inclined plain falls first because the HRB has just experienced a wet season and then rises according to future climate change. The groundwater level of the inclined plain east of the Heihe River and Yanchi basin decreases continuously because of the change in water cycle caused by human activities. The erosion accumulation plain is located in the groundwater discharge zone, and the water level is basically stable. Under the conditions of water resource development and utilization, the runoff of Zhengyixia hydrological station cannot meet the requirements of the "97 Water Dividing Plan" of the State Council in most years in the future, and the ecological and production water in the lower reaches of HRB cannot be effectively guaranteed. With the implementation of water-saving irrigation under the RCP4.5 and RCP8.5 scenarios, the runoff of Zhengyixia can meet the "97 Water Diversion Plan". It is suggested to further improve the level of agricultural water savings in the middle reaches of the HRB and control the reasonable scale of cultivated land in order to reduce water consumption in the middle reaches of the HRB and implement sustainable utilization of water resources in the HRB.

Keywords: sustainable utilization of water resources; coupled numerical simulation of surface water and groundwater; water cycle and water balance; climate change; Heihe River Basin

Citation: Qi, X.; Li, W.; Zheng, Y.; Cui, H.; Kang, W.; Liu, Z.; Shao, X. Coupling Simulation and Prediction of Sustainable Utilization of Water Resources in an Arid Inland River Basin under Climate Change. *Water* **2023**, *15*, 3232. https://doi.org/10.3390/w15183232

Academic Editor: Yuanzheng Zhai

Received: 24 May 2023
Revised: 15 August 2023
Accepted: 15 August 2023
Published: 11 September 2023

1. Introduction

The arid region of northwest China is short on water resources and has a fragile ecological environment [1]. As an active natural resource, water resources play an important role in supporting the economic, social, and ecological civilization construction of arid areas. Water resources in arid areas have their own unique formation and evolution modes, mainly forming in mountainous areas, dissipating in plains, and existing in various forms such as precipitation, glacial meltwater, surface water, groundwater, etc. The groundwater and surface water in the basin show a huge amount of mutual transformation [2–4]. It is of great significance to scientifically understand the water cycle mechanism of the basin, make full use of the function of large-thickness aquifer regulation and storage, and realize the joint storage and utilization of groundwater and surface water for the sustainable utilization of water resources in arid areas [1–5].

The Heihe River Basin (HRB) is the second-largest arid inland river basin in China, and its water cycle and water resource characteristics have always been the research hotspots of domestic and foreign scholars [6,7]. Numerical simulation is an effective means to carry out relevant research, and groundwater models, including groundwater models, soil water and groundwater coupling models, and surface water and groundwater coupling models in different regions of the HRB, have been established. Based on the groundwater model, the relationship between base flow and recharge in mountainous areas in the upper reaches of the Heihe River [8], groundwater equilibrium and water cycle evolution under water resource management scenarios such as the history and future of agricultural irrigation and water conservation in the middle reaches of Zhangye Basin [9,10], and the eco-hydrological effect of Ejina Banner under environmental flow management of the Heihe River downstream were studied [11]. Based on the SWAT model and intelligent optimization, the optimal allocation of agricultural water and soil resources in the middle reaches of the HRB and Zhangye areas was studied [12]. The coupling model of soil water and groundwater was used to study the water cycle process in Zhangye area under historical and future water-saving scenarios [13], and the soil water balance model and numerical model of farmland gas wrapping zone based on empirical assumptions were used to study the soil water balance in Zhangye agricultural irrigation area [14]. The eco-hydrological model is used to evaluate the water cycle flux in the middle and lower basins of the HRB [15], and the role of groundwater in the eco-hydrological system of the middle and lower basins [16]. Based on the coupling model of surface water and groundwater, the water use contradiction and water cycle flux of agriculture and ecosystems in the middle reaches of the Heihe River Basin and Yanchi Basin were studied [17], and the relationship between surface water and groundwater transformation was studied [18]. The coupling model of eco-hydrology and socio-economy at the basin scale was used to study multiple eco-hydrological processes in the basin and deepen the understanding of the coupling relationship between water, soil, atmosphere, vegetation, and humans in the arid basin [19].

The accuracy of water recycling resources is key to the effectiveness of numerical models. In the middle reaches of the HRB, the water cycle mainly includes river runoff, water diversion, and extraction, among which river runoff can be used to calculate river seepage and groundwater discharge [3,4]. The amount of river leakage depends on the amount of water coming from the river, the length of the riverbed, the leakage capacity, etc. Through the preliminary geological survey work, empirical functions such as the power function regression equation of the leakage rate per unit length of the Heihe River channel and the power function of the incoming water have been established, and the curves of the cumulative discharge of groundwater and the discharge per unit length of the midstream basin have also been obtained [3,4,20–22]. Water diversion and extraction quantities can be used to calculate irrigation channel leakage and field infiltration quantities. The preliminary work summarized the relationship between irrigation channel leakage and the lithology of the unsaturated zone, canal lining, groundwater depth, and the relationship between field infiltration and stratum lithology, irrigation water quantity, and burial depth of water level.

Climate change is a global issue of universal concern in today's society. In the past century, major changes have taken place in the global climate and environment, whose main feature is global warming [23], which has had an important impact on the global water cycle [24]. Due to the fragile ecological environment in northwest China, climate change prediction is extremely important for maintaining water security, ecological security, and food security in the region. Global climate models (GCMs) are currently an effective tool for predicting climate change [25]. GCMs can better simulate the characteristics of large-scale meteorological elements, but they are difficult to directly apply to meteorological and water resource prediction at the basin scale because of their low spatial resolution [26]. The downscaling method can transform large-scale, low-resolution global climate models into small-scale, high-resolution regional climate information [25], which has the advantages of a simple method, a small calculation amount, and no need to consider boundary conditions. It is widely used in China and abroad [27].

In the past 33 years, the HRB has experienced the most fluctuating hydrological change process in the dry period and the most abundant period since observation data was obtained. The cultivated land area has continued to expand, the water-saving level of agricultural irrigation has increased, the water consumption has changed significantly, and the water cycle has new characteristics. In order to deepen the understanding of the mechanism of the long hydrological cycle and future changes in water resources in the basin of the middle reaches of the HRB under the conditions of climate change and human activities, based on the latest 1:50,000 hydrogeological survey, National Natural Science Foundation projects, groundwater monitoring, and other multi-source data, this paper further analyzes the characteristics of the river–aquifer system in the middle reaches of the HRB. Based on the new observation data at the turning point of river leakage and groundwater discharge to the river, the law of river leakage in the middle reaches of the Heihe River is recognized, and a coupled numerical model of surface water and groundwater on a long time scale is constructed based on multi-source data. The process and trend of the water cycle in the study area are studied, and the reasons are analyzed. The statistical downscaling model is used to predict the future precipitation in both the upper and middle reaches of the HRB, and then the mountain runoff is estimated. Three future climate change scenarios are constructed, and the future changes in water resources in the study area are predicted by numerical models. The dynamic changes in groundwater levels, groundwater storage, and Zhengyixia station runoff are analyzed. The rationality of the development and utilization of water resources in the basin is judged, and the problems that should be paid attention to in the sustainable utilization of water resources are put forward so as to provide reference for the sustainable utilization of water resources in the arid inland river basins.

2. Study Area

Located in the middle section of the Qilian Mountains and Hexi Corridor, The Heihe River Basin is the second-largest basin of the most arid inland river basin in China, bounded by the Dahuangshan Mountain in Shandan County of Zhangye City in the east and the Black Mountain in Jiayuguan City in the west, bordering the Shiyang River Basin and the Shule River Basin. It covers the Qilian Mountain Watershed in the south to the terminal Juyanhai Lake in the north, with a basin area of 22.25×10^4 km^2 (Figure 1).

From 1944 to 2022, the total runoff of the Heihe River Basin was about 37.87×10^8 m^3/a, of which the main stream of the Heihe River was 16.86×108 m^3/a and the Liyuanhe River was 2.48×10^8 m^3/a. The water in the basin is mainly used for agricultural irrigation, production, life, and natural oases. Water resources are mainly formed in the Qilian Mountains in the Southern part of the basin and dissipated in the Southern and Northern basins. Water resources generally flow from south to north, eventually converging in the East Yanhai Sea and the West Yanhai Sea in the north. The main ecological and geological environmental problems in the Heihe River Basin are ecological degradation and land desertification [2].

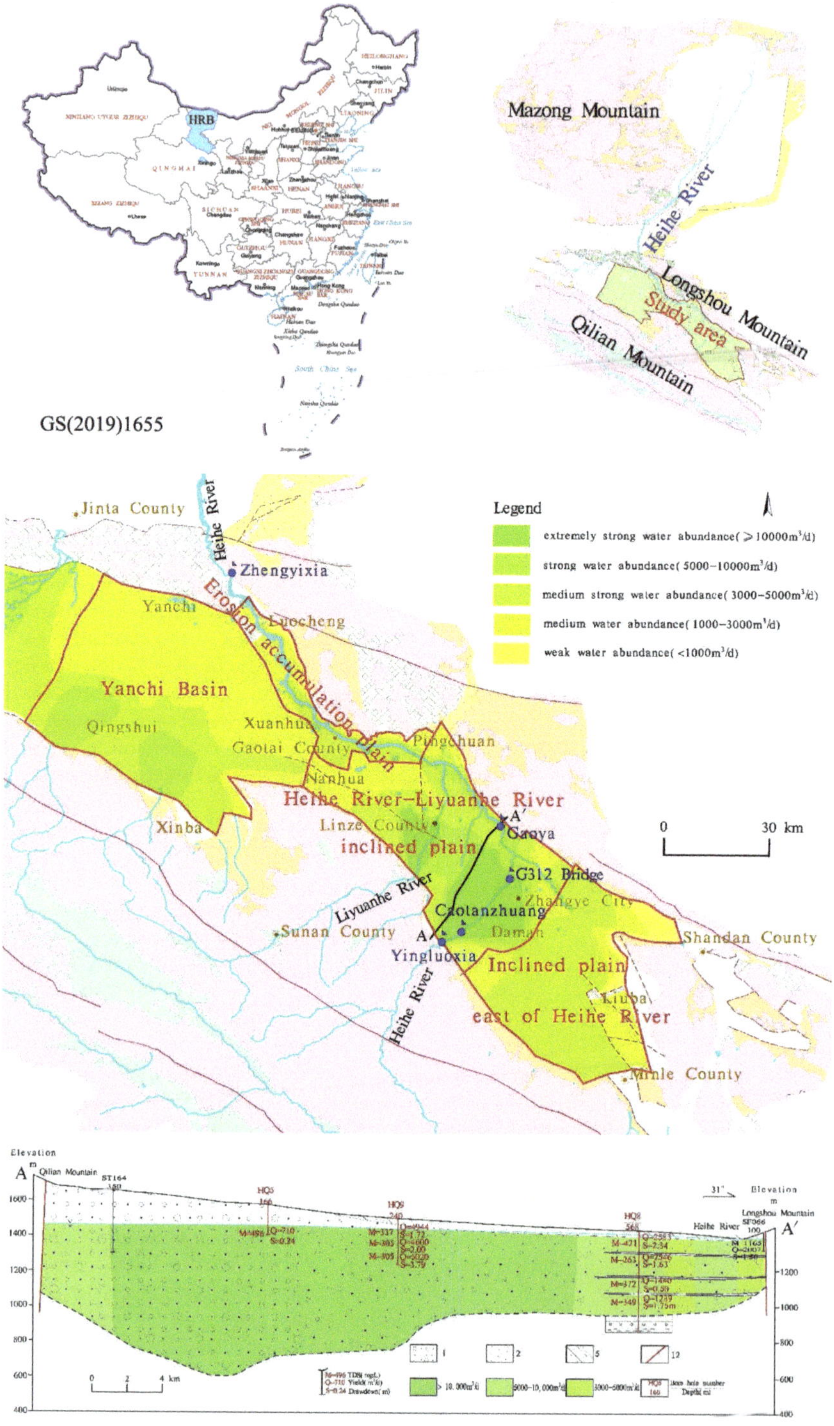

Figure 1. Sketch map of the study area.

The study area is located in the middle reaches of the Heihe River Basin, including Yanchi Basin and Zhangye Basin. Zhangye Basin consists of the Heihe River–Liyuanhe River inclined plain, the inclined plain east of the Heihe River, the erosion accumulation plain in the lower middle reaches of the HRB, and Yanchi Basin, with a study area of 0.93×10^4 km^2 (Figure 1).

3. Materials and Methods

3.1. Data Source

Over the years, the natural resources department, the water resources department, the meteorological department, and the Chinese Academy of Sciences have carried out a lot of work in the Heihe River Basin. Among them, the representative research projects in recent years include the 2014–2018 "Hydrogeological Survey of Key Areas in the Heihe River Basin" and "1:50,000 Hydrogeological Survey of the Heihe River Basin in the Hexi Corridor" by the China Geological Survey, the 2010–2017 Major Research Plan "Integrated Study of Ecological-Hydrological Processes in the Heihe River Basin" funded by the National Natural Science Foundation and the National Groundwater Monitoring Project launched in 2015. Based on the above-mentioned work, a groundwater, water conservancy, and remote sensing monitoring network for the Heihe River Basin has been established, constituting rich groundwater, surface water, water resource development and utilization, land use, and evapotranspiration data. On the basis of various types of hydrogeological drilling and geophysical prospecting results, the relationship between the mountains and the basin, as well as the hydrogeological conditions of the basin, are deeply investigated. A multi-layer, layered pumping system a layered monitoring scientific test base, and a test site for hydrological parameters such as infiltration coefficient are built to update the key hydrological parameters of the basin [28,29]. The above resources provide abundant basic data for this study.

The monthly hydrological data from 1990 to 2022 on the runoff of the Yingluoxia hydrological station, the Zhengyixia station of the Heihe River, and the Yinggezui station of the Liyuanhe River comes from Zhangye Hydrological Station (Zhangye, China). The monthly data from 1990 to 2020 on the development and utilization of water resources, such as surface water diversion and groundwater extraction quantity in each irrigation area, are from the Zhangye Water Bureau (Zhangye, China). Groundwater level dynamics and groundwater flow fields are from the China Institute of Geo-Environment Monitoring (Beijing, China).

The water resources forecast period is 2023–2055, including three future climate change scenarios. The first one is a historical reappearance scenario, that is, natural conditions such as mountain runoff and rainfall use data from 1990 to 2022 to reproduce the dry-to-flood season process in the HRB. The other two are climate prediction scenarios, namely the CNRM-CM5 model RCP4.5 and RCP8.5 pathways scenarios of CMIP5. The development and utilization of water resources, such as water diversion and groundwater exploitation in irrigation areas, are all reappearances of current data in 2020.

The climate prediction data includes three kinds of data: meteorological station measurement data (China Meteorological Data Network, Beijing, China), global atmospheric NCEP reanalysis data (NOAA, Washington, DC, USA) and CMIP5 climate model prediction data (Lawrence Livermore National Laboratory, Livermore, CA, USA).

3.1.1. Heihe Mountain Runoff

The observation section of the runoff from the Heihe River is the Yingluoxia hydrological station (Figure 2), and the multi-year runoff dynamics of this station show the characteristics of alternating dry-flat-dry-flat-dry-abundant (Figure 2). The period from 1990 to 2022 is a dry period up to 33 years, of which 1990–2001 is a dry period and 2002–2022 is a flood period, which is very typical. The multi-year dynamic changes in the difference between the runoff of Yingluoxia and the runoff of Zhengyixia in the Heihe River roughly reflect the change in water consumption scale in the middle basin, dominated by agricultural irrigation.

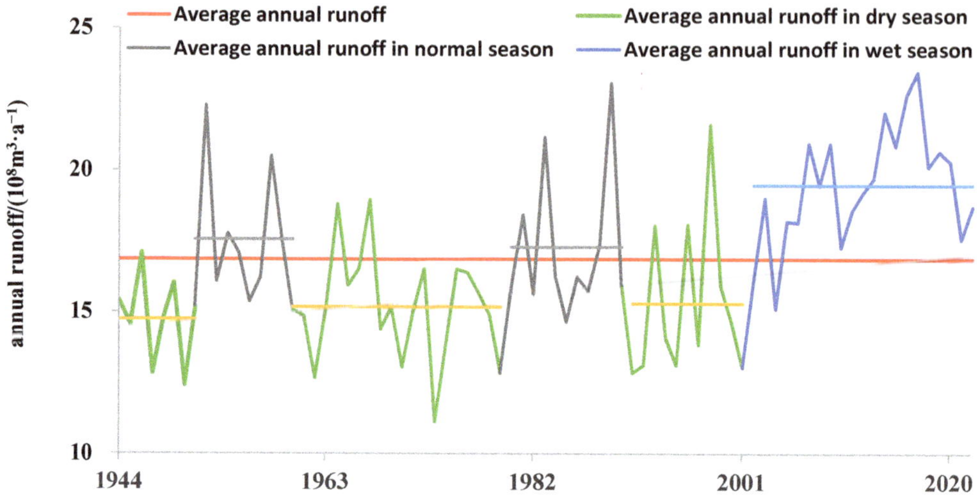

Figure 2. Annual runoff of Yingluoxia.

3.1.2. Heihe River Leakage

In the middle basin of the Heihe River Basin, there is a huge amount of groundwater and surface water conversion, of which the leakage of the Heihe River and the discharge of groundwater through the river channel are the main manifestations. The infiltration section and discharge section of the main stream of the Heihe River in the Zhangye Basin are roughly bounded by the G312 bridge, also known as the turning point of groundwater and surface water conversion, which is located about 1 km upstream of the G312 bridge. The Yingluoxia–G312 bridge section is the seepage section of the hanging river, including the Yingluoxia–Caotanzhuang section and the Caotanzhuang–G312 bridge section, and the river infiltration and replenishment are mainly controlled by the actual water flow into the river. The G312 bridge–Zhengyixia section is a groundwater discharge section. Based on such observational data as hydrological station runoff, irrigation channel diversion, G312 section runoff, and evaporation in the Heihe River–Zhangye Basin, the water leakage law of the Heihe River is analyzed by the river water balance method in order to obtain the river water leakage amount over the years [30].

In 1990–2020, the runoff from Yingluoxia was 17.96×10^8 m^3/a. Under the current conditions of water resource development and utilization, the annual average river water leakage of the Yingluoxia–G312 Heihe bridge section was 7.11×10^8 m^3/a, accounting for 39.59% of the total runoff of Yingluoxia. The annual average infiltration of Yingluoxia–Caotanzhuang was 2.48×10^8 m^3/a, and the annual average infiltration of Caotanzhuang–G312 bridge was 4.63×10^8 m^3/a. The annual average water intake was 6.68×10^8 m^3/a, accounting for 37.19% of the runoff from Yingluoxia. The discharge under the transformation section was 4.02×10^8 m^3/a, accounting for 22.38% of the runoff from Yingluoxia. Compared with the natural conditions, the amount of water flowing through the river is reduced under human activity conditions, and the leakage and recharge rate of the river channel increase.

3.1.3. Groundwater Lateral Runoff Discharge to the Plain of the Southern Qilian Mountain

The amount of lateral runoff recharge to plain areas by mountain groundwater in the arid inland basin of northwest China has long been controversial, and this amount has been estimated to be small or large. Combined with the latest geo-hydrogeological survey, geophysical prospecting, drilling, pumping test, groundwater dynamic observation, water chemistry test, basin groundwater level measurement, and comprehensive research, the multi-year average lateral runoff recharge of groundwater in the plain area of Qilian Mountain is estimated to be 0.77×10^8 m^3/a, the recharge of runoff in the shallow mountain zone is 0.40×10^8 m^3/a, the lateral recharge of bedrock in the valley section is 0.07×10^8 m^3/a, and the recharge of the fourth series in the valley section is 0.30×10^8 m^3/a. The total recharge accounts for about 3.0% of the basin's groundwater resources [31].

3.1.4. Infiltration Replenishment Coefficient

Irrigation water and rainfall infiltration recharge are important sources of groundwater recharge in arid endorheic basins. For a long time, infiltration recharge was calculated using empirical parameter methods, lacking mature monitoring methods and empirical data. Based on the artificial bromine tracing method, the infiltration and recharge coefficients of three irrigation conditions and natural precipitation conditions in the middle reaches of the Heihe River were determined. The infiltration recharge coefficients under large-flood irrigation, small-flood irrigation, and drip irrigation were 0.16, 0.07, and 0.11, respectively, and that is 0.1 under natural precipitation conditions [32].

3.1.5. Hydrogeological Structure and Parameters

Among the 314 boreholes in the middle reaches of the HRB, 98 core boreholes were selected, and the total drilling advance was 17,632.27 m, of which the deepest Quaternary borehole was 570.34 m. The hydrogeological structure of the basin was identified based on boreholes and geophysical data [28,33]. The hydrogeological parameters of the basin were updated based on the pumping test data [23,24]. At the top of the alluvial fan of the Qilian Mountain and the Northern Longshou Mountain, etc., the aquifer is a single phreatic aquifer, and the fine soil plain area is a multi-layer aquifer. The main lithology of the basin is pebbles and gravel, with a little sand, sub-sandy, and sub-clay. The groundwater storage space in the midstream basin is huge.

3.1.6. Distribution of Cultivated Land Area

There are a total of 35 irrigated areas in the study area. In the inclined plain north of the Qilian Mountain, the groundwater is buried deeply, and irrigation mainly relies on surface water; the buried depth of groundwater decreases along the north direction, and the amount of groundwater extraction increases, which constitutes the agricultural irrigation water source together with surface water. At the spring discharge area, the proportion of spring irrigation increased, and the Wujiang irrigation area was totally a spring irrigation area.

3.1.7. Climate Prediction Data

The meteorological data comes from the China Meteorological Data Network, and the precipitation data from Qilian, Yeniugou, Shandan, Zhangye, and Gaotai stations is selected. The global atmospheric NCEP reanalysis data are from the global atmospheric NCEP reanalysis jointly released by the American Center for Environmental Forecast (NCEP) and the National Center for Atmospheric Research (NCAR). The alternative predictors of NCEP reanalysis data are land surface temperature (air), surface relative humidity (rhum), surface pressure (pres), sea level pressure (slp), surface wind speed (wspd), surface zonal wind speed (uwnd), surface meridional wind speed (vwnd), as well as temperature, relative humidity, wind speed, zonal wind speed, meridional wind speed, vertical wind speed (omega), and geopotential height (hgt) at 500/850 hPa. On the one hand, these factors affect the longitude and latitude of climate zonality by affecting the meridional and zonal distribution of atmospheric circulation. On the other hand, meteorological circulation factors such as wind speed, sea level pressure, convective temperature field, potential height field and surface temperature affect the diffusion and transport processes of water and heat and then affect the change in precipitation. The CNRM-CM5 model, which has less absolute error in precipitation simulation in the arid region of northwest China, was chosen [34].

3.2. Hydrogeological Conceptual Model

The middle basin of the HRB is a fault basin, and the north and south sides of the model are bounded by the piedmont reverse fault; the Eastern boundary of the model is bounded by faults; and the Western boundary of the model is the Western boundary of the Yanchi Basin. Vertically, the top boundary is the groundwater surface, and the bottom boundary is the Quartenary basal boundary. The model boundaries are all flow boundaries. Horizontally, the model area is split using a 500 × 500 m finite difference grid. Vertically, the model is divided into 20 layers (Figure 3). The hydrogeological parameters, such as hydraulic conductivity, are given initial values first and then adjusted during model identification and validation.

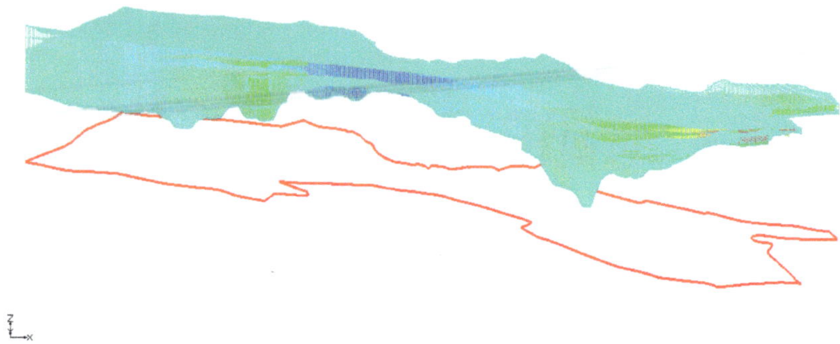

Figure 3. Numerical model space discretization.

The model recharge items included 8 items, namely rainfall infiltration recharge quantity (RCH), river seepage recharge quantity (RIV), irrigation canal seepage recharge quantity (CHA), canal irrigation (field) infiltration recharge quantity (CIR), well irrigation (field) recharge quantity (WIR), undercurrent lateral recharge quantity of the piedmont river valley (LRP), lateral recharge quantity of the piedmont non-river valley (LR), and spring water irrigation recharge quantity (SIR). The discharge items include three items, namely diving evaporation quantity (EVT), groundwater extraction quantity (WEL), and groundwater discharge quantity (GD).

The initial flow field of the numerical model is the flow field in January 1990. The simulation period is from 1990 to 2020, and the stress period is 1 month. The sources and sinks, water resources, and other elements of the water cycle are constant during each stress period.

3.3. Mathematical Model

The numerical model of groundwater flow in this study is a heterogeneous anisotropic transient three-dimensional flow model that is simulated using the MODFLOW-2005 simulation program issued by the United States Geological Survey [35]. The mathematical model of groundwater flow is:

$$\begin{cases} \frac{\partial}{\partial x}\left(K_h\frac{\partial H}{\partial x}\right)+\frac{\partial}{\partial y}\left(K_h\frac{\partial H}{\partial y}\right)+\frac{\partial}{\partial z}\left(K_z\frac{\partial H}{\partial z}\right)-p=S_s\frac{\partial H}{\partial t}, & (x,y,z)\in\Omega, t>0 \\ H(x,y,z,t)|_{t=0}=H_0(x,y,z), & (x,y,z)\in\Omega, t=0 \\ -K_n\frac{\partial H}{\partial n}\Big|_{(x,y,z)\in\Gamma_2}=q(x,y,z), & (x,y,z)\in\Gamma_2, t>0 \\ K_h\left(\frac{\partial H}{\partial x}\right)^2+K_h\left(\frac{\partial H}{\partial y}\right)^2-K_z\frac{\partial H}{\partial z}+w=\mu\frac{\partial H}{\partial t}, & (x,y,z)\in S_1 \\ K_z\frac{\partial H}{\partial z}=0, & (x,y,z)\in S_2 \end{cases} \quad (1)$$

In the equation: Ω—seepage area: H—aquifer head (m); S_1—diving surface; S_2—bulkhead; μ—yield of water (−); S_s—water storage rate (1/m); K_h—horizontal permeability coefficient; K_z—vertical permeability coefficient (m/d); w—diving surface source sink term (m/d); p—groundwater pumping source sink term (1/d); H_0—initial water level of the aquifer (m); Γ_2—second boundary; n—the outer normal direction of the second boundary; and q—specific discharge of the second boundary (m/d).

The mathematical model of river flow in the study area is:

$$\begin{cases} \frac{\partial Q_R(L,t)}{\partial L} = E(L,Q_R) + \sum\limits_j Q_{Cj}(t)\delta(L-L_j) & L>0, t>0 \\ Q_R(L,t)|_{L=0} = Q_{R0}(t) & t>0 \\ Q_R(L,t) \geq 0 & L>0, t>0 \end{cases} \tag{2}$$

In the equation: $Q_R(L,t)$—the flow of the Heihe River, which varies with time and distance; $Q_{R0}(t)$—inbound flow of Heihe River (Yingluoxia); $E(L,Q_R)$—the exchange intensity between Heihe River and aquifer (gain from groundwater is positive, loss to groundwater is negative); L—distance along the Heihe River; Q_{Cj}—inflows of tributaries along the Heihe River (inflow is positive, outflow is negative); L_j—the distance of the tributaries along the Heihe River.

3.4. Statistical Downscaling Model

Firstly, the statistical downscaling model is used to predict the future precipitation of the basin, including the upstream precipitation that determines the mountain runoff, as well as the precipitation in the study area. Secondly, mountain runoff is predicted. The essence of the statistical downscaling model is to establish a significant relationship between climate forecasting factors and regional meteorological elements. The process of establishing the future climate change prediction of the CMIP5 model through the statistical downscaling model is shown in Figure 4. The data period used to establish the model is from 1960 to 2014, including 1960–2005 for model calibration and 2006–2014 for model verification.

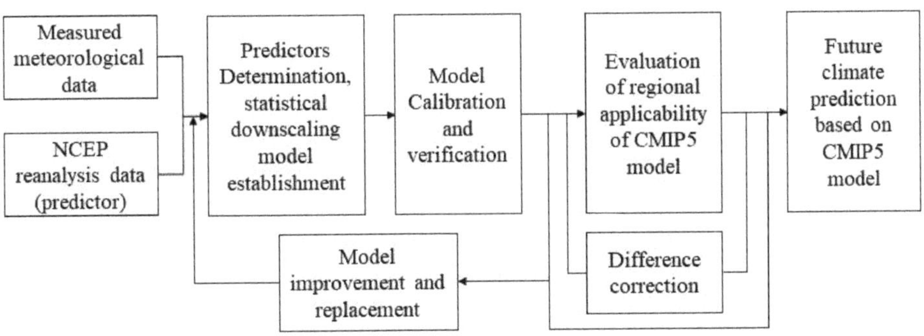

Figure 4. Prediction process of future climate change based on statistical downscaling and the CMIP5 model.

3.4.1. Prediction Factor Selection

The steps of selecting forecasting factors are as follows: first, all candidate factors are screened by the stepwise regression method, and the collinearity of the included factors in the regression equation is checked. Secondly, for similar factors with significant collinearity, according to their correlation and partial correlation with the predicted quantity, the factors with greater correlation are selected. Thirdly, after collinear exclusion, if there are still too many prediction factors in the regression equation, the standardized coefficient in the equation is small, i.e., the factors that have little influence on the prediction.

3.4.2. Multiple Regression Downscaling Model

The multiple regression downscale model of forecast factor and forecast quantity is established, and the sub-rate is predicted periodically and in the verification period, in which the prediction threshold of precipitation is set at 0.2 mm, i.e., no precipitation occurs in that month when the monthly precipitation is less than 0.2 mm. The degree of fitting of the monthly value and the relative error of the annual value are selected as indicators to evaluate the simulation effect.

3.4.3. Supplementary Multiple Regression Downscaling Model

The residuals of simulated and measured precipitation at periodic stations are calculated, and it is found that each residual can be well fitted by linear regression. Therefore, the deviation correction is carried out, and a supplementary multiple regression model based on residual analysis is established to improve the simulation effect of the downscale model of the HRB [34].

3.4.4. Evaluation of Regional Applicability of CMIP5 Model

The simulation process of the Global Climate Model (GCM) is uncertain, so it is necessary to evaluate its regional applicability before using the GCM to predict the future climate. The starting time of CMIP5′s CNRM-CM5 model prediction is January 2006, so the prediction factors of the period from January 2006 to December 2014 are used to simulate the precipitation and temperature of each station according to the two models established above. Before simulation, the spatial resolution of the CNRM-CM5 model is adjusted to 2.5×2.5 by using the distance reciprocal weight interpolation method, which is consistent with the NCEP reanalysis data. Regional applicability evaluation is divided into two paths: RCP4.5 and RCP8.5.

3.4.5. Prediction of Future Precipitation

According to the downscaling model determined by regional applicability evaluation, the future precipitation of the basin is predicted under the two paths of RCP4.5 and RCP8.5.

3.4.6. Prediction of Future Runoff

According to the regression model of mountain runoff and precipitation in the upper reaches of the basin, the mountain runoff in the study area is estimated.

4. Results and Discussion

4.1. Calibration and Verification of Numerical Model

Using a variety of data, such as groundwater level dynamics, groundwater flow field, groundwater discharge along the river, annual groundwater discharge to the river, go runoff, etc., the model is calibrated and verified by manual and automatic parameter adjustments [36]. The model calibration period is from 1990 to 2015, and the model verification period is from 2016 to 2020. Using Nash efficiency coefficients [4], Taylor diagrams [37], Box plots [38], Radar charts [39], and other statistics, the numerical model is considered effective in simulation [30].

4.2. Groundwater Budget in the Midstream Basin

From 1990 to 2020, the average annual groundwater recharge in the study area was 17.98×10^8 m^3/a, the average annual discharge was 18.62×10^8 m^3/a, and the storage change was -0.58×10^8 m^3/a. From 1990 to 2001, the average annual groundwater recharge in the basin was 16.39×10^8 m^3/a, the average annual discharge was 17.85×10^8 m^3/a, and the storage change was -1.46×10^8 m^3/a. From 2002 to 2020, the average annual groundwater recharge in the basin was 18.98×10^8 m^3/a, the average annual discharge was 19.11×10^8 m^3/a, and the storage change was -0.13×10^8 m^3/a (Table 1). The total groundwater storage change during the simulation period was -19.99×10^8 m^3, of which the groundwater storage change was -17.52×10^8 m^3 from 1990 to 2001 and -2.47×10^8 m^3 from 2002 to 2020.

Table 1. Groundwater budget in the middle reaches of the HRB.

Recharge Item	Recharge/(10^8 m^3·a^{-1})						Discharge Item	Discharge/(10^8 m^3·a^{-1})					
	1990	2001	1990–2001	2014	2020	2002–2020		1990	2001	1990–2001	2014	2020	2002–2020
RCH	1.2	0.98	1.09	1.02	1.58	1.21	EVT	2.21	1.92	2.17	2.07	2.09	2.02
RIV	8.29	6.36	7.66	11.30	10.45	10.08	WEL	0.91	3.57	1.95	4.36	5.31	4.45
CHA	6.28	5.03	5.86	5.91	5.85	5.64	GD	15.11	12.27	13.73	13.23	12.96	12.64
CIR	1.19	1.01	1.04	1.10	1.09	1.00							
WIR	0.10	0.40	0.21	0.45	0.55	0.47							
LRP	0.21	0.18	0.20	0.26	0.24	0.23							
LR	0.26	0.22	0.26	0.32	0.30	0.29							
SIR	0.08	0.07	0.07	0.06	0.06	0.06							
Total in.	17.61	14.25	16.39	20.42	20.12	18.98	Total out.	18.23	17.76	17.85	19.66	20.36	19.11
Storage change								−0.62	−3.51	−1.46	0.76	−0.24	−0.13

The interannual fluctuation of groundwater recharge in the middle basin of the Heihe River Basin is large, while the interannual fluctuation of discharge is small (Figure 5). This reflects the regulation and storage effect of the large-thickness aquifer system in the basin, and the interannual fluctuation of discharge is more stable than that of recharge after the adjustment of the aquifer. At the same time, groundwater discharge has a self-adaptation feature. For example, when the amount of groundwater extraction in the shallow buried area on both sides of the Heihe River increases, the groundwater discharge decreases accordingly, and the sum of the two reflects the macroscopic change in groundwater discharge in the midstream basin.

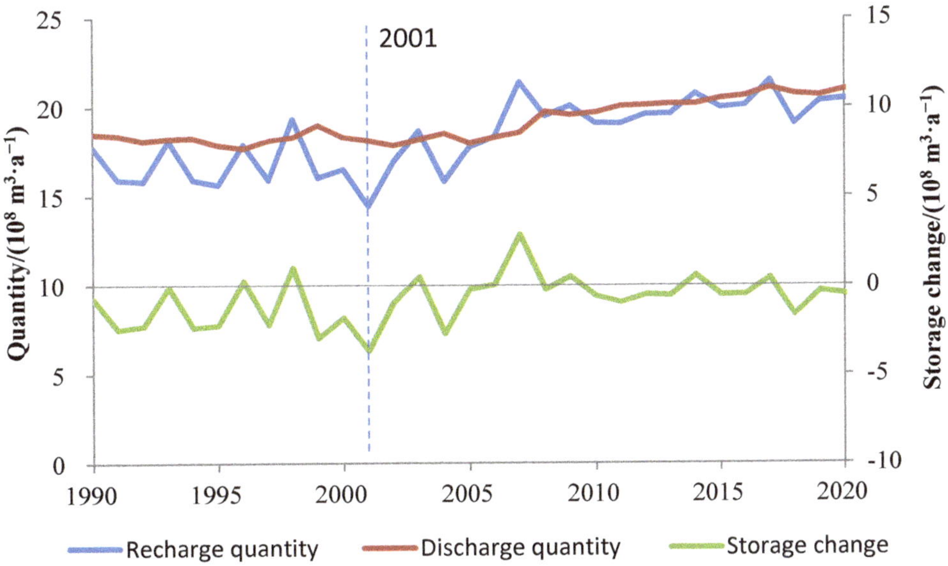

Figure 5. Difference between groundwater recharge and recharge discharge in the midstream basin over the years.

4.3. Influence Area of the Heihe River

Based on the groundwater flow field in typical years, the influence area of the Heihe River was identified (Figure 6). From 1990 to 2001, due to the continuous drought in the HRB and the increase in groundwater extraction, the area affected by the recharge of the Heihe River decreased. From 2001 to 2014, the river basin turned from dry to wet, and the influence area increased. From 2014 to 2020, the HRB continued to be abundant in precipitation, and under the combined effects of mountain runoff, irrigation water diversion, and groundwater extraction, the area affected by the leakage of the Heihe River decreased slightly.

1990 (Median water year) 2001 (Dry year) 2014 (Wet Year) 2020 (Median water year)

Figure 6. Influence area of the Heihe River in typical years.

4.4. Proportions of Groundwater Discharge to the Heihe River

Four typical years were selected, and the multi-year variation in the proportions of discharge of the Heihe River from recharge of the Heihe River was analyzed by streamlining tracking (Figure 7, Table 2). Compared with the median water year in 1990, the proportion of leakage recharge of the Heihe River to the total recharge in the dry year of 2001 decreased from 36.92% to 34.22%, which was mainly affected by the decrease in mountain runoff. The ratio of discharge of the Heihe River to recharge of the Heihe River increased from 63.37% to 67.80%. The reason may be that, with the significant increase in groundwater exploitation in the irrigation area, more groundwater comes from storage nearby since the recharge area of the Heihe River is relatively far away. Compared with the dry year in 2001, the proportion of leakage recharge of the Heihe River in the 2014 wet year increased significantly, the discharge of the Heihe River increased, and the ratio of discharge in the Heihe River from recharge of the Heihe River also increased significantly. Compared with 2014, the mountain runoff in 2020 decreased, and the proportion of leakage recharge of the Heihe River to the total recharge also fell. In general, the larger the mountain runoff, the greater the proportion of leakage recharge of the Heihe River to the total recharge, and the greater the ratio of discharge of the Heihe River from recharge of the Heihe River.

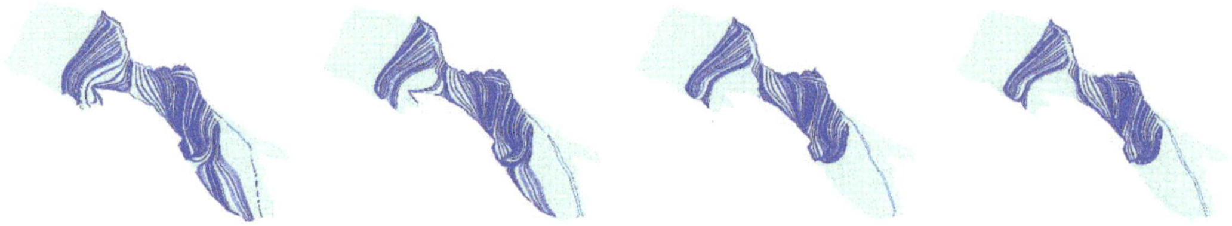

1990 (Median water year) 2001 (Dry year) 2014 (Wet Year) 2020 (Median water year)

Figure 7. Streamline diagram of Heihe River overflow in a typical year.

Table 2. Statistical values of leakage and discharge characteristics of the Heihe River.

Year	Heihe River Leakage Recharge /(10^8 $m^3 \cdot a^{-1}$)	The Proportion of Leakage Recharge of the Heihe River to the Total Recharge (%)	Discharge of the Heihe River /(10^8 $m^3 \cdot a^{-1}$)	Discharge of the Heihe River from Recharge of the Heihe River/ (10^8 $m^3 \cdot a^{-1}$)	Ratio of Discharge of the Heihe River from Recharge of the Heihe River(%)
1990	6.50	36.92	11.15	7.07	63.37
2001	4.88	34.22	9.78	6.63	67.80
2014	8.80	43.10	10.42	7.63	73.24
2020	8.15	40.51	10.26	7.20	70.13

4.5. Prediction of Mountain Runoff in the Watershed

According to the principle of prediction factor selection, the precipitation prediction factors of the Qilian station are air850, rhum500, and omega850, and the prediction factors of the Yeniugou station are air, rhum500, and omega850.

Based on the multiple regression downscaling model, the fitting degree of the monthly value of the precipitation rate is 0.536 to 0.837, and the verification period is 0.454 to 0.837. The relative error of the annual value of the precipitation rate is −22.17 percent, and the verification period is −43.47 percent. The overall fitting rate of the verification period is poor. For different stations, with the decrease in average annual precipitation, the fitting degree of the monthly value of the rate periodicity and the verification period becomes smaller, and the relative error of the annual value becomes larger. Generally speaking, there is a big error in the simulated precipitation values of each meteorological station during the verification period, and the simulated values are all smaller than the measured values, which cannot meet the prediction requirements.

The simulation effect of the precipitation supplementary regression model is obviously better than that of the multiple regression model, and the annual mean relative error of the simulation is obviously reduced. The annual mean relative error of the supplementary regression model rate of precipitation is −1.63~0.01%, and the verification period is −18.88~2.30%, which are significantly smaller than those of the multiple regression model.

The evaluation of regional applicability shows that under the two pathways, the simulation annual mean relative error of the supplementary regression models of precipitation is generally smaller than the multiple regression models. Under the two pathways, the simulated precipitation is higher than the measured value from January to April and from October to December and lower than the measured value from May to September. The simulation effect of the model for the annual distribution of precipitation is slightly poor, but the simulation effect for annual precipitation is better.

The mountain runoff of the Heihe River and other rivers is predicted according to the prediction of precipitation in the upper reaches of the HRB. Under the RCP4.5 scenario, the predicted annual runoff of Yingluoxia from 2023 to 2055 is 21.17 × 10^8 m^3/a, and that of Yingluoxia from 2023 to 2055 is 21.46 × 10^8 m^3/a under the RCP8.5 scenario (Figure 8).

4.6. Groundwater Budget from 2023 to 2055

Under the predicted historical reappearance scenario, the average annual recharge of groundwater in the study area from 2023 to 2055 is 18.93 × 10^8 m^3/a, the average annual discharge is 19.59 × 10^8 m^3/a, and the difference between recharge and discharge is −0.66 × 10^8 m^3/a. Under the RCP4.5 scenario, the average annual recharge of groundwater in the study area from 2023 to 2055 is 19.67 × 10^8 m^3/a, the average annual discharge is 20.23 × 10^8 m^3/a, and the difference between recharge and discharge is −0.56 × 10^8 m^3/a. Under the RCP8.5 scenario, the average annual recharge of groundwater in the study area from 2023 to 2055 is 19.77 × 10^8 m^3/a, the average annual discharge is 20.30 × 10^8 m^3/a, and the difference between recharge and discharge is −0.53 × 10^8 m^3/a (Figure 9).

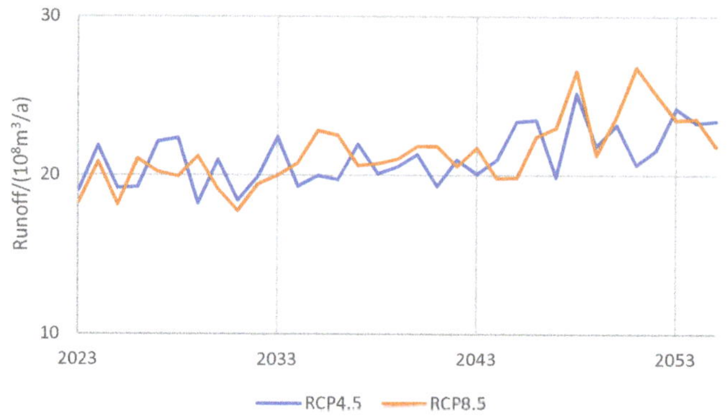

Figure 8. Predicted Yingluoxia runoff from 2023 to 2055.

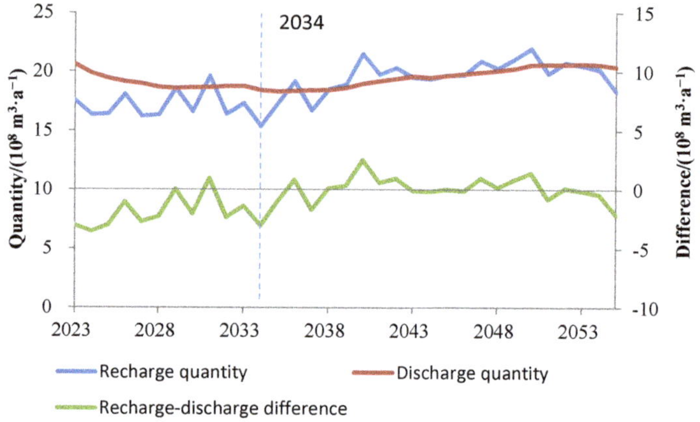

(**a**) Historical reappearance scenario.

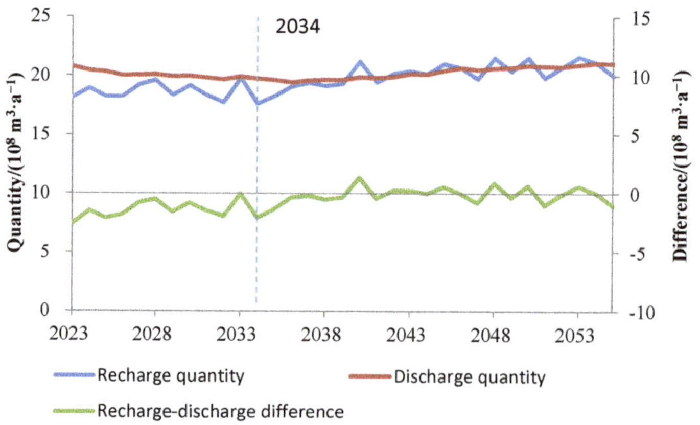

(**b**) CNRM-CM5 model RCP4.5 pathway scenario

Figure 9. *Cont.*

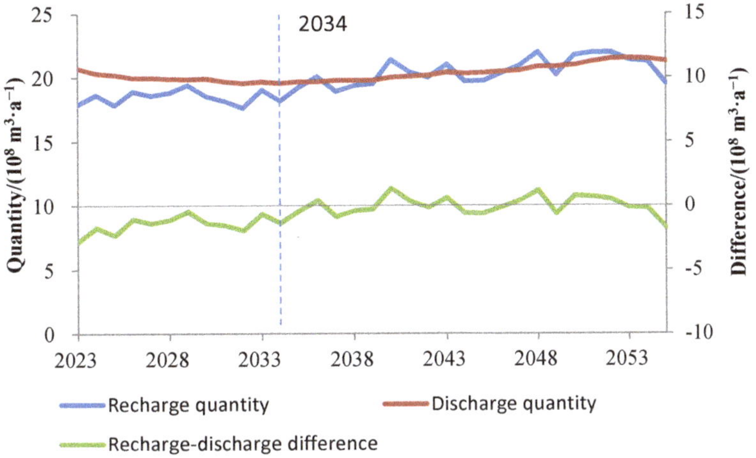

(c) CNRM-CM5 model RCP8.5 pathway scenario

Figure 9. The groundwater recharge, discharge, and storage changes in the study area.

4.7. Prediction of Groundwater Levels in the Middle Reaches of the HRB

Under the three scenarios, the groundwater level in the inclined plain east of the Heihe River is in a state of continuous decline from 2023 to 2055, such as in the H65 observation hole, which is due to excessive surface water diversion and reduced groundwater recharge. Under the RCP4.5 and RCP8.5 scenarios, the water resources in the future basin increased, and the decline in groundwater level was significantly lower than that of the historical reappearance scenario. The groundwater level dynamics in the Heihe River–Liyuanhe River inclined plain are mainly affected by mountain runoff, and there is a certain trend of falling first and then rising in the future (H13-0). The falling process reflects the rise of the groundwater level during the strongest flood period from 2002 to 2022, and the rising process reflects the future increase in water resources in the basin. The erosion accumulation plain is located in the groundwater discharge zone of the midstream basin, and the groundwater table is basically stable, such as in the H77-1 observation hole. The groundwater level in the Yanchi Basin continued to decline, most typically in the Southern mountainous area, mainly affected by overexploitation in the irrigation area. For example, the continuous decline of the groundwater level in the H11-0 observation hole is due to the continuous exploitation of the Luotuocheng County irrigation area. Over time, the downward trend of H11-0 water level slowed down, which was due to the increase in lateral runoff recharge of groundwater caused by the decline of the groundwater table and the increase in the hydraulic gradient around the irrigation area. The downward trends of the groundwater levels in the Yanchi basin under the RCP4.5 and RCP8.5 scenarios are slower than those in the historical reappearance scenario (Figures 10 and 11).

4.8. Groundwater Budget in the Midstream Basin

In the future, the strong groundwater runoff zone in the middle reaches of the HRB will change with the amount of water resources and human activity in the basin. In the flood year, the groundwater runoff is large, and the recharge characteristics of Heihe River leakage to groundwater are significant; in the dry year, the groundwater runoff is small, and the groundwater recharge caused by Heihe River leakage is relatively weak (Figure 12).

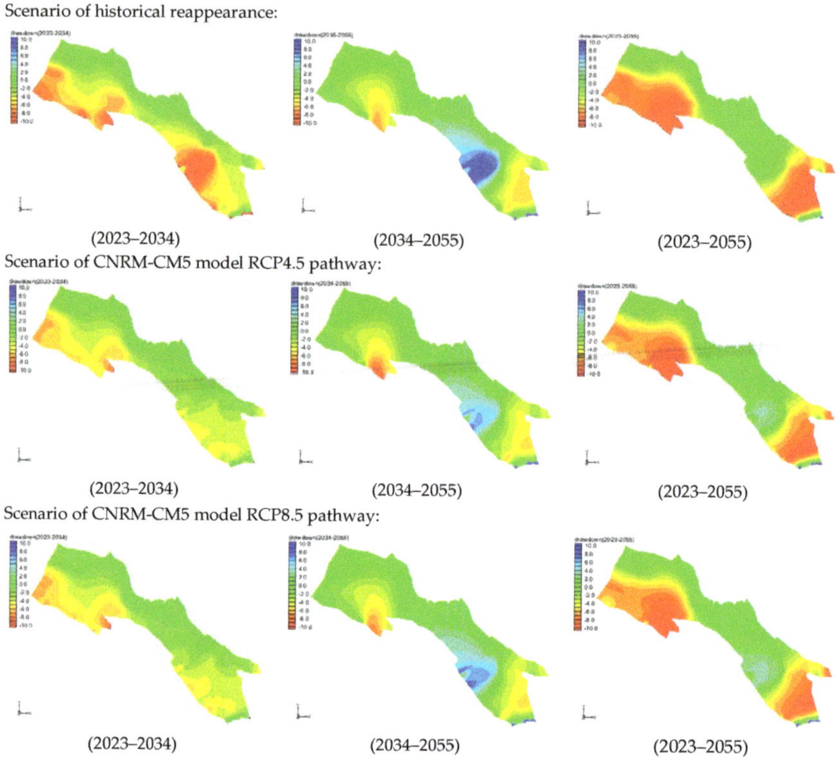

Figure 10. Future groundwater regime variations.

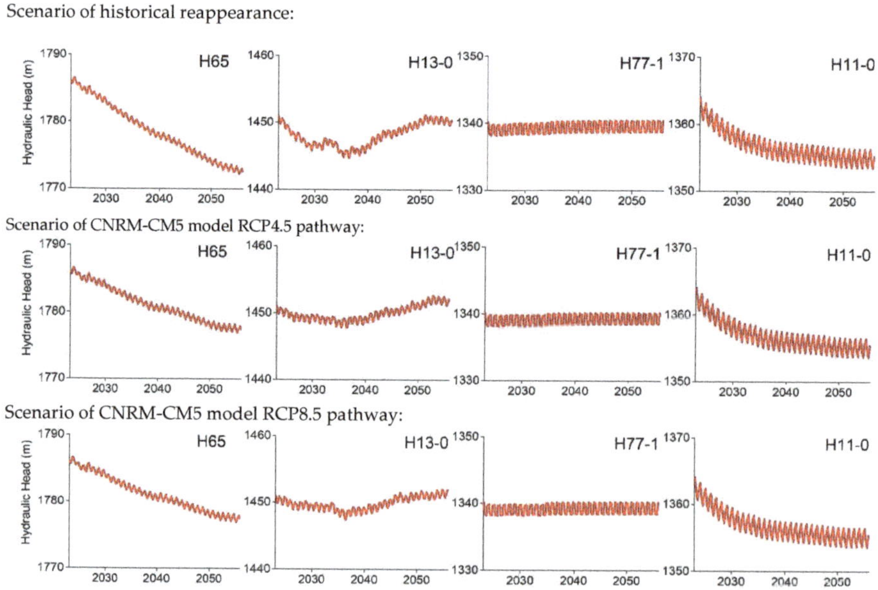

Figure 11. Groundwater level variations.

Scenario of historical reappearance:

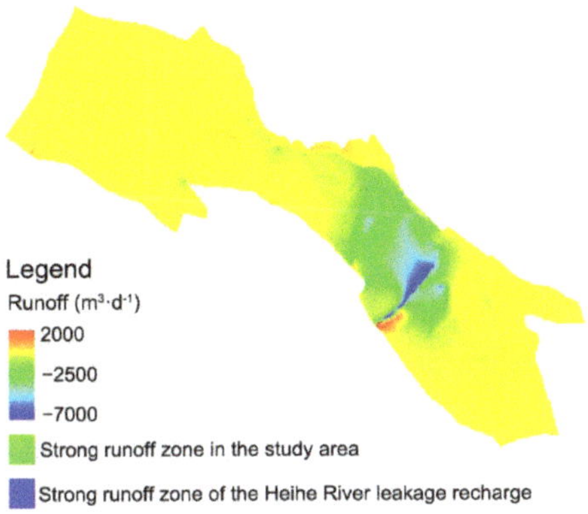

2023

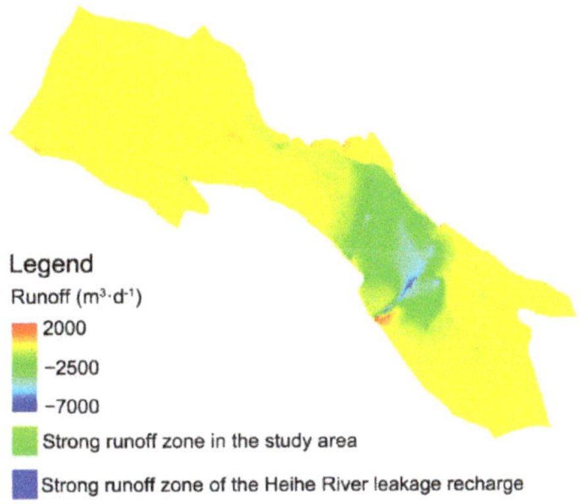

2034

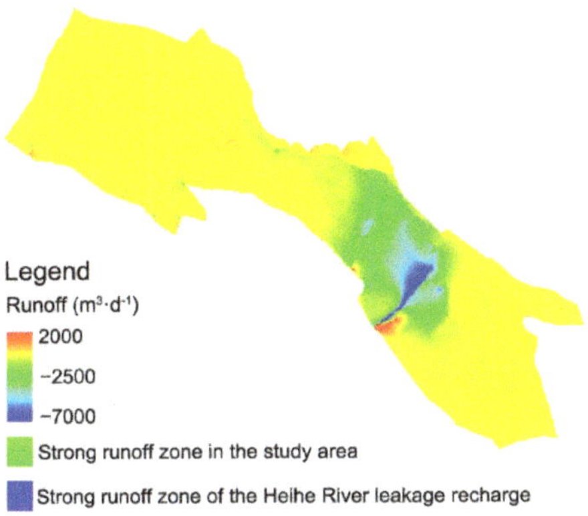

2050

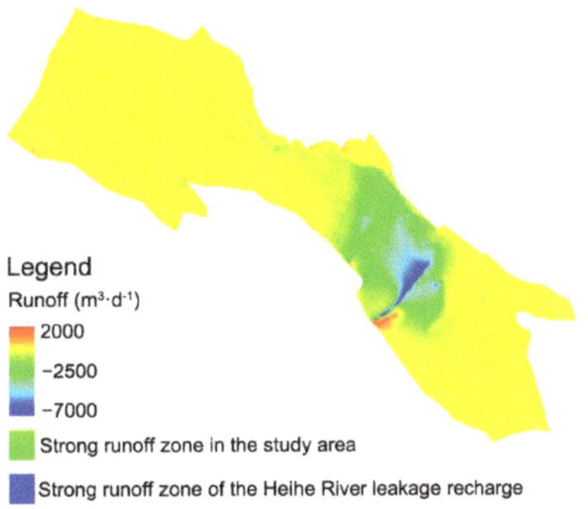

2055

Scenario of CNRM-CM5 model RCP4.5 pathway:

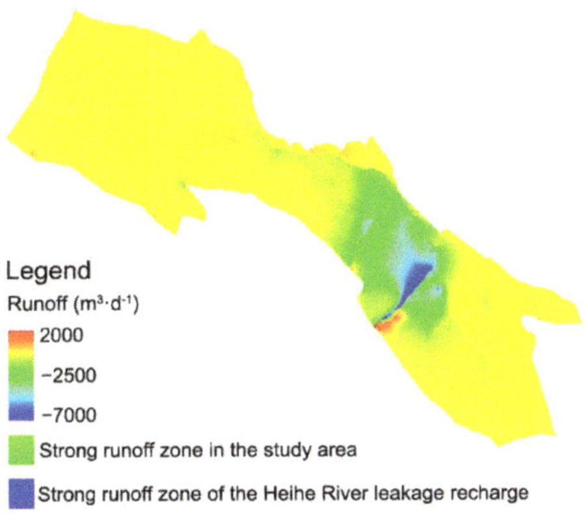

2023

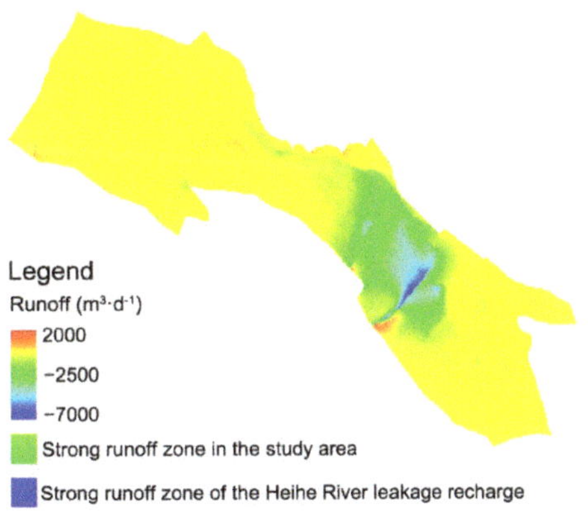

2034

Figure 12. *Cont.*

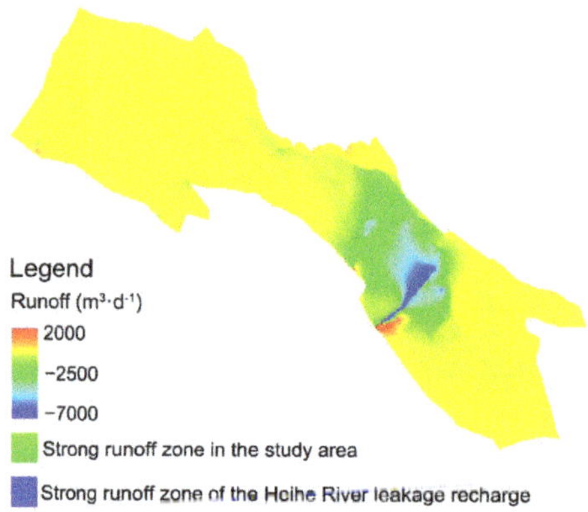

2050

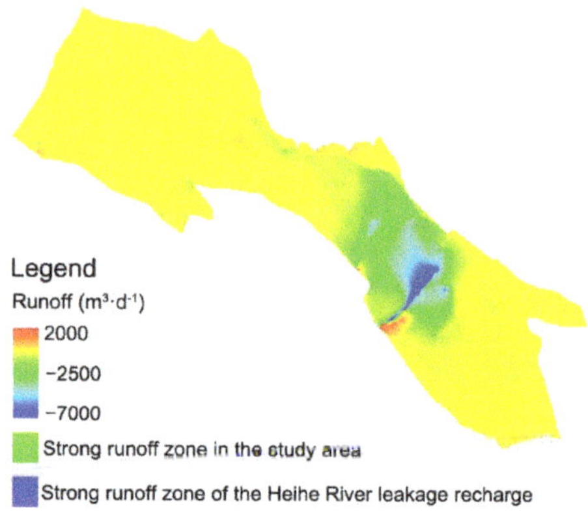

2055

Scenario of CNRM-CM5 model RCP8.5 pathway:

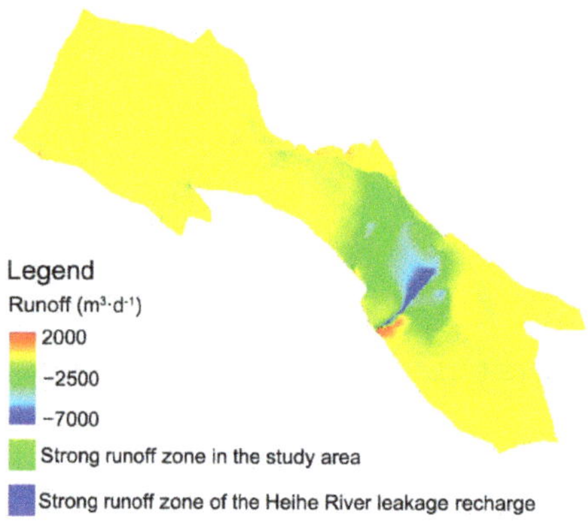

2023

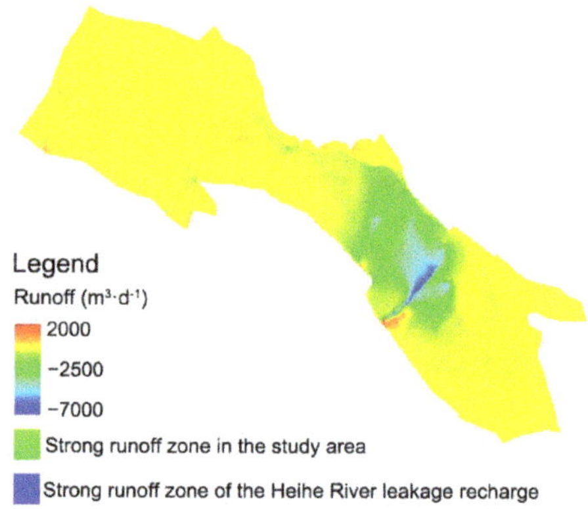

2034

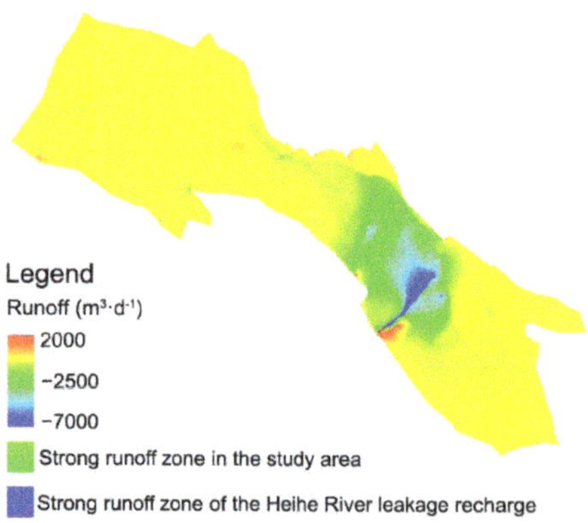

2050

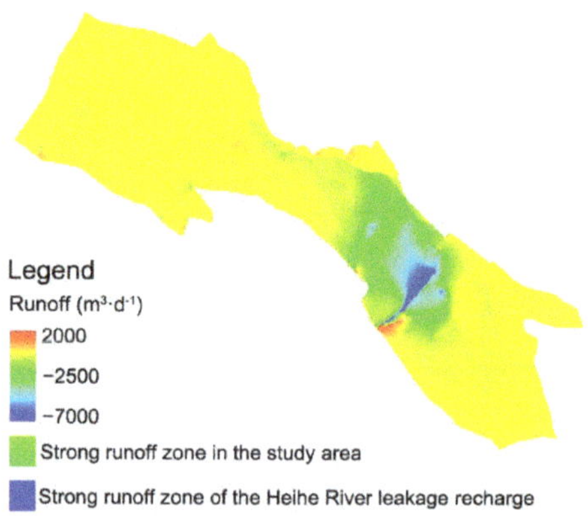

2055

Figure 12. Strong groundwater runoff zones of typical years in the middle reaches of HRB.

4.9. Future Runoff of Zhengyixia

The "97 Water Diversion Plan" of the State Council stipulates the functional relationship between the runoff of Zhengyixia and Yingluoxia, so as to ensure a certain amount of runoff from Zhengyixia to the lower reaches of the HRB and safeguard the downstream ecology. Under the historical reappearance scenario, the annual average difference between the runoff of Zhengyixia and the dividing runoff amount in the future is predicted to be -1.35×10^8 m^3/a. Under the RCP4.5 and RCP8.5 scenarios, the differences between the two are -3.74×10^8 m^3/a and -3.48×10^8 m^3/a, respectively (Figure 13). In the future, the basin will be in a wet period and will not be able to meet the "97 Water Diversion Plan" for many more years. In the previous study, with the implementation of the water-saving irrigation under the RCP4.5 and RCP8.5 scenarios, the runoff of Zhengyixia can meet the "97 Water Diversion Plan" from 2023 to 2033 continuously [4]. It is suggested that the middle reaches of the HRB further improve the level of agricultural water saving and control the reasonable scale of cultivated land to reduce the water consumption of agricultural irrigation and ensure ecological water use in the lower reaches, aiming to realize the sustainable utilization of water resources in the basin.

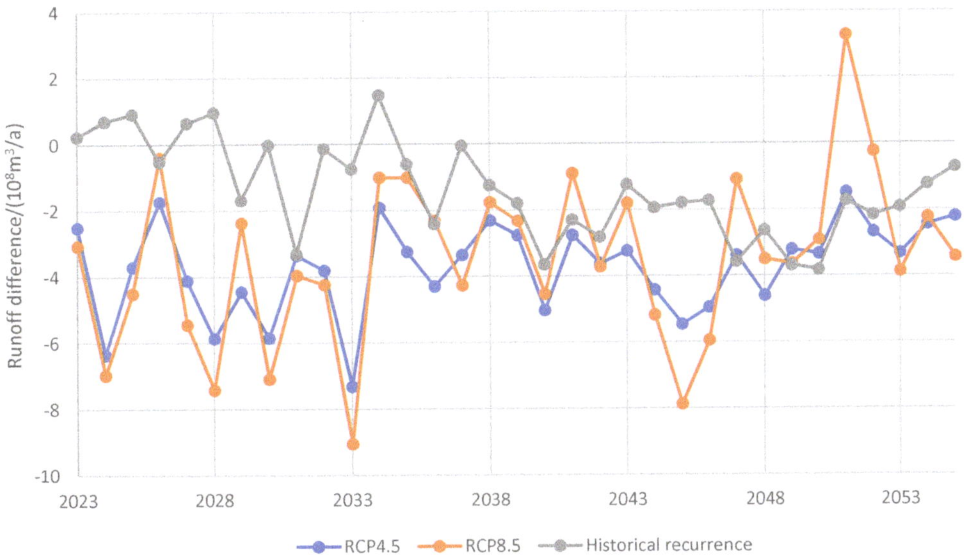

Figure 13. Annual differences between the amount of the predicted Zhengyixia runoff and the water dividing amount in the future.

5. Conclusions

Based on the latest hydrogeological survey and other multi-source data, this paper constructs a coupled numerical model of surface water and groundwater in the middle reaches of the HRB and analyzes the changing process of the water cycle. The future precipitation in the upper and middle reaches of the HRB is predicted by the statistical downscaling model, and the mountain runoff is predicted. Three future climate change scenarios are constructed, and the numerical model is used to predict the future change in water resources and judge the rationality of the development and utilization of water resources in the basin.

Under the combined influence of natural conditions and human activities, the average annual recharge of groundwater in the study area from 1990 to 2020 was 17.98×10^8 m^3/a, the average annual discharge was 18.62×10^8 m^3/a, and the difference between recharge and discharge was -0.64×10^8 m^3/a. The total groundwater storage change was -19.99×10^8 m^3, of which the groundwater storage change from 1990 to 2001 was -17.52×10^8 m^3 and from 2002 to 2020 was -2.47×10^8 m^3.

Under the scenarios of historical reappearance and RCP4.5 and RCP8.5 pathways of the CMIP5 CNRM-CM5 model, the groundwater level in the Heihe River–Liyuanhe River inclined plain falls first and then rises; the groundwater levels in the inclined plain east of the Heihe River and the Yanchi basin continue to decline; and the water level in the erosion accumulation plain is basically stable. In the future, the wet period will significantly improve the groundwater deficit in the basin. The future runoff of Zhengyixia cannot meet the requirements of the State Council's "97 Water Diversion Plan" in most years, and the ecological and production water usage in the lower reaches of the HRB cannot be effectively guaranteed. With the implementation of the water-saving irrigation under the RCP4.5 and RCP8.5 scenarios, the runoff of Zhengyixia can meet the "97 Water Diversion Plan".

The water resources in the inland river basin have multiple functions, such as resources, environment, ecology, etc. The middle reaches of the HRB should maintain a reasonable scale of land development, develop water-saving agriculture, appropriately return farmland to forest and grassland, protect wetlands and natural vegetation, reduce water consumption in the basin, and achieve water diversion and exploitation in an orderly manner to promote the coordinated and healthy development of hydrology, ecology, economy, and society in the inland river basins.

Author Contributions: Conceptualization, W.L.; Investigation, H.C. and X.Q.; Writing—original draft, X.Q.; Writing—review & editing, X.Q. and W.L.; Supervision, W.L., Y.Z., W.K., Z.L. and X.S.; Project administration, W.L. and Y.Z.; Funding acquisition, W.L. and Y.Z. All authors have read and agreed to the published version of the manuscript.

Funding: This work was supported by Projects of the National Natural Science Foundation of China (Grant Nos. 91025019; 91325107); the China Geological Survey Project (DD20160292; DD20190322; DD20230075); the China Geological Survey Administration Project (10220222018000009060); and the Xinjiang Uygur Autonomous Region Water Conservancy Department Financial Special Project (2022.B-004).

Data Availability Statement: The data presented in this study are available within the article.

Acknowledgments: The authors are very grateful to all anonymous reviewers, chief editors and associate editors for their valuable comments and suggestions, which improved the quality of this paper.

Conflicts of Interest: The authors declare no conflict of interest.

References

1. Li, W.; Zhou, H.; Zhou, Y.; Jiao, P.; Hao, A. *Groundwater Flow Systems of Typical Arid Areas in Northwest China*; Earthquake Press: Beijing, China, 1995. (In Chinese)
2. Li, W.; Kang, W.; Liu, Z.; Hao, A.; Li, Z.; Li, S.; Zhao, C.; Zheng, Y.; Yang, X.; Li, J.; et al. *Water Resources Regulation and Optimal Utilization Models of Typical Inland Basins in Northwest China—A Case Study of Heihe River Basin*; Geology Press: Beijing, China, 2010. (In Chinese)
3. Cui, H. Water Cycle Mechanism and Co-Evolution of Water Cycle and Oasis in Heihe River Basin. Ph.D. Thesis, Northwest University, Xian, China, 2016; p. 143. (In Chinese)
4. Qi, X. *Study on Sustainable Utilization of Water Resources in an Arid Inland Basin under Conditions of Climate Change*; China University of Geosciences: Beijing, China, 2018. (In Chinese)
5. Zhou, Y.; Nonner, J.C.; Meekes, S.; Griffieon, J.; Su, Z.; Li, W.; Hao, A.; Zhou, Y.; Li, X.; Shao, X.; et al. *Strategies and Techniques for Groundwater Resources Development in Northwest China*; China Land Press: Beijing, China, 2007.
6. Zhang, M.; Wang, S.; Fu, B.; Gao, G.; Shen, Q. Ecological effects and potential risks of the water diversion project in the Heihe River Basin. *Sci. Total Environ.* **2017**, *619–620*, 794–803. [CrossRef]
7. Cai, W.; Jiang, X.; Sun, H.; Lei, Y.; Nie, T.; Li, L. Spatial scale effect of irrigation efficiency paradox based on water accounting framework in Heihe River Basin, Northwest China. *Agric. Water Manag.* **2023**, *277*, 108118. [CrossRef]
8. Yao, Y.; Zheng, C.; Andrews, C.; Zheng, Y.; Zhang, A.; Liu, J. What controls the partitioning between baseflow and mountain block recharge in the Qinghai-Tibet Plateau? *Geophys. Res. Lett.* **2017**, *44*, 8352–8358. [CrossRef]
9. Wen, X.; Wu, Y.; Lee, L.; Su, J.; Wu, J. Groundwater flow modeling in the Zhangye Basin, Northwestern China. *Environ. Geol.* **2007**, *53*, 77–84. [CrossRef]
10. Chen, S.; Yang, W.; Huo, Z.; Huang, G. Groundwater simulation for efficient water resources management in Zhangye Oasis, Northwest China. *Environ. Earth Sci.* **2016**, *75*, 647. [CrossRef]
11. Yao, Y.; Zheng, C.; Tian, Y.; Li, X.; Liu, J. Eco-hydrological effects associated with environmental flow management: A case study from the arid desert region of China. *Ecohydrology* **2017**, *11*, e1914. [CrossRef]

12. Zhang, Z.; Wang, Q.; Guan, Q.; Xiao, X.; Mi, J.; Lv, S. Research on the optimal allocation of agricultural water and soil resourcesin the Heihe River Basin based on SWAT and intelligent optimization. *Agric. Water Manag.* **2023**, *279*, 108177. [CrossRef]
13. Li, J.; Mao, X.; Li, M. Modeling hydrological processes in oasis of Heihe River Basin by landscape unit-based conceptual models integrated with FEFLOW and GIS. *Agric. Water Manag.* **2017**, *179*, 338–351. [CrossRef]
14. Li, J.; Mao, X.; Shang, S.; Steenhuis, T.S. Modeling Regional Soil Water Balance in Farmland of the Middle Reaches of Heihe River Basin. *Water* **2017**, *9*, 847. [CrossRef]
15. Yao, Y.; Zheng, C.; Liu, J.; Cao, G.; Xiao, H.; Li, H.; Li, W. Conceptual and numerical models for groundwater flow in an arid inland river basin. *Hydrol. Process.* **2015**, *29*, 1480–1492. [CrossRef]
16. Yao, Y.; Tian, Y.; Andrews, C.; Li, X.; Zheng, Y.; Zheng, C. Role of Groundwater in the Dryland Ecohydrological System: A Case Study of the Heihe River Basin. *J. Geophys. Res. Atmos.* **2018**, *123*, 6760–6776. [CrossRef]
17. Wu, X.; Zheng, Y.; Wu, B.; Tian, Y.; Han, F.; Zheng, C. Optimizing conjunctive use of surface water and groundwater for irrigation to address human-nature water conflicts: A surrogate modeling approach. *Agric. Water Manag.* **2015**, *163*, 380–392. [CrossRef]
18. Hu, L.; Xu, Z.; Huang, W. Development of a river-groundwater interaction model and its application to a catchment in Northwestern China. *J. Hydrol.* **2016**, *543*, 483–500. [CrossRef]
19. Li, X.; Zhang, L.; Zheng, Y.; Yang, D.; Wu, F.; Tian, Y.; Han, F.; Gao, B.; Li, H.; Zhang, Y.; et al. Novel hybrid coupling of ecohydrology and socioeconomy at river basin scale: A watershed system model for the Heihe River basin. *Environ. Model. Softw.* **2021**, *141*, 105058. [CrossRef]
20. Fan, X.; Ao, S.; Jin, L.; Jiang, Z.; Hou, X.; Wan, J.; Chen, S.; Zuo, M.; Xu, Z.; Dang, Y.; et al. *Study on Groundwater Distribution and Rational Development and Utilization in Hexi Corridor of Gansu Province*; Geological Science Institute of Gansu Geological and Mineral Bureau: Lanzhou, China, 1984. (In Chinese)
21. Zhou, X.; Zhao, J.; Zhang, A.; Zhao, Y.; Chen, X. *Two-Water Conversion and Comprehensive Development and Utilization of Water Resources in the Middle and Lower Reaches of Heihe River in Western Gansu and Inner Mongolia*; Second Hydrogeological Engineering Geological Team of Gansu Geological and Mineral Bureau: Zhangye, China, 1990. (In Chinese)
22. Zhou, X.; Zhao, J.; Wang, Z.; Zhao, Y.; Zhang, Y.; Hua, J.; Zhang, A.; Wang, F.; Ding, H. *Investigation on Groundwater Resources and Their Rational Development and Utilization in the Middle Reaches of Heihe River in Gansu Province*; Second Hydrogeological Engineering Geological Team of Gansu Geological and Mineral Bureau: Zhangye, China, 1990. (In Chinese)
23. Mora, C.; Frazier, A.G.; Longman, R.J.; Dacks, R.S.; Walton, M.M.; Tong, E.J.; Sanchez, J.J.; Kaiser, L.R.; Stender, Y.O.; Anderson, J.M.; et al. The projected timing of climate departure from recent variability. *Nature* **2013**, *502*, 183–187. [CrossRef] [PubMed]
24. Core Writing Team; Pachauri, R.K.; Meyer, L.A. *Climate Change 2014: Synthesis Report. Contribution of Working Groups I, II and III to the Fifth Assessment Report of the Intergovernmental Panel on Climate Change*; IPCC: Geneva, Switzerland, 2014.
25. Chang-Ming, L.; Wen-Bin, L.; Guo-Bin, F.; Ru-Lin, O. A discussion of some aspects of statistical downscaling climate impacts assessment. *Adv. Water Sci.* **2012**, *23*, 427–437.
26. Wilby, R.L.; Harris, I. A framework for assessing uncertainties in climate change impacts: Low-flow scenarios for the River Thames, UK. *Water Resour. Res.* **2006**, *42*, W02419. [CrossRef]
27. Acharya, N.; Kar, S.C.; Mohanty, U.C.; Kulkarni, M.A.; Dash, S.K. Performance of GCMs for seasonal prediction over India—A case study for 2009 monsoon. *Theor. Appl. Clim.* **2011**, *105*, 505–520. [CrossRef]
28. An, Y.; Wang, W.; Wu, X.; Xie, W.; Gong, L.; Yin, D.; Sun, Y.; Wang, X.; Zhang, J.; Yang, J.; et al. *Report on Hydrogeological Survey Results in Key Areas of Heihe River Basin*; Center for Hydrogeology and Environmental Geology, CGS: Baoding, China, 2016. (In Chinese)
29. Wu, X.; An, Y.; Yin, D.; Cui, H.; Wang, X.; Wu, B.; Wang, X.; Wang, W.; Xie, W.; Sun, Y.; et al. *Report of 1:50,000 Hydrogeological Survey Secondary Project in Heihe River Basin, Hexi Corridor*; Center for Hydrogeology and Environmental Geology, CGS: Baoding, China, 2020. (In Chinese)
30. Qi, X.; Li, W.; Cui, H.; Kang, W.; Liu, Z.; Shao, X. Study on the conversion mechanism of surface water and groundwater in the middle reaches of the Heihe River Basin. *Hydrogeol. Eng. Geol.* **2022**, *49*, 29–43. (In Chinese)
31. Li, W.; Shao, X.; Qi, X. Estimation of groundwater lateral flow in the southern mountainous area of the middle Heihe River Basin. *Hydrogeol. Eng. Geol.* **2022**, *49*, 1–10. (In Chinese)
32. Cui, H.; Li, W.; Kang, W. A study of groundwater recharge under different irrigation conditions in the middle reaches of the Heihe River. *Hydrogeol. Eng. Geol.* **2022**, *49*, 22–28. (In Chinese)
33. Ma, N. *Construction of 3D Hydrogeological Structure Model for the Middle Reaches of Heihe River Basin*; China University of Geosciences: Beijing, China, 2019. (In Chinese)
34. Qi, X.; Li, W.; Li, H.; Liu, H. Future climate change prediction of arid inland riverbasin based on CMIP5 model. *Arid. Land Geol.* **2017**, *40*, 987–996. (In Chinese)
35. Harbaugh, A.W. MODFLOW-2005, the U.S. Geological Survey modular ground-water model—The Ground-Water Flow Process: U.S. Geological Survey Techniques and Methods 6-A16. Available online: https://pubs.usgs.gov/publication/tm6A16 (accessed on 6 March 2015).
36. Doherty, J. *PEST: Model-Independent Parameter Estimation, User Manual*, 5th ed.; Watermark Computing: Brisbane, QL, Australia, 2005.
37. Elbeltagi, A.; Di Nunno, F.; Kushwaha, N.L.; de Marinis, G.; Granata, F. River flow rate prediction in the Des Moines watershed (Iowa, USA): A machine learning approach. *Stoch. Environ. Res. Risk Assess.* **2022**, *36*, 3835–3855. [CrossRef]

38. Di Nunno, F.; Granata, F. Future trends of reference evapotranspiration in Sicily based on CORDEX data and Machine Learning algorithms. *Agric. Water Manag.* **2023**, *280*, 108232. [CrossRef]
39. Jiang, L.; Wu, Y.; Wang, J.; Wang, Z.; Jiang, Q. Scenario simulation and comprehensive assessment of water footprint sustainability system in Heilongjiang Province, China: A model combining system dynamics with improved radar chart approach. *Ecol. Indic.* **2023**, *154*, 110527. [CrossRef]

water

Review

A Review of Advances in Groundwater Evapotranspiration Research

Xianglong Hou [1], Hui Yang [2], Jiansheng Cao [2,*], Wenzhao Feng [2,3] and Yuan Zhang [1,*]

1 Institute of Geographical Sciences, Hebei Academy of Sciences, Hebei Technology Innovation Center for Geographic Information Application, Shijiazhuang 050011, China
2 Key Laboratory of Agricultural Water Resources, Hebei Key Laboratory of Agricultural Water-Saving, Center for Agricultural Resources Research, Institute of Genetics and Developmental Biology, Chinese Academy of Sciences, Shijiazhuang 050001, China
3 School of Land Science and Space Planning, Hebei GEO University, Shijiazhuang 050021, China
* Correspondence: caojs@sjziam.ac.cn (J.C.); 83zhangyuan@sohu.com (Y.Z.)

Abstract: Groundwater evapotranspiration (ET_g) is an important component of the hydrological cycle in water-scarce regions and is important for local ecosystems and agricultural irrigation management. However, accurate estimation of ET_g is not easy due to uncertainties in climatic conditions, vegetation parameters, and the hydrological parameters of the unsaturated zone and aquifers. The current methods for calculating ET_g mainly include the WTF method and the numerical groundwater model. The WTF method often requires data supplementation from the numerical unsaturated model to reduce uncertainty; in addition, it relies on point-monitoring data and cannot solve the spatial heterogeneity of ET_g. The ET_g calculation module of the numerical groundwater model is set up too simply and ignores the influence from the unsaturated zone and surface cover. Subsequent research breakthroughs should focus on the improvement of WTF calculation theory and the setting up of an aquifer water-table fluctuation monitoring network. The numerical groundwater model should couple the surface remote sensing data with the unsaturated zone model to improve the accuracy of ET_g calculation.

Keywords: groundwater evapotranspiration; water-table fluctuation method; unsaturated zone; aquifer; MODFLOW; HYDRUS

Citation: Hou, X.; Yang, H.; Cao, J.; Feng, W.; Zhang, Y. A Review of Advances in Groundwater Evapotranspiration Research. *Water* **2023**, *15*, 969. https://doi.org/10.3390/w15050969

Academic Editor: Adriana Bruggeman

Received: 13 January 2023
Revised: 26 February 2023
Accepted: 28 February 2023
Published: 2 March 2023

1. Introduction

Water scarcity is already the biggest challenge for global agricultural development [1], with one third of the population in developing countries living in water-scarce areas and fifty-four percent of agricultural land also being located in water-scarce areas [2,3]. In water-scarce areas, groundwater evapotranspiration is an important part of the hydrological cycle; it is one of the main sources of regional evapotranspiration and the main consumer of groundwater in areas with a shallow water table [4,5]. Arid and semiarid regions occupy approximately 30% of the land surface of the Earth [6], including the majority of northern and southern Africa; the Middle East; western USA and southern South America; most of Australia; large parts of central Asia; and parts of Europe [7]. Vegetation provides natural protection against desertification and dust storms in these regions. Some vegetation, known as phreatophyte, is groundwater dependent [8]. Phreatophyte transpiration consumes groundwater and causes diurnal fluctuations of groundwater levels [9]. On the other hand, surface water is scarce, and groundwater is often the only reliable water resource for socio-economic development in arid regions [10]. Irrigation water for crops is usually provided by the abstraction of groundwater. The over-exploitation of groundwater resources has caused decreasing groundwater levels and resulted in desertification in many parts of arid regions [11]. The sustainable management of groundwater resources must consider water use both from human activities and by nature. The starting point to develop a sustainable groundwater use plan is the assessment of groundwater balance. In arid environments, an important component of the groundwater balance is groundwater evapotranspiration (ET_g) [12]. Accurate estimation of groundwater evapotranspiration is essential for understanding hydrological cycle processes and sustainable groundwater resource use and management [13–15], and it is useful for natural ecosystem conservation and restoration [16]. The quantification of groundwater evapotranspiration is particularly important in areas dependent on groundwater ecosystems [17–20] and is important for water management of crops and the investigation of soil salinization processes [21–23].

Groundwater evapotranspiration (ET_g) can result in significant loss of groundwater storage. ET_g is generated when water moves from the unsaturated zone to replenish soil storage depleted by surface evapotranspiration and root water uptake. The shallow water table allows groundwater to be used to directly supply crop growth [24–29], so an accurate estimation of ET_g for farmland can help to improve irrigation management. Any efforts toward improving ET_g estimation methods are worthwhile for agricultural water management and land and water environmental protection [15].

However, accurate estimation of groundwater evapotranspiration remains a challenge because it is often influenced by uncertainties associated with climatic variables, vegetation parameters, geological variables, and hydrologic parameters [30,31].

Numerous studies have been conducted on evapotranspiration estimation models worldwide, and these studies have mainly included empirical statistical models, energy balance models, remote sensing models based on Penman's formula, complementary correlation models, and hydrological models. The evapotranspiration model (later called the Food and Agricultural Organization Penman–Monteith (FAO56-PM) model) proposed by Allen [32] is the most suitable model for estimating international evapotranspiration at present, and this model has been widely used worldwide. Evapotranspiration (ET) includes surface evaporation (E_s), evaporation of water from below the ground surface (E_{ss}), and transpiration of water by plants (T_{ss}). The latter two were together defined as subsurface evapotranspiration (ET_{ss}) [33], which includes groundwater evapotranspiration (ET_g) and unsaturated water evapotranspiration (ET_u) [34].

There are many computational methods applied to different scales to calculate evapotranspiration, including the method based on evapotranspiration lysimeter weighing [35–38], the method based on field water balance equations [39–42], micrometeorological methods [16], vorticity covariance methods [43–46], the Bowen ratio method [47], and regional-scale ET calculation models (for example, the TSEB model [48], the SEBAL model [49], the S-SEBI model [50], the SEBS model [51], and the LandSAF model [52]), as well as mapping of evapotranspiration using the internal scale method (METRIC) [53], the STSEB model [54], the GLEAM model [55], the MODIS-ET model [56], and the ETwatch product [57]. However, these methods are usually used more for the measurement and calculation of earth surface evapotranspiration and are not able to directly measure the value of groundwater evapotranspiration, mainly because the hydraulic connection between the earth surface and the aquifer is blocked by the unsaturated zone, and the uncertainty of the unsaturated zone increases the difficulty of calculating groundwater evapotranspiration. Therefore, the calculation of groundwater evapotranspiration often needs to consider the variation of moisture content in the unsaturated zone. In addition to this, the accurate estimation of ET_g needs to consider the variation of local atmospheric conditions [58] and groundwater table variation [59], and is also influenced by the spatial heterogeneity of land use [60]. Moreover, many factors such as lateral inflow at the recharge boundary, vertical recharge at the surface, and complex geological structure and soil composition in the unsaturated zone can affect ET_g [61–64]. In addition, intensive anthropogenic measures, such as water diversions and irrigation in water-scarce areas, also have a significant impact on groundwater evapotranspiration. External water transfers directly change the water storage capacity of inland lakes, and overall raise the groundwater table in the wetlands around the lakes and diversion channels. This change in groundwater is continuous, unlike the transfer of water for irrigation in agricultural areas. The shallower water table makes the water storage capacity and regulation capacity of the unsaturated zone weaker, the capillary zone is closer to the surface, and thus groundwater evapotranspiration becomes stronger; in addition, the shallower water table allows more channels for groundwater to rise into the air, and the root systems of some deep-rooted plants can act directly on the unsaturated zone and even the aquifer. The branches and roots of plants establish the hydraulic connection between air and aquifer, and driven by the transpiration of plants, a large amount of groundwater enters the air directly without passing through the unsaturated zone. The shallower unsaturated zone channels and the newly added plant channels change the proportion of groundwater evapotranspiration (ET_g) in total evapotranspiration (ET_a). This combination of multiple factors makes ET_g difficult to calculate.

2. Methodology

2.1. Advances in Research on Groundwater Evapotranspiration

Historical studies of groundwater evapotranspiration by researchers date back to the 1920s. The early researchers found that groundwater is constantly supplied with evapotranspiration through "capillary rise," which is evident in groundwater when the water table is less than 3 m. Remson and Fox [65] proposed a method for estimating groundwater discharge by evapotranspiration from the water capillary rise of the water table. "Potential capillary water loss" is defined as a measure of the ability of the soil capillary gap to raise water from the groundwater to the earth surface, and they consider the depth of the water table as the most important factor affecting the magnitude of groundwater evapotranspiration. Gardner's [66] analysis showed that the evapotranspiration rate depends on the depth of the water table and the capillary flux. Subsequently, Gardner [66] and Willis [67] proposed the calculation of ET_g as a function of water table depth, and although their assumption of the constant moisture content in the unsaturated zone is inaccurate today, this idea has had a profound impact on subsequent studies of groundwater evapotranspiration. Schoeller [68] introduced the concept of ultimate evapotranspiration depth, and he theorized that groundwater evapotranspiration occurs only when the groundwater table depth is smaller than the ultimate evapotranspiration depth. The magnitude of the groundwater evapotranspiration value is determined by the groundwater table depth together with the potential evapotranspiration value, and this method is widely used because of its simple form. In the classic groundwater numerical model MODFLOW, the groundwater package EVT still follows this idea in the calculation of evapotranspiration. Doorenbos and Pruitt [69] proposed a method to calculate ET_g by quantifying the variation of soil water content in the root zone. They were the first to elaborate on the relationship between ET_g and soil moisture in the root zone of crops. This method has been used in soil water balance models for calculating the magnitude of ET_g. The calculation of soil water balance is a specific application of the water balance equation theory, which is also a common method for groundwater evapotranspiration calculations and is applicable to different scales [4,70–72].

Wang et al. [15] used the water balance equation to calculate the soil water content in the root zone under irrigation and rainfall conditions, and then proposed a new equation that integrates multiple influencing factors to estimate ET_g during the growing season based on the methods proposed by Doorenbos and Pruitt [69] and the Averianov equation [68]. Groundwater evapotranspiration was calculated using the following equation.

$$ET_g = K_c \times ET_0 \times \left(1 - \frac{H}{Hmax}\right)^n \times \frac{\theta_{fc} - \theta}{\theta_{fc} - \theta_r} \quad (1)$$

where K_c is crop coefficient, ET_0 is reference crop evapotranspiration in mm·day^{-1}, H is the actual water table depth in m, H_{max} is the potential maximum depth in m, beyond which no ET_g occurs; n is the soil characteristics parameter, θ is the actual averaged soil water content in the root zone in cm^3·cm^{-3} (usually about 60 cm below the soil surface), θ_{fc} is the field capacity of the soil in the root zone in cm^3·cm^{-3}, θ_r is the soil water content close to permanent wilting point in $cm^3 \cdot cm^{-3}$ (in this paper, a constant value 0.05 is used [73]). The method was tested and validated against the data from the lysimeter experiments.

Wang et al. [74] then used the following water balance equation to calculate ET_g based on ET_a obtained in a subsequent study.

$$ET_g = (P + I) + \Delta SWC - ET_a - P_c \quad (2)$$

where P and I are precipitation and irrigation in mm, respectively, ΔSWC is the variation in soil water storage up to 90 cm depth where most of the maize root system is concentrated [75], and P_c is the deep percolation to shallow groundwater. Since the daily ΔSWC was sometimes too small to accurately determine using Hydra probe measurements, the two-day average was used to calculate the water balance. In any event, it was assumed that either P_c or ET_g was zero. Lai et al. [76] considered quantifying the contribution of shallow groundwater to evapotranspiration (ET_g) as an important topic that has been extensively studied [77,78]. They used lysimeters to calculate groundwater evapotranspiration values for wheat fields in the lower Yellow River basin at different groundwater level conditions. He concluded that reasonable groundwater level control can help increase yields while reducing the risk of soil salinization and is important for sustainable management of the lower Yellow River basin.

2.2. *Using the WTF Method to Calculate the ET_g*

Among the methods for calculating groundwater evapotranspiration considering the influence of crops or plant roots, the groundwater table fluctuation (WTF) method is one of the most commonly used methods. In recent decades, there has been an increasing emphasis on using the WTF method to quantify ET_g. Calculating ET_g based on the WTF method is a relatively straightforward, simple and inexpensive method [9,79,80], Users can measure water loss due to evapotranspiration directly from groundwater table changes and therefore does not require additional measurements at the soil surface [73,81–83]. The use of the groundwater table fluctuation method assumes that groundwater table changes in a shallow aquifer are caused by evapotranspiration only [84], and this method has been widely used to estimate ET_g in riparian zones and wetlands [13,58,85–87].

The method of groundwater table fluctuation exploits the law of daily variation of the water table in riparian zones or wetlands. During the daytime, plant transpiration makes the groundwater table lower, and at night, the intensity of plant transpiration decreases and a significant rebound of the groundwater table level occurs. Based on the discovery of this pattern, White [81] proposed a classic method for calculating groundwater evapotranspiration, namely the White method. This method has long been used for groundwater evapotranspiration rates in arid and semi-arid areas [83,88–90]; it has been continuously improved by many researchers and has been applied to wetlands [8], riparian zones [79,91–93], prairies [94], and forests [95] in a variety of ecosystems.

2.2.1. White Method

The White method is the most classic method of the water-table fluctuation methods. White theorized that groundwater evapotranspiration is mainly composed of two parts—the recovery, and the storage variation in 24 h—based on the regular fluctuation of groundwater, which can be calculated by the following equation.

$$ET_g = S_y(24r \pm s) \quad (3)$$

where S_y is the specific yield of the shallow aquifer, r is the groundwater recharge rate from 00:00 a.m. to 04:00 a.m., and s is the 24-h shallow groundwater table variation. However, the White method is not universally applicable, and his application is based on four assumptions: (1) the day and night dynamics of groundwater table fluctuation are caused by plant evapotranspiration; (2) the evapotranspiration of plants from 0:00 a.m. to 4:00 a.m. at night is 0; (3) the groundwater recovery rate at night is a constant value; and (4) the specific yield is a constant value and representative. Due to the simplicity and ease of operation of White method, this method is also the most widely used for evapotranspiration calculation at the site scale. However, some of the elements in the assumptions are subjective and limited, and researchers have made corresponding improvements in subsequent studies based on the shortcomings of the assumptions of the White method. Inspired by White method, many researchers have proposed other methods based on water-table fluctuation information, including the Dolan method [96], the Hays method [13], the Gribovszki method [85], the Loheide method [62], the Soylu method [86], etc. These methods are based on the White method and classed as approaches that improve upon the White method (Figure 1).

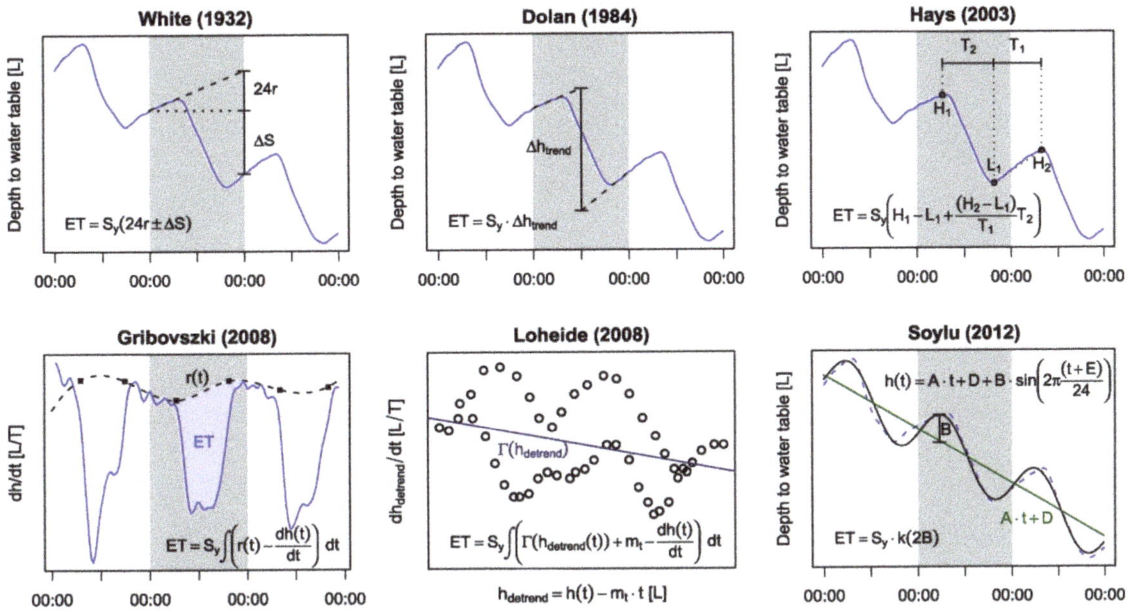

Figure 1. Computational schematic of the White method and its improved approaches [88].

2.2.2. Dolan Method

Dolan et al. [96] utilized continuous records of water-table elevation in the marsh soil, and related a drop in the water table to evapotranspiration loss. The observed rise or fall in the groundwater table at night represents the net inflow or outflow of water into or out of the marsh due to hydraulic forces alone. The rate of change of each night's groundwater table rise was extrapolated to noon the next day and back to noon the day before. The midday elevation represents the altitude of the groundwater table if no evapotranspiration had occurred during the 24 h period. The cycle is centered on each successive night. Thus, the difference between the elevation extrapolated from the previous night and the elevation extrapolated from the next night represents the water loss due to evapotranspiration on that day.

2.2.3. Hays Method

Hays [97] developed a new method for estimating ET_g. The biggest advantage of this method is the flexibility to determine the time of the water-table rise and fall according to the water-table waveform, instead of fixing it at a certain definite time period. Its calculation equation is:

$$ET_g = [(H_1 - H_L) + \frac{H_2 - H_L}{T_1} T_2] \times S_y \tag{4}$$

where H_1 is the maximum water table depth on the morning of the calculation day in m, H_2 is the maximum water table depth on the next day in m, H_L is the minimum water table depth on the day of the calculation in m, T_2 is the duration of the groundwater table decline period, and T_1 is the duration of the groundwater table rise period. The prerequisites for the use of the Hays method are similar to those of the White method.

2.2.4. Loheide Method

The Loheide method [62] is an improved method based on the White method. The main idea of the Loheide method is that it is first assumed that the trend of the water table change in the recharge source is consistent with the general trend of observed water table change, so that the measured groundwater table can be detrended. The detrending analysis of the water table included in this method not only improves the calculation accuracy of vegetation evapotranspiration, but also reduces the uncertainty in the calculation process. The change in groundwater storage near the water table observation wells is expressed as the change in water table with time (dWT/dt) and is controlled by the net inflow or outflow of groundwater in the vicinity (r(t)(L/T)) together with the ET_g.

$$S_y{}^*\frac{dWT}{dt} = r(t) - ET_g(t) \quad (5)$$

When ET_g is 0, Equation (3) can be simplified as

$$S_y{}^*\frac{dWT}{dt} = r(t) \quad (6)$$

It has been clarified that the daily recharge rate is a function of time, and Loheide [52] assumes a constant head in the recharge source area and that the recharge rate can be obtained from water table changes, so Equation (5) can be expressed as

$$r(WT) = S_y{}^*\frac{dWT}{dt} \quad (7)$$

The Loheide method considers the trend of the groundwater recharge zone water table to be included in the water table change at the observation point, so this trend is therefore removed from the water table change at the observation point.

$$WT_{DT}(t) = WT(t) - m_T \times t - b_T \quad (8)$$

where $WT_{DT}(t)$ is the detrended water table, WT(t) is the observed water table, m_T is the trend line slope, and b_T is the trend line intercept. dWT_{DT}/dt and $WT_{DT}(t)$ using the water table of ET_g as 0 in the early morning of the current day and the next day, established the functional relationship $\Gamma(WT_{DT})$, and thus the recharge rate function is obtained, which can be expressed as

$$r(t) = S_y^* \times [\Gamma(WT_{DT}(t)) + mT] \quad (9)$$

Finally, the available ET_g is expressed as:

$$ET_g(t) = r(t) - S_y^* \times \frac{dWT}{dt} \quad (10)$$

$S_y{}^*$ in the above equation is the complete specific yield instead of the traditional specific yield concept. The reason for the proposed concept of complete specific yield is because Loheide [62] theorized that the error of specific yield is an important factor in causing the error in the calculation of groundwater evapotranspiration. Therefore, he proposed the concept of complete specific yield, and the equation of complete specific yield is as follows

$$S_y^* = S_{yu} - \frac{S_{yu}}{\left[1+\alpha\left(\frac{z_i+z_f}{2}\right)^n\right]^{1-\frac{1}{n}}}, S_{yu} = \theta_s - \theta_r \tag{11}$$

where θ_s is the saturated soil water content, θ_r is the residual water content, and z_i and z_f are the initial and end values for the variation of groundwater table depth. The parameters α and n are in the van Genuchten model. The average values are taken when the groundwater table varies in the soil layer.

2.2.5. Yin Method

The Loheide method assumes that the rate of change of the detrended water table is linearly related to the detrended water table. In response to this assumption, Yin et al. [12] further refined the Loheide method in their study, and found that applying the exponential equation to fit the relationship curve between the rate of change of the detrended water table and the detrended water table, can yield more accurate calculation results. Inspired by the Loheide method, Yin et al. [12] proposed an improved calculation of hourly-scale evapotranspiration based on the White method, as Equation (12)

$$\mathrm{ET}_g = S_y \times (r + (H_{i-1} - H_i)) \tag{12}$$

where S_y is the specific yield, r is the water level recovery rate (LT^{-1}), i is the moment value (T), and H is the water table value at the ith moment (T). Compared with the Loheide method, the calculation process of this method is simpler.

2.2.6. Gribovszki Method

Gribovszki et al. [85] also concluded that the rate of water table recovery is not constant throughout the day and proposed the use of empirical interpolation methods to calculate a non-constant rate of water table recovery throughout the day. The main idea is to estimate the rate by using hydraulic derivation and empirical methods. The maximum (positive) rate of water table changes between midnight and 6 a.m. and the average of dh/dt are chosen to obtain the maximum and minimum recharge rates, respectively. These values were assigned to the times of maximum and minimum groundwater table rates of recovery, respectively. Therefore, two points (sections) were defined for each day. Based on the points for several days, a spline interpolation was performed to describe the recharge rate r(t) (L/T) over time. Subdaily evapotranspiration was then calculated (e.g., dt = 1 h).

$$\mathrm{ET(t)} = S_y\left(r(t) - \frac{dh}{dt}\right) \tag{13}$$

However, although the empirical interpolation method obtains the variation of water table recovery rate for each hour of the day, there may be some error in calculating the hour-by-hour water level recovery rate because the empirical interpolation method is based on only two pieces of daily data.

2.2.7. Soylu Method

Soylu et al. [86] proposed to use a new 'Fourier method' to calculate groundwater evapotranspiration. They found that the amplitude of the fluctuations of the water table after detrending can be used to calculate evapotranspiration directly by combining the water balance equation of the White method and the Fourier equation proposed by Czikowsky and Fitzjarrald [98]. The equations are as follows.

$$\mathrm{H(t)} = A \times t + D + B\sin\left(2\pi\frac{(t+E)}{24}\right) \tag{14}$$

$$\mathrm{ET}g = Sy \times k(2B) \tag{15}$$

where H is the water table depth (L), A is the water table depth change trend for multiple days (LT^{-1}), t is the time (T); D is the average deviation of water table depth change (L), B is the diurnal fluctuation amplitude (L), E is the day and night fluctuation phase (T), and k is the empirical correction factor for correcting the water table recovery and evapotranspiration components included in $A \times t$.

2.2.8. Wang Method

Wang et al. [13] used the statistical method of day and night fluctuations of the water table to analyze the characteristics of de-trended groundwater table fluctuations, and then calculated the evapotranspiration at different time scales based on the relevant parameters. This method can effectively deal with continuously changing groundwater table fluctuation data.

$$ETg = Sy \times \frac{\sigma}{\lambda} \tag{16}$$

where σ is the variance (L) of the variation of the de-trended water table, and λ is the evapotranspiration cycle correlation coefficient. It can be seen that the subsequent improvements in the water table fluctuation methods—both the Fourier method and the water table statistics method—are mostly improvements in the fluctuation characteristics about the detrended water table, and this series of improvement ideas is derived from the detrending theory proposed by Loheide.

2.2.9. Other Improvements and Applications

Wang et al. [99] concluded that one of the limitations of the White method is the large uncertainty in quantifying the daily groundwater recovery rate (r). Since ET_g is highly dependent on the shape and duration of the diurnal clear-sky solar radiation curve, using the groundwater recovery rate at short nighttime intervals to represent daily r may lead to large uncertainties in the ET_g estimates. They analyzed the dependence of the estimated daily r on sunset and sunrise times. and found that the estimated r is highly sensitive to the duration between sunset and sunrise and varies with the season. Instead of using fixed time spans (TS_s), they suggest using a more universal method for determining TS_s for estimating daily r. They tested this dynamic T of S method at a *Tamarix ramosissima*-dominated riparian site in northwestern China. The results proved that their improved method was better and less subjective than the traditional White method. Subsequently, many researchers have used these representative methods mentioned above in different study areas for comparison to see their applicability. Three water table fluctuation methods [13,69,74] were used by Su et al. [100] in 2017 to calculate daily ET_g in a riparian forest area in northwestern China. The purpose of comparing these three methods was to evaluate and compare their performance under various groundwater table conditions in the wild. The results indicated that the White method is applicable to the period of declining groundwater table. In addition, the selected time period may affect the estimation of ET_g. The Soylu and Hays methods performed well under various groundwater table conditions. Therefore, it appears that the Hays and Soylu methods are more suitable for long-term ET_g estimation in the wild. In addition, it was found that the percentage of water transpired by plants from groundwater varies during the growing season, and that riparian plants mainly use soil water in early growth and tend to use groundwater in late growth. Yin et al. [12] also compared three frequently used water table fluctuation methods: the White, Hayes, and Loheide methods. These three methods use the water table generated by the model to calculate the ET_g. The comparison of actual and estimated ET_g reveals the accuracy of each method and determines the applicability of the methods. When the recovery branch of the groundwater table process line is nonlinear, these methods underestimate daily ET_g. The Loheide method is relatively good, and all three methods can accurately estimate daily ET_g. The modified White method can provide hourly ET_g estimates and is recommended for general use. It was found that in practical applications, analysis of the shape of the groundwater table recovery branch and the difference in groundwater head between the upper and lower gradients can determine the appropriate estimation method of ET_g. Fahle and Dietrich [101] compared the existing water table fluctuation method with their field measurements of evapotranspiration. They used 85 days of rain-free data from a weighable groundwater solution meter located in the wetland meadows of the Spreewald in northeastern Germany. It can be shown that some researchers have used multiple water table fluctuation methods in different study areas around the world and have continuously made targeted improvements. These case studies continue to promote the development and improvement of water table fluctuation methods.

However, there are still some uncertainties regarding the White method and its improvements [102]. These uncertainties are mainly in the quantification of specific yield (S_y), the choice of recharge time, the high heterogeneity of surface vegetation, and the effect of rainfall or irrigation on the water table. The largest uncertainty of White method mainly lies in the estimation of the S_y. The S_y was defined as the volume of water released by gravity from a unit area of rock column extending from the water table to the ground surface as the water table decreases by one unit depth (head) [103]. Errors in the S_y estimates translate directly into ET_g estimates [62,69,70,104]; both variables are involved in the ET_g calculation in the White method, since storage variability and groundwater recharge variability need to be multiplied by the S_y. However, there are many difficulties in estimating S_y, as this parameter is not constant over time. The specific yield is highly variable in shallow aquifers, and its magnitude depends on factors such as soil texture, water table depth, and the state of drainage or recharge [42,62,70,85,103,105–107]. S_y is usually used only for groundwater drainage, but for rising water table conditions, the presence of air within the pore space is likely to reduce the value of S_y, i.e., the equivalent change in water table corresponds to a different change in water volume for the recharge and drainage states [42].

In addition, the White method for calculating the recharge rate (the underlying assumption for r) uses the average rate of groundwater table rise between 0:00 a.m. and 04:00 a.m. for a total of 4 h to equal the average groundwater recharge rate for the entire day (Table 1); therefore, this method is not applicable to variable groundwater fluctuations. The recharge rate is typically a function of the head difference between the observation well and the recharge source [62]. The transient recharge rate will vary with time. Replacing the average daily recharge rate with the average daily recharge rate determined over a range of time can cause errors in the estimation of ET_g [106]. Therefore, many researchers usually modify the White method according to the time period used for recharge estimation [62,94,108]. Some other methods even avoid using varying recharge rates [13,86].

Table 1. The time periods selected for water table recovery rate calculation in different studies.

Method	Experimental Period
White [81]	0:00~4:00
Dolan et al. [96]	0:00~4:00
Hays [97]	0:00~4:00
Loheide [62]	0:00~6:00
Yin et al. [12]	the previous day 21:00~5:00
Rushton [108]	the previous day 18:00~6:00
Miller et al. [94]	the previous day 22:00~7:00

In addition to the degree of water availability and the timing of recharge, the magnitude of ET_g is also influenced by surface vegetation cover and the growth state of vegetation [109–111]. Of course, vegetation effects are not limited to groundwater evapotranspiration ET_g; for total surface evapotranspiration ET, the contribution of plant transpiration (T) to evapotranspiration (T/ET) is estimated to range from 38 to 77% at the global scale, with an average of 64 ± 13% [112–114]. Thus, terrestrial vegetation is an important force in the global water cycle [115]. In arid areas or areas with a low leaf area index (LAI), the average contribution of T to ET can reach 70% to 95% [113,116,117]. In arid or semi-arid zones, plants can only extract water from deeper soils or directly from the shallow water table for transpiration during the growing season due to the low water content of the topsoil.

Multiple indications suggest a strong correlation between the distribution of plant species and the depth of the groundwater table [118,119]. Vegetation has a strong adaptive capacity, especially in water-scarce ecosystems, to make full use of water in the unsaturated zone and aquifers through root growth [120,121]. Nepstad et al. [122] suggest that the semi-enclosed forests of the Brazilian Amazon rely on deep root systems to maintain a green canopy during the dry season. Evergreen forests can meet evapotranspiration requirements during droughts of up to 5 months by absorbing water from the soil at depths greater than 8 m. Maraux and Lafolie [123] found, in a maize-sorghum field in Nicaragua, that upward infiltration of water fluxes into the root zone reached 2 mm per day during drought, while actual evapotranspiration ranged from 2 to 4 mm per day. Kleidon and Heimann [124] similarly found that water uptake from deep soil or groundwater plays an important role in dry season transpiration in Amazonia. More recently, Saleska et al. [125] used Moderate Resolution Imaging Spectroradiometer (MODIS) satellite data to find that the greenness of Amazonian forests increased even during the 2005 drought, and concluded that trees were able to use deep roots to access groundwater to survive extreme drought periods. Considering the influence of crop growth processes on ET_g, Liu and Luo [126] combined two different methods, proposed by Doorenbos and Pruitt [59] and Schmid et al. [127], to calculate the values of groundwater evapotranspiration in areas with burial depths of less than 1.5 m for rainy periods (presence of rainfall or irrigation) and non-rainy periods, respectively. This method used a negative linear relationship between the water content of the unsaturated zone (root zone) and ET_g. However, when the groundwater depth is greater than 1.5 m, the variation in ET_g and the effect of irrigation can cause the relationship to deviate from the linear relationship. Under field conditions, the calculation of ET_g is influenced by multiple factors. Even during non-rainy periods (when there is no rainfall or irrigation) ET_g is still affected by multiple factors such as soil properties, crop water requirements, available soil water content, and groundwater table depth [128]. During rainy periods, a mixed upward and downward water potential gradient is formed in the soil profile [126], and downward fluxes caused by rainfall or irrigation may lead to a gradual development of local downward water potential gradients toward the bottom of the root zone. Some researchers, such as Yuan et al. [129], observed a significant positive correlation between evapotranspiration ET_g and potential evapotranspiration (PET). In addition, the results of Carlson Mazur [58] recognized a significant positive correlation between the two. Some studies have also reported that ET_g and potential PET have a weak positive correlation. According to some previous calculations, the R^2 range of ET_g and PET was 0.02–0.43 [130]. Lautz [79] also reported a similar correlation between groundwater evapotranspiration (ET_g) and potential evapotranspiration (PET) in semi-arid riverfront areas. Zhang et al. [131] reported that the strength of the correlation between ET_g and PET was significantly different in different research regions or conditions. The above complexity of the relationship between PET and ET_g is mainly due to their different controlling factors. Compared with PET, which is affected by climate conditions, there are many influencing factors for groundwater evapotranspiration. including plant species in the study area, root depth, local meteorological conditions, groundwater level variation patterns, lateral recharge and discharge of groundwater, and so on. Under the combined effect of these factors, groundwater evapotranspiration exhibits a large spatial and temporal heterogeneity.

The water table fluctuation (WTF) method, as a typical point-scale ET_g calculation method, is not a good choice in solving problems in terms of spatial heterogeneity. Due to some restrictive nature of its own application conditions and influenced by the uncertainty of the envelope, many researchers use numerical models as an auxiliary method to the water table fluctuation method to solve the problems of envelope uncertainty and spatial heterogeneity. Using an arid desert environment in northwestern China as his study area, Wang et al. [13] found that groundwater evapotranspiration is an important factor controlling hydrological processes in arid riparian zones, while the accuracy of estimating groundwater evapotranspiration values is influenced by the groundwater flow rate in the aquifer, the redistribution of water within the riparian aquifer during river flow [132], and the specific yield [84]. He also used HYDRUS as a complement to understand the different seasonal variations in groundwater evapotranspiration values, but the point-scale HYDRUS+WTF approach still cannot address the uncertainties associated with time-varying lateral flow velocities, spatial variations in groundwater dynamic patterns, and specific yields in the riparian corridor, and he argued that monitoring networks rather than individual monitoring of point settings is necessary. Jia et al. [133] also used HYDRUS as a complement to the WTF method and found that in areas with shallow groundwater depths (<1 m), the groundwater replenishment of ET-induced depleted soils during nighttime is significant, and the use of the traditional White method would seriously underestimate groundwater evapotranspiration values because this factor is neglected; he therefore used HYDRUS to correct the omission of White method and used an improved method to estimate ET_g. Diouf et al. [134] also used HYDRUS and they found that in urban or shallow depth groundwater areas with agricultural irrigation, groundwater table fluctuation methods may be influenced by urban water use or agricultural irrigation, so in order to analyze the applicability of the WTF method to this area, he found that the WTF method achieved better accuracy in the evapotranspiration values calculated in the dry season using the water table simulated by HYDUS; however, the accuracy was not guaranteed for different surfaces. The accuracy could be guaranteed only under the vegetation type. Therefore, the feasibility of using the simulation results of the numerical model as the source data for the WTF method to calculate the evapotranspiration needs to be further demonstrated. By studying historical data on water content and pressure level fluctuations on soil profiles. Zhao et al. [135] found that in a study area in northwest China, direct groundwater recharge does not occur from mid-June to mid-September when evapotranspiration is high, and soil water content changes in the upper unsaturated zone are mainly controlled by atmospheric conditions; however, in deeper parts, they are controlled by pressure surface fluctuations. Accordingly, a one-dimensional model (HYDRUS-1D) with variable head for the lower boundary condition (BC) was developed to explain the response of groundwater flow to changes in atmospheric and groundwater conditions, and the lower BC model with variable head reproduced the observed variation in soil moisture content, with a much smaller total evapotranspiration value obtained for the variable head below BC compared with the fixed head corresponding to the mean water table depth, which is similar to Zhu et al.'s model results [136].

2.3. Using the Numerical Model to Calculate the Groundwater Evapotranspiration

Unlike HYDRUS, another commonly used groundwater flow numerical model, MODFLOW [137], is often used without the need to incorporate the WTF method because it has a well-established ET_g calculation package. The calculation of ET_g is based on the fact that potential evapotranspiration shows a linear variation depending on the depth of the water table, with ET_g reaching a maximum when the water table is near the surface [138]. When the water table is below a fixed depth (ultimate evapotranspiration depth), ET_g is zero. In addition, MODFLOW includes the segmented linear evapotranspiration (ETS) package, which considers the segmented linear relationship between the depth of groundwater burial and the ET_g rate [138], and the ETS package more accurately represents the numerical variation of ET_g at different burial depth stages [139]. In the ETS package, the relation of ET rate to head is conceptualized as a segmented line (a piecewise linear function) in the variable interval. The segments that determine the shape of the function in the variable interval are defined by intermediate points where adjacent segments join. The ends of the segments at the top and bottom of the variable interval are defined by the ET surface and the extinction depth. For the simplest case, where a single ET segment is used (equivalent to the EVT package in MODFLOW2005), the ET rate is calculated as:

$$\mathrm{RET}_{\mathrm{nb}} = EVTR_{\mathrm{nb}},\ h_n > SURF_{nb} \tag{17}$$

$$RET_{\text{nb}} = EVTR_{\text{nb}} \frac{h_n - (SURF_{nb} - EXDP_{nb})}{EXDP_{nb}}, \ (SURF_{nb} - EXDP_{nb}) \le h_{\text{n}} \le SURF_{nb} \quad (18)$$

$$\text{RET}_{\text{nb}} = 0, \ h_{\text{n}} < (SURF_{nb} - EXDP_{nb}) \quad (19)$$

where RET_{nb} is the rate of loss per unit surface area of water table due to ET, in units of volume of water per unit area per unit time (LT $^{-1}$); hn is the head, or water-table elevation in cell n from which the ET occurs (L); $EVTR_{\text{nb}}$ is the maximum possible value of RET (LT^{-1}); $SURF_{\text{nb}}$ is the ET surface elevation, or the water table elevation at which this maximum value of ET loss occurs (L); and $EXDP_{\text{nb}}$ is the cutoff or extinction depth (L), such that when the distance between h_{n} and $SURF_{\text{nb}}$ exceeds $EXDP_{\text{nb}}$, ET ceases.

Although the ETS package is an improvement in computational accuracy over the traditional EVT package, the calculations of the ETS package ignore other influences on ET_g, such as plant canopy cover, leaf area, community composition, and water content in the air pocket.

Pozdniakov et al. [140] argues that groundwater vapor emission has an important influence on the water balance in river valley zones, and he uses MODFLOW with the RIV and EVT packages to calculate ET_g. Based on the MODFLOW evapotranspiration package (EVT), Baird and Maddock [141] developed the Riparian Evapotranspiration for MODFLOW (RIP-ET) package to simulate riparian and wetland ET_g using a nonlinear ET_g curve that accounts for reduced ET_g rates due to hypoxic conditions. The spatial and temporal variability of riparian ET_g is controlled by climate, vegetation structure, water content of the unsaturated zone, and groundwater table depth (DTWT). To address the spatial heterogeneity of riparian zone vegetation ET_g, Ajami et al. [142] implemented a GIS tool in conjunction with RIPGIS-NET, which creates data input files in the MODFLOW-2000 or RIP-ET packages and visualizes MODFLOW results. Combining RIP-ET in MODFLOW with the GIS functionality of RIPGIS-NET can be used to calculate ET_g at different scales. This relationship was later modified in the evapotranspiration (ETS1) package to a segmented linear relationship, to include the nonlinearity between DTWT and ET_g rates. El-Zehairy et al. [143] calculated groundwater evapotranspiration ET_g and unsaturated zone evapotranspiration ET_{uz} for reservoir areas with large stage fluctuations in the water table by using MODFLOW with the addition of the packages SFR7, UZF1, and LAK7. Sergey et al. [144] analyzed the role of groundwater evapotranspiration in the water balance by using the MODFLOW-2005 hydrogeological model with the STR package; the annual variation of recharge was obtained with the codes from Surfbal and HYDRUS. The program code SurfBal is a surface water balance model, which was developed to simulate the processes of precipitation and heat energy transformations on the land surface in order to generate the upper boundary meteorological conditions for HYDRUS. Hou et al. [145] used the remote sensing evapotranspiration data ETwatch with the UZF1 package and MODFLOW to calculate the region-scale ET_g of the shallow depth groundwater zone in the eastern North China Plain.

Some other numerical models have also become common in recent years to calculate groundwater evaporation ET_g. Researchers conducting climate simulations have also started to calculate groundwater evapotranspiration [146–157]. Other researchers have combined numerical simulations with experiments to estimate groundwater evapotranspiration from agricultural fields [69,158]. Blin et al. [159] used evapotranspiration (ET) from the Earth Engine Evapotranspiration Flux (EEFlux) tool as calibration data, then used Parameter Estimation Software (PEST) as a tool to calibrate ET_g from MODFLOW for an undeveloped basin located in the arid Chilean highlands. Liu et al. [160] proposed an alternative approach to estimate ET through the lower boundary of the root zone (<1.0 m depth). Askri et al. [161] developed the hydrological model OASIS-MOD to study the effects of irrigation management on groundwater table fluctuations and soil salinity. The model involves evapotranspiration processes in the unsaturated zone, crop transpiration, and groundwater evapotranspiration processes.

3. Conclusions

Based on the above discussion, it is evident that accurate estimation of groundwater evapotranspiration is still a challenging task at present. The progress and breakthroughs in the technical methods related to this problem are mainly focused on two directions; the first is based on the theory of the WTF method, for which further improvement of the key parameters is required (such as specific yield S_y and groundwater recovery rate, etc.) There are various methods for improving the accuracy of the parameters, such as using numerical models or measuring in sample fields with the help of various instruments. It should be noted that the improvement of accuracy should not be limited only to the methods of obtaining parameters, and the discovery of new computational theories should not be ignored. The second type of method to calculate ET_g mainly relies on the numerical groundwater model. The distributed groundwater model can better solve the problem of spatial heterogeneity, and remote sensing technology can provide more accurate hydrological and meteorological parameters for the model, but the uncertainty of the unsaturated zone blocks the combination of remote sensing technology and the groundwater model, so the coupled model of surface remote sensing data and the unsaturated zone model and groundwater model, should subsequently be constructed for calculating the groundwater evapotranspiration at a regional scale.

Author Contributions: Writing—original draft, X.H.; writing—review and editing, H.Y. and J.C.; conceptualization, W.F. and Y.Z. All authors have read and agreed to the published version of the manuscript.

Funding: This research was financially supported by the Science & Technology Fundamental Resources Investigation Program (grant numbers 2022FY100104), National Natural Science Foundation of China (grant number 42001009), the Key Research and Development Plan Project of Hebei Province (grant numbers 20324201D, 22324202D, 20536001D), the Open Project of Water Environment Science Laboratory in Hebei Province, China (grant number HBSHJ202101) and the Science and Technology Program of Hebei Academy of Sciences, China (22106).

Institutional Review Board Statement: Not applicable.

Informed Consent Statement: Not applicable.

Data Availability Statement: All data, models, and code generated or used during the study appear in the submitted article.

Conflicts of Interest: The authors declare no conflict of interest.

References

1. Xue, J.Y.; Guan, H.D.; Huo, Z.L.; Wang, F.X.; Huang, G.H.; Boll, J. Water saving practices enhance regional efficiency of water consumption and water productivity in an arid agricultural area with shallow groundwater. *Agric. Water Manag.* **2017**, *194*, 78–89. [CrossRef]
2. Rijsberman, F.R. Water scarcity: Fact or fiction? *Agric. Water Manag.* **2006**, *80*, 5–22. [CrossRef]
3. Vörösmarty, C.J.; Green, P.; Salisbury, J.; Lammers, R. Global water resources:vulnerability from climate change and population growth. *Science* **2000**, *289*, 284–288. [CrossRef]
4. Yeh, P.J.F.; Famiglietti, J.S. Regional groundwater evapotranspiration inillinois. *J. Hydrometeorol.* **2009**, *10*, 464–478. [CrossRef]
5. Babajimopoulos, C.; Panoras, A.; Georgoussis, H.; Arampatzis, G.; Hatzigiannakis, E.; Papamichail, D. Contribution to irrigation from shallow water table under field conditions. *Agric. Water Manag.* **2007**, *92*, 205–210. [CrossRef]
6. Dregne, H.E. Global status of desertification Ann. *Arid Zone* **1991**, *30*, 179–185.
7. NOAA, JetStream: Online School for Weather. Available online: http://www.srh.noaa.gov/jetstream/global/climate_max.htm (accessed on 26 February 2023).
8. Bulter, J.J., Jr.; Kluitenberg, G.J.; Whittemore, D.O.; Loheide, S.P., II; Jin, W.; Billinger, M.A.; Zhan, X.Y. A field investigation of phreatophyte-induced fluctuations in the water–table. *Water Resoureces Res.* **2007**, *43*, W02404.
9. Gribovszki, Z.; Szilágyi, J.; Kalicz, P. Diurnal fluctuations in shallow groundwater levels and streamflow rates and their interpretation-A review. *J. Hydrol.* **2010**, *385*, 371–383. [CrossRef]
10. Scanlon, B.R.; Keese, K.E.; Flint, A.L.; Flint, L.E.; Gaye, C.B.; Edmunds, W.M.; Simmers, I. Global synthesis of groundwater recharge in semiarid and arid regions. *Hydrol. Process.* **2006**, *20*, 3335–3370. [CrossRef]

11. Qi, S.Z.; Luo, F. Hydrological indicators of desertification in the Heihe River Basin of arid northwest China. *AMBIO A J. Hum. Environ.* **2006**, *35*, 319–321.
12. Yin, L.H.; Zhou, Y.X.; Ge, S.M.; Wen, D.G.; Zhang, E.Y.; Dong, J.Q. Comparison and modification of methods for estimating evapotranspiration using diurnal groundwater level fluctuations in arid and semiarid regions. *J. Hydrol.* **2013**, *496*, 9–16. [CrossRef]
13. Wang, P.; Grinevsky, S.O.; Pozdniakov, S.P. Application of the water table fluctuation method for estimating evapotranspiration at two phreatophyte-dominated sites under hyper-arid environments. *J. Hydrol.* **2014**, *519*, 2289–2300. [CrossRef]
14. Goodrich, D.C.; Scott, R.; Qi, J.; Goff, B.; Unkrich, C.L.; Moran, M.S.; Williams, D.; Schaeffer, S.; Snyder, R.K.; MacNish, R.; et al. Seasonal estimates of riparian evapotranspiration using remote and in situ measurements. *Agric. For. Meteorol.* **2000**, *105*, 281–309. [CrossRef]
15. Wang, X.W.; Huo, Z.L.; Feng, S.Y.; Guo, P.; Guan, H.D. Estimating groundwater evapotranspiration from irrigated cropland incorporating root zone soil texture and moisture dynamics. *J. Hydrol.* **2016**, *543*, 501–509. [CrossRef]
16. Drexler, J.Z.; Snyder, R.L.; Spano, D.; Tha Paw, U.K. A review of models and micrometeorological methods used to estimate wetland evapotranspiration. *Hydrol. Process.* **2004**, *18*, 2071–2101. [CrossRef]
17. Cooper, D.J.; Sanderson, J.S.; Stannard, D.I.; Groeneveldd, D.P. Effects of long-term water table drawdown on evapotranspiration and vegetation in an arid region phreatophyte community. *J. Hydrol.* **2006**, *325*, 21–34. [CrossRef]
18. Naumburg, E.; Mata-Gonzalez, R.; Hunter, R.G.; Mclendon, T.; Martin, D.W. Phreatophytic vegetation and groundwater fluctuations: A review of current research and application of ecosystem response modeling with an emphasis on great basin vegetation. *Environ. Manag.* **2005**, *35*, 726–740. [CrossRef]
19. Orellana, F.; Verma, P.; Loheide, S.P., II; Daly, E. Monitoring and modeling water-vegetation interactions in groundwater-dependent ecosystems. *Rev. Geophys.* **2012**, *50*, 3. [CrossRef]
20. Yuan, G.F.; Zhang, P.; Shao, M.A.; Luo, Y.; Zhu, X.C. Energy and water exchanges over a riparian Tamarix spp. stand in the lower Tarim River basin under a hyper-arid climate. *Agric. For. Meteorol.* **2014**, *194*, 144–154. [CrossRef]
21. Salama, R.B.; Fitzpatrick, O. Contributions of groundwater conditions to soil and water salinization. *Hydrogeol. J.* **1999**, *7*, 46–64. [CrossRef]
22. Northey, J.E.; Christen, E.W.; Ayars, J.E.; Jankowski, J. Occurrence and measurement of salinity stratification in shallow groundwater in the Murrumbidgee Irrigation Area, south-eastern Australia. *Agric. Water Manag.* **2006**, *81*, 23–40. [CrossRef]
23. Zipper, S.C.; Soylu, M.E.; Booth, E.G.; Loheide, S.P., II. Untangling the effects of shallow groundwater and soil texture as drivers of subfield-scale yield variability. *Water Resour. Res.* **2015**, *51*, 6338–6358. [CrossRef]
24. Ayars, J.E.; Schoneman, R.A. Use of saline water from a shallow water table by cotton. *Trans ASAE* **1986**, *29*, 1674–1678. [CrossRef]
25. Yang, F.; Zhang, G.; Yin, X.; Liu, Z.; Huang, Z. Study on capillary rise from shallow groundwater and critical water table depth of a saline-sodic soil in western Songnen plain of China. *Environ. Earth Sci.* **2011**, *64*, 2119–2126. [CrossRef]
26. Pratharpar, S.A.; Qureshi, A.S. Modelling the effects of deficit irrigation on soil salinity, depth to water table and transpiration in semi-arid zones with monsoonal rains. *Int. J. Water Resour. Dev.* **1999**, *15*, 141–159. [CrossRef]
27. Yang, J.F.; Li, B.Q.; Liu, S.P. A large weighing lysimeter for evapotranspiration and soil-water-groundwater exchange studies. *Hydrol. Process.* **2000**, *14*, 1887–1897. [CrossRef]
28. Kahlown, M.A.; Ashraf, M. Zia-ul-Haq, Effect of shallow groundwater table on crop water requirements and crop yields. *Agric. Water Manag.* **2005**, *76*, 24–35. [CrossRef]
29. Wu, Y.; Liu, T.; Paredes, P.; Duan, L.; Pereira, L.S. Water use by a groundwater dependent maize in a semi-arid region of Inner Mongolia: Evapotranspiration partitioning and capillary rise. *Agric. Water Manag.* **2015**, *152*, 222–232. [CrossRef]
30. Gou, S.; Miller, G. A groundwater-soil-plant-atmosphere continuum approach for modelling water stress, uptake, and hydraulic redistribution in phreatophytic vegetation. *Ecohydrology* **2014**, *7*, 1029–1041. [CrossRef]
31. Newman, B.D.; Wilcox, B.P.; Archer, S.R.; Breshears, D.D.; Dahm, C.N.; Duffy, C.J.; McDowell, N.G.; Phillips, F.M.; Scanlon, B.R.; Vivoni, E.R. Ecohydrology of water-limited environments: A scientific vision. *Water Resour. Res.* **2006**, *42*, W06302. [CrossRef]
32. Allen, R.G. Assessing integrity of weather data for reference evapotranspiration estimation. *J. Irrig. Drain. Eng.* **1996**, *122*, 97–106. [CrossRef]
33. Balugani, E.; Lubczynski, M.W.; Reyes-Acosta, L.; Van Der, T.C.; Francés, A.P.; Metselaar, K. Groundwater and unsaturated zone evaporation and transpiration in a semi-arid open wood land. *J. Hydrol.* **2017**, *547*, 54–66. [CrossRef]
34. Lubczynski, M.W.; Gurwin, J. Integration of various data sources for transient groundwater modeling with spatio-temporally variable fluxes—Sardon study case, Spain. *J. Hydrol.* **2005**, *306*, 71–96. [CrossRef]
35. Gavilán, P.; Berengena, J.; Allen, R.G. Measuring versus estimating net radiation and soil heat flux: Impact on penman–monteith reference et estimates in semiarid regions. *Agric. Water Manag.* **2007**, *89*, 275–286. [CrossRef]
36. Gazal, R.M.; Scott, R.L.; Goodrich, D.C.; Williams, D.G. Controls on transpiration in a semiarid riparian cottonwood forest. *Agric. For. Meteorol.* **2006**, *137*, 56–67. [CrossRef]
37. Hultine, K.R.; Bush, S.E.; West, A.G.; Ehleringer, J.R. Effect of gender on sap-flux-scaled transpiration in a dominant riparian tree species: Box elder (acer negundo). *J. Geophys. Res. Biogeosci.* **2015**, *112*, 87–101. [CrossRef]
38. Moore, D.J.P.; Hu, J.; Sacks, W.J.; Schimel, D.S.; Monson, R.K. Estimating transpiration and the sensitivity of carbon uptake to water availability in a subalpine forest using a simple ecosystem process model informed by measured net CO_2 and H_2O fluxes. *Agric. For. Meteorol.* **2008**, *148*, 1467–1477. [CrossRef]

39. Lenka, S.; Singh, A.K.; Lenka, N.K. Water and nitrogen interaction on soil profile water extraction and ET in maize-wheat cropping system. *Agric. Water Manag.* **2009**, *96*, 195–207. [CrossRef]
40. Richard, G.; Benyon, S.; Theiveyanathan Doody, T.M. Impacts of tree plantations on groundwater in south-eastern Australia. *Aust. J. Bot.* **2006**, *54*, 181–192.
41. Doody, T.; Benyon, R. Quantifying water savings from willow removal in Australian streams. *J. Environ. Manag.* **2011**, *92*, 926–935. [CrossRef]
42. Nachabe, M.; Shah, N.; Ross, M.; Vomacka, J. Evapotranspiration of two vegetation covers in a shallow water table environment. *Soil Sci. Soc. Am. J.* **2005**, *69*, 492–499. [CrossRef]
43. Devitt, D.A.; Fenstermaker, L.F.; Young, M.H.; Conrad, B.; Baghzouz, M.; Bird, B.M. Evapotranspiration of mixed shrub communities in phreatophytic zones of the Great Basin region of Nevada (USA). *Ecohydrology* **2011**, *4*, 807–822. [CrossRef]
44. Moore, G.W.; Cleverly, J.R.; Owens, M.K. Nocturnal transpiration in riparian Tamarix thickets authenticated by sap flux, eddy covariance and leaf gas exchange measurements. *Tree Physiol.* **2008**, *28*, 521–528. [CrossRef]
45. Scott, R.L.; Edwards, E.A.; Shuttleworth, W.J.; Huxman, T.E.; Watts, C.; Goodrich, D.C. Interannual and seasonal variation in fluxes of water and carbon dioxide from a riparian woodland ecosystem. *Agric. For. Meteorol.* **2004**, *122*, 65–84. [CrossRef]
46. Scott, R.L.; Huxman, T.E.; Williams, D.G.; Goodrich, D.C. Ecohydrological impacts of woody-plant encroachment: Seasonal patterns of water and carbon dioxide exchange within a semiarid riparian environment. *Glob. Change Biol.* **2006**, *12*, 311–324. [CrossRef]
47. Scott, R.L.; Shuttleworth, W.J.; Goodrich, D.C. The water use of two dominant vegetation communities in a semiarid riparian ecosystem. *Agric. For. Meteorol.* **2000**, *105*, 241–256. [CrossRef]
48. Norman, J.M.; Kustas, W.P.; Humes, K.S. Source approach for estimating soil and vegetation energy fluxes in observations of directional radiometric surface temperature. *Agric. For. Meteorol.* **1995**, *77*, 263–293. [CrossRef]
49. Bastiaanssen, W.G.M.; Pelgrum, H.; Wang, J.; Ma, Y.; Moreno, J.F.; Roerink, G.J.; van der Wal, T. A remote sensing surface energy balance algorithm for land (SEBAL).: Part 2: Valiation. *J. Hydrol.* **1998**, *212*, 213–229. [CrossRef]
50. Roerink, G.J.; Su, Z.; Menenti, M. S-SEBI: A simple remote sensing algorithm to estimate the surface energy balance. *Phys. Chem. Earth Part B Hydrol. Oceans Atmos.* **2000**, *25*, 147–157. [CrossRef]
51. Su, Z. The surface energy balance system (SEBS) for estimation of turbulent heat fluxes. *Hydrol. Earth Syst. Sci.* **2002**, *6*, 85–99. [CrossRef]
52. Gellens-Meulenberghs, F.; Arboleda, A.; Ghilain, N. Towards a continuous monitoring of evapotranspiration based on MSG data. *Environ. Sci.* **2007**, *316*, 228–234.
53. Allen, R.G.; Tasumi, M.; Trezza, R. Satellite-based energy balance for mapping evapotranspiration with internalized calibration (METRIC)-model. *J. Irrigat. Drain. Eng.* **2007**, *133*, 380–394. [CrossRef]
54. Sánchez, J.M.; Kustas, W.P.; Caselles, V.; Anderson, M.C. Modelling surface energy fluxes over maize using a two-source patch model and radiometric soil and canopy temperature observations. *Remote Sens. Environ.* **2008**, *112*, 1130–1143. [CrossRef]
55. Miralles, D.G.; Holmes, T.R.H.; DeJeu, R.A.M.; Gash, J.H.; Meesters, A.G.C.A. Global land-surface evaporation estimated from satellite-based observations. *Hydrol. Earth Syst. Sci. Discuss.* **2011**, *7*, 453–469. [CrossRef]
56. Mu, Q.; Zhao, M.; Running, S.W. Improvements to a MODIS global terrestrial evapotranspiration algorithm. *Remote Sens. Environ.* **2011**, *115*, 1781–1800. [CrossRef]
57. Wu, B.F.; Yan, N.N.; Xiong, J.; Bastiaanssen, W.G.M.; Zhu, W.W.; Stein, A. Validation of ETWatch using field measurements at diverse landscapes: A case study in Hai basin of China. *J. Hydrol.* **2012**, *436*, 67–80. [CrossRef]
58. Mazur, M.C.; Wiley, M.J.; Wilcox, D.A. Estimating evapotranspiration and groundwater flow from water-table fluctuations for a general wetland scenario. *Ecohydrology* **2014**, *7*, 378–390. [CrossRef]
59. Ridolfi, L.; Odorico, P.D.; Laio, F. Vegetation dynamics induced by phreatophyte–aquifer interactions. *J. Theor. Biol.* **2007**, *248*, 301–310. [CrossRef]
60. Sanderson, J.S.; Cooper, D.J. Ground water discharge by evapotranspiration in wetlands of an arid intermountain basin. *J. Hydrol.* **2008**, *351*, 344–359. [CrossRef]
61. Healy, R.W.; Scanlon, B.R. *Estimating Groundwater Recharge: Groundwater Recharge*; Cambridge University Press: Cambridge, UK, 2010.
62. Loheide II, S.P. A method for estimating subdaily evapotranspiration of shallow groundwater using diurnal water table fluctuations. *Ecohydrology* **2008**, *1*, 59–66. [CrossRef]
63. Simmers, I.; Hendrickx, J.; Kruseman, G.P.; Rushton, K.R. *Recharge of Phreatic Aquifers in (Semi-) Arid Areas*; Balkema: Rotterdam, The Netherlands, 1997.
64. Yu, L.Y.; Zeng, Y.J.; Su, Z.B.; Cai, H.J.; Zheng, Z. The effect of different evapotranspiration methods on portraying soil water dynamics and ET partitioning in a semi-arid environment in Northwest China. *Hydrol. Earth Syst. Sci.* **2016**, *20*, 975–990. [CrossRef]
65. Remson, I.; Fox, G.S. Capillary losses from ground water. *Trans. Am. Geophys. Union* **1955**, *36*, 304–310. [CrossRef]
66. Gardner, W.R. Some steady-state solutions of the unsaturated moisture flow equation with application to evaporation from a water table. *Soil Sci.* **1958**, *85*, 228–232. [CrossRef]
67. Willis, W.O. Evaporation from layered soils in the presence of a water table. *Soil Sci. Soc. Am. J.* **1960**, *24*, 239–242. [CrossRef]
68. Schoeller, H. *Les Eaux Souterraines*; Mason and Cie: Paris, France, 1962; Volume 642.

69. Doorenbos, J.; Pruitt, W.O. *FAO Guidelines for Predicting Crop Water Requirements*; FAO Irrigation & Drainage Paper; Food and Agriculture Organization of the United Nations: Rome, Italy, 1975.
70. Steinwand, A.L.; Harrington, R.F.; Or, D. Water balance for great basin phreatophytes derived from eddy covariance, soil water, and water table measurements. *J. Hydrol.* **2006**, *329*, 595–605. [CrossRef]
71. Stuff, R.G.; Dale, R.F. A soil moisture budget model accounting for shallow water table influences. *Soil Sci. Soc. Am. J.* **1978**, *42*, 637–643. [CrossRef]
72. Wallender, W.W.; Grimes, D.W.; Henderson, D.W.; Stromberg, L.K. Estimating the contribution of a perched water table to the seasonal evapotranspiration of cotton. *Agron. J.* **1979**, *71*, 1056–1060. [CrossRef]
73. Loheide, S.P., II; Butler, J.J.; Gorelick, S.M. Estimation of groundwater consumption by phreatophytes using diurnal water table fluctuations: A saturated-unsaturated flow assessment. *Water Resour. Res.* **2005**, *41*, W07030. [CrossRef]
74. Wang, X.W.; Huo, Z.L.; Shukla, M.K.; Wang, X.H.; Guo, P.; Xu, X.; Huang, G.H. Energy fluxes and evapotranspiration over irrigated maize field in an arid area with shallow groundwater. *Agric. Water Manag.* **2020**, *228*, 105922. [CrossRef]
75. Ren, D.Y.; Xu, X.; Hao, Y.Y.; Huang, G.H. Modeling and assessing field irrigation water use in a canal system of Hetao, upper Yellow River basin: Application to maize, sunflower and watermelon. *J. Hydrol.* **2016**, *532*, 122–139. [CrossRef]
76. Lai, J.; Liu, T.; Luo, Y. Evapotranspiration Partitioning for Winter Wheat with Shallow Groundwater in the Lower Reach of the Yellow River Basin. *Agric. Water Manag.* **2022**, *266*, 107561. [CrossRef]
77. Karimov, A.K.; Šimůnek, J.; Hanjra, M.A.; Avliyakulov, M.; Forkutsa, I. Effects of the shallow water table on water use of winter wheat and ecosystem health: Implications for unlocking the potential of groundwater in the Fergana Valley (Central Asia). *Agric. Water Manag.* **2014**, *131*, 57–69. [CrossRef]
78. Nirjhar, S.; Mahmood, N.; Mark, R. Extinction depth and evapotranspiration from ground water under selected land covers. *Ground Water* **2007**, *45*, 329–338.
79. Lautz, L.K. Estimating groundwater evapotranspiration rates using diurnal water-table fluctuations in a semi-arid riparian zone. *Hydrogeol. J.* **2008**, *16*, 483–497. [CrossRef]
80. Mould, D.J.; Frahm, E.; Salzmann Th Miegel, K.; Acreman, M.C. Evaluating the use of diurnal groundwater fluctuations for estimating evapotranspiration in wetland environments: Case studies in southeast England and northeast Germany. *Ecohydrology* **2010**, *3*, 294–305. [CrossRef]
81. White, W.N. *Method of Estimating Ground-Water Supplies Based on Discharge by Plants and Evaporation from Soil: Results of Investigation in Escalante Valley, Utah (R)*; Geological Survey Water Supply Paper, 659-A; United States Department of the Interior, Government Printing Office: Washington, DC, USA, 1932; Volume 115.
82. Meyboom, P. Three observations on streamflow depletion by phreatophytes. *J. Hydrol.* **1965**, *2*, 248–261. [CrossRef]
83. Gerla, P.J. The Relationship of Water-Table Changes to the Capillary Fringe, Evapotranspiration, and Precipitation in Intermittent Wetlands. *Wetlands* **1992**, *12*, 91–98. [CrossRef]
84. Healy, R.W.; Cook, P.G. Using groundwater levels to estimate recharge. *Hydrogeol. J.* **2002**, *10*, 91–109. [CrossRef]
85. Gribovszki, Z.; Kalicz, P.; Szilagyi, J.; Kucsara, M. Riparian zone evapotranspiration estimation from diurnal groundwater level fluctuations. *J. Hydrol.* **2008**, *349*, 6–17. [CrossRef]
86. Soylu, M.E.; Lenters, J.D.; Istanbulluoglu, E.; Loheide, S.P., II. On evapotranspiration and shallow groundwater fluctuations:\na fourier-based improvement to the white method. *Water Resources Res.* **2012**, *48*, W06506. [CrossRef]
87. Wang, P.; Yu, J.J.; Pozdniakov, S.P.; Grinevsky, S.O.; Liu, C.M. Shallow groundwater dynamics and its driving forces in extremely arid areas: A case study of the lower Heihe river in northwestern China. *Hydrol. Process.* **2014**, *28*, 1539–1553. [CrossRef]
88. Troxell, H.C. The diurnal fluctuation in the ground-water and flow of the santa ana river and its meaning. *Trans. Am. Geophys. Union* **1936**, *17*, 496–504. [CrossRef]
89. Gatewood, J.S. Use of water by bottom-land vegetation in lower safford valley, Arizona. *US Geol. Surv. Water Supply Pap.* **1950**, *1103*, 210.
90. Laczniak, R.J.; Demeo, G.A.; Reiner, S.R.; Smith, J.L.; Nylund, W.E. *Estimates of Ground-Water Discharge as Determined from Measurements of Evapotranspiration, Ash Meadows area, Nye County, Nevada*; USGS US Geological Survey Water Resources Investigations Report 99-4079; Geological Survey: Las Vegas, NV, USA, 1999.
91. Mclaughlin, D.L.; Cohen, M.J. Ecosystem specific yield for estimating evapotranspiration and groundwater exchange from diel surface water variation. *Hydrol. Process.* **2013**, *28*, 1495–1506. [CrossRef]
92. Schilling, K.E. Water table fluctuations under three riparian land covers, Iowa (USA). *Hydrol. Process.* **2010**, *21*, 2415–2424. [CrossRef]
93. Martinet, M.C.; Vivoni, E.R.; Cleverly, J.R.; Thibault, J.R.; Dahm, C.N. On groundwater fluctuations, evapotranspiration, and understory removal in riparian corridors. *Water Resour. Res.* **2009**, *45*, 207–213. [CrossRef]
94. Miller, G.R.; Chen, X.; Rubin, Y.; Ma, S.; Baldocchi, D.D. Groundwater uptake by woody vegetation in a semiarid oak savanna. *Water Resour. Res.* **2010**, *46*, 2290–2296. [CrossRef]
95. Vincke, C.; Thiry, Y. Water table is a relevant source for water uptake by a Scots pine (*Pinus sylvestris* L.) stand: Evidences from continuous evapotranspiration and water table monitoring. *Agric. For. Meteorol.* **2008**, *148*, 1419–1432. [CrossRef]
96. Dolan, T.J.; Hermann, A.J.; Bayley, S.E.; Zoltek, J.J. Evapotranspiration of a Florida, U.S.A. freshwater wetland. *J. Hydrol.* **1984**, *74*, 355–371. [CrossRef]

97. Hays, K.B. Water Use by Saltcedar (*Tamarix* sp.) and Associated Vegetation on the Canadian, Colorado and Pecos Rivers in Texas. Master's Thesis, Texas A & M University, Texas, TX, USA, 2003.
98. Czikowsky, M.J.; Fitzjarrald, D.R. Evidence of seasonal changes in evapotranspiration in eastern u.s. hydrological records. *J. Hydrometeorol.* **2009**, *5*, 974–988. [CrossRef]
99. Wang, T.Y.; Wang, P.; Yu, J.J.; Pozdniakov, S.P.; Zhang, Y. Revisiting the white method for estimating groundwater evapotranspiration: A consideration of sunset and sunrise timings. *Environ. Earth Sci.* **2019**, *78*, 412.1–412.7. [CrossRef]
100. Su, Y.H.; Feng, Q.; Zhu, G.F.; Zhang, Q. Evaluating the different methods for estimating groundwater evapotranspiration using diurnal water table fluctuations. *J. Hydrol.* **2022**, *607*, 127508.
101. Fahle, M.; Dietrich, O. Estimation of evapotranspiration using diurnal groundwater level fluctuations: Comparison of different approaches with groundwater lysimeter data. *Water Resour. Res.* **2014**, *50*, 273–286. [CrossRef]
102. Acharya, S.; Jawitz, J.W.; Mylavarapu, R.S. Analytical expressions for drainable and fillable porosity of phreatic aquifers under vertical fluxes from evapotranspiration and recharge. *Water Resour. Res.* **2012**, *48*, W11526. [CrossRef]
103. Sophocleous, M. The role of specific yield in ground-water recharge estimations: A numerical study. *Groundwater* **1985**, *23*, 52–58. [CrossRef]
104. Cheng, D.H.; Li, Y.; Chen, X.; Wang, W.K.; Hou, G.C.; Wang, C.L. Estimation of groundwater evaportranspiration using diurnal water table fluctuations in the mu us desert, northern China. *J. Hydrol.* **2013**, *490*, 106–113. [CrossRef]
105. Duke, H.R. Capillary properties of soils–influence upon specific yield. *Amer Soc Agr Eng Trans Asae* **1972**, *15*, 688–691. [CrossRef]
106. Alley, W.M.; Healy, R.W.; Labaugh, J.W.; Reilly, T.E. Flow and storage in groundwater systems. *Science* **2002**, *296*, 1985–1990. [CrossRef]
107. Scanlon, B.R.; Faunt, C.C.; Longuevergne, L.; Reedy, R.C.; Mcmahon, P.B. Groundwater depletion and sustainability of irrigation in the us high plains and central valley. *Proc. Natl. Acad. Sci. USA* **2012**, *109*, 9320–9325. [CrossRef]
108. Rushton, B. Hydrologic budget for a freshwater marsh in Florida. *Water Resour. Bull.* **1996**, *32*, 13–21. [CrossRef]
109. Bethenod, O.; Katerji, N.; Goujet, R.; Bertolini, J.M.; Rana, G. Determination and validation of corn crop transpirationby sap flow measurement under field conditions. *Theor. Appl. Climatol.* **2000**, *67*, 153–160. [CrossRef]
110. Hsiao, B.T.; Xu, L. Evapotranspiration and Relative Contribution by the Soil and the Plant. *Calif. Water Plan Update* **2005**, *4*, 129–160.
111. Wilson, K.B.; Hanson, P.J.; Baldocchi, D.D. Factors controlling evaporation and energy partitioning beneath a deciduous forest over an annual cycle. *Agric. For. Meteorol.* **2000**, *102*, 83–103. [CrossRef]
112. Coenders-Gerrits, A.M.J.; van der Ent, R.J.; Bogaard, T.A.; Wang, E.L.; Harchowitz, M.; Savenijie, H.H.G. Uncertainties in transpiration estimates. *Nature* **2014**, *506*, E1–E2. [CrossRef]
113. Wang, L.; Good, S.P.; Caylor, K.K. Global synthesis of vegetation control on evapotranspiration partitioning. *Geophys. Res. Lett.* **2014**, *41*, 6753–6757. [CrossRef]
114. Good, S.P.; Noone, D.; Bowen, G. Hydrologic connectivity constrains partitioning of global terrestrial water fluxes. *Science* **2015**, *349*, 175–177. [CrossRef]
115. Schlesinger, W.H.; Jasechko, S. Transpiration in the global water cycle. *Agric. For. Meteorol.* **2014**, *189*, 115–117. [CrossRef]
116. Cavanaugh, M.L.; Kurc, S.A.; Scott, R.L. Evapotranspiration partitioning in semiarid shrubland ecosystems: A two-site evaluation of soil moisture control on transpiration. *Ecohydrology* **2011**, *4*, 671–681. [CrossRef]
117. Schlaepfer, D.R.; Ewers, B.E.; Shuman, B.N.; Williams, D.G.; Frank, J.M.; Massman, W.J.J.; Lauenroth, W.K. Terrestrial water fluxes dominated by transpiration: Comment. *Ecosphere* **2016**, *5*, 1–9. [CrossRef]
118. Nichols, W.D. Estimating discharge of shallow groundwater by transpiration from greasewood in the northern great basin. *Water Resour. Res.* **1993**, *29*, 2771–2778. [CrossRef]
119. Nichols, W.D. Groundwater discharge by phreatophyte shrubs in the great basin as related to depth to groundwater. *Water Resour. Res.* **1994**, *30*, 3265–3274. [CrossRef]
120. Oliveira, R.S.; Dawson, T.E.; Burgess, S. Hydraulic redistribution in three Amazonian trees. *Oecologia* **2005**, *145*, 354–363. [CrossRef]
121. Amenu, G.G.; Kumar, P. A model for hydraulic redistribution incorporating coupled soil-root moisture transport. *Hydrol. Earth Syst. Sci.* **2008**, *12*, 55–74. [CrossRef]
122. Nepstad, D.C.; Carvalho, C.D.; Davidson, E.A.; Jipp, P.H.; Lefebvre, P.A.; Negreiros, G.H.; Silva, E.D.D.; Stone, T.A.; Trumbore, S.E.; Vieira, S. The role of deep roots in the hydrological and carbon cycles of Amazonian forests and pastures. *Nature* **1994**, *372*, 666–669. [CrossRef]
123. Maraux, F.; Lafolie, F. Modeling Soil Water Balance of a Maize-Sorghum Sequence. *Soil Sci. Soc. Am. J.* **2011**, *62*, 75–82. [CrossRef]
124. Kleidon, A.; Heimann, M. Assessing the role of deep rooted vegetation in the climate system with model simulations: Mechanism, comparison to observations and implications for Amazonian deforestation. *Clim. Dyn.* **2000**, *16*, 183–199. [CrossRef]
125. Saleska, S.R.; Didan, K.; Huete, A.R.; da Rocha, H.R. Amazon forests green-up during 2005 drought. *Science* **2007**, *318*, 612. [CrossRef]
126. Liu, T.; Luo, Y. An empirical approach simulating evapotranspiration from groundwater under different soil water conditions. *Environ. Earth Sci.* **2012**, *67*, 1345–1355. [CrossRef]

127. Schmid, W.; Hanson, R.T.; Iii, T.M.; Leake, S. User guide for the Farm Process (FMP1) for the U.S. Geological Survey's modular three-dimensional finite-difference ground-water flow model, MODFLOW-2000. In *Book 6: Modeling Techniques, Section A. Ground-Water*; The U.S. Geological Survey Office of Ground Water, Ground-Water Resources Program: Reston, VA, USA, 2006; Chapter 17.
128. Mermoud, A.; Morelseytoux, H.J. Modélisation et observation du flux hydriquevers la surface du sol depuisune nappe peu profonde. *J. Cardiovasc. Pharmacol.* **1989**, *4*, 11–23.
129. Yuan, G.F.; Luo, Y.; Shao, M.A.; Zhang, P.; Zhu, X.C. Evapotranspiration and its main controlling mechanism over the desert riparian forests in the lower Tarim river basin. *Sci. China Earth Sci.* **2015**, *58*, 1032–1042. [CrossRef]
130. Yue, W.; Wang, T.; Franz, T.E. Spatiotemporal patterns of water table fluctuations andevapotranspiration induced by riparian vegetation in a semiarid area. *Water Resour. Res.* **2016**, *52*, 1948–1960. [CrossRef]
131. Zhang, W.J. Study on the Evapotranspiration and Water Balance of Lake Basin in the Hinterland of Badain Jaran Desert. Master's Thesis, Lanzhou University, Lanzhou, China, 2020. (In Chinese with English abstract)
132. Wondzell, S.M.; Gooseff, M.N.; Mcglynn, B.L. An analysis of alternative conceptual models relating hyporheic exchange flow to diel fluctuations in discharge during baseflow recession. *Hydrol. Process.* **2010**, *24*, 686–694. [CrossRef]
133. Jia, W.H.; Yin, L.H.; Zhang, M.S.; Zhang, J.; Zhang, X.X.; Gu, X.F.; Dong, J.Q. Modified method for the estimation of groundwater evapotranspiration under very shallow water table conditions based on diurnal water table fluctuations. *J. Hydrol.* **2021**, *597*, 126193. [CrossRef]
134. Diouf, O.A.; Weihermüller, L.; Diedhiou, M.; Vereecken, H.; CisséFaye, S.C.; Faye, S.; Sylla, S.N. Modelling groundwater evapotranspiration in a shallow aquifer in a semi-arid environment. *J. Hydrol.* **2020**, *587*, 124967. [CrossRef]
135. Zhao, K.Y.; Jiang, X.W.; Wang, X.S.; Wan, L. Restriction of groundwater recharge and evapotranspiration due to a fluctuating water table: A study in the Ordos Plateau, China. *Hydrogeol. J.* **2020**, *2*, 1–11. [CrossRef]
136. Zhu, Y.; Ren, L.; Skaggs, T.H.; Lue, H.; Yu, Z.; Wu, Y.; Fang, X. Simulation of Populus euphratica root uptake of groundwater in an arid woodland of the Ejina Basin, China. *Hydrol. Process.* **2010**, *23*, 2460–2469. [CrossRef]
137. Mcdonald, M.G. *A Modular Three-Dimensional Finite-Difference Groundwater Flow Model*; Techniques of Water-Resources; U. S. Geological Survey: Reston, VA, USA, 1988.
138. Banta, E.R. *MODFLOW-2000, the U.S. Geological Survey Modular Ground-Water Model-Documentation of Packages for Simulating Evapotranspiration with a Segmented Function (EST1) and Drains with Return Flow (DRT1)*; The U.S. Geological Survey in cooperation with the Colorado Water Conservation Board and the Colorado Division of Water Resources: Reston, VA, USA, 2000.
139. Warrick, W.A. Additional solutions for steady-state evaporation from a shallow water table. *Soil Sci.* **1988**, *146*, 63–66. [CrossRef]
140. Pozdniakov, S.P.; Vasilevsky, P.Y.; Wang, P. Analysis of a steady-state model of groundwater discharge in a river valley without and with evapotranspiration Adv. *Water Resour. Res.* **2022**, *168*, 104306. [CrossRef]
141. Baird, K.J.; Maddock, T. III. Simulating riparian evapotranspiration: A new methodology and application for groundwater models. *J. Hydrol.* **2015**, *312*, 176–190. [CrossRef]
142. Ajami, H. Maddock III, T. Meixner, T. Hogan, J.F. Guertin, D.P. RIPGIS-NET: A GIS tool for riparian groundwater evapotranspiration in MODFLOW. *Groundwater* **2012**, *50*, 154–158. [CrossRef]
143. El-Zehairy, A.A.; Lubczynski, M.W.; Gurwin, J. Interactions of artificial lakes with groundwater applying an integrated MODFLOW solution. *Hydrogeol. J.* **2017**, *26*, 109–132. [CrossRef]
144. Sergey, G. Elena, F. Victor, S. Vsevolod, S. Sergey, P. Evapotranspiration capture and stream depletion due to groundwater pumping under variable boreal climate conditions: Sudogda River basin, Russia. *Hydrogeol. J.* **2018**, *26*, 2753–2767.
145. Hou, X.L.; Wang, S.Q.; Jin, X.R.; Li, M.L.; Lv, M.Y.; Feng, W.Z. Using an ETWatch (RS)-UZF-MODFLOW coupled model to optimize joint use of transferred water and local water sources in a saline water area of the North China Plain. *Water* **2020**, *12*, 3361. [CrossRef]
146. Famiglietti, J.S.; Wood, E.F. Multiscale modeling of spatially variable water and energy balance processes. *Water Resour. Res.* **1994**, *30*, 3061–3078. [CrossRef]
147. Stieglitz, M.; Rind, D.; Famiglietti, J. An efficient approach to modeling the topographic control of surface hydrology for regional and global climate modeling. *J. Clim.* **1997**, *10*, 118–137. [CrossRef]
148. Koster, R.D.; Ducharne, A.; Stieglitz, M.; Kumar, P.; Suare, M.J. A catchment-based approach to modeling land surface processes in a general circulation model 1. Model structure. Journal of Geophysical Research. *Biogeosciences* **2000**, *105*, 24809–24822.
149. Gutowski, W.J., Jr.; Vorosmarty, C.J.; Person, M.; Otles, Z.; Fekete, B.; York, J. A coupled land-atmosphere simulation program (CLASP): Calibration and validation. *J. Geophys. Res. Atmos.* **2002**, *107*, 1–17. [CrossRef]
150. York, J.P.; Person, M.; Gutowski, W.J.; Winter, T.C. Putting aquifers into atmospheric simulation models: An example from the mill creek watershed, northeastern Kansas. *Adv. Water Resour.* **2002**, *25*, 221–238. [CrossRef]
151. Liang, X.; Xie, Z.H.; Huang, M.Y. A new parameterization for groundwater and surface water interactions and its impact on water budgets with the variable infiltration capacity (VIC) land surface model. *J. Geophys. Res. Atmos.* **2003**, *108*, 1–17. [CrossRef]
152. Miller, N.L.; Maxwell, R.M. Development of a coupled land surface and groundwater model. *J. Hydrometeorol.* **2005**, *6*, 233–247.
153. Yeh, P.J.F.; Eltahir, E.A.B. Representation of water table dynamics in a land surface scheme. Part I: Model development. *J. Clim.* **2005**, *18*, 1861–1880. [CrossRef]
154. Yeh, P.J.F.; Eltahir, E.A.B. Representation of water table dynamics in a land surface scheme: Observations, models, and analyses. Representation of Water Table Dynamics in a Land Surface Scheme. *Part II Subgrid Variability. J. Clim.* **2005**, *18*, 1881–1901.

155. Cohen, D.; Person, M.; Daannen, R.; Locke, S.; Dahlstrom, D.; Zabielski, V.; Winter, T.C.; Rosenbery, D.O.; Wright, H.; Ito, E.; et al. Groundwater-supported evapotranspiration within glaciated watersheds under conditions of climate change. *J. Hydrol.* **2006**, *320*, 484–500. [CrossRef]
156. Niu, G.Y.; Yang, Z.L.; Dickinson, R.E.; Gulden, L.E.; Su, H. Development of a simple groundwater model for use in climate models and evaluation with gravity recovery and climate experiment data. *J. Geophys. Res.* **2007**, *112*, D07103. [CrossRef]
157. Fan, Y.; Miguez-Macho, G.; Weaver, C.P.; Walko, R.; Robock, A. Incorporating water table dynamics in climate modeling: 1. Water table observations and equilibrium water table simulations. *J. Geophys. Res. Atmos.* **2007**, *112*, D10125. [CrossRef]
158. Jorenush, M.H.; Sepaskhah, A.R. Modelling capillary rise and soil salinity for shallow saline water table under irrigated and non-irrigated conditions. *Agric. Water Manag.* **2003**, *61*, 125–141. [CrossRef]
159. Blin, N.; Suárez, F. Evaluating the contribution of satellite-derived evapotranspiration in the calibration of numerical groundwater models in remote zones using the EEFlux tool. *Sci. Total Environ.* **2023**, *858*, 15. [CrossRef]
160. Liu, Y.; Pereira, L.S.; Fernando, R.M. Fluxes through the bottom boundary of the root zone in silty soils: Parametric approaches to estimate groundwater contribution and percolation. *Agric. Water Manag.* **2006**, *84*, 27–40. [CrossRef]
161. Askri, B.; Bouhlila, R.; Job, J.O. Development and application of a conceptual hydrologic model to predict soil salinity within modern Tunisian oases. *J. Hydrol.* **2010**, *380*, 45–61. [CrossRef]

MDPI AG
Grosspeteranlage 5
4052 Basel
Switzerland
Tel.: +41 61 683 77 34

Water Editorial Office
E-mail: water@mdpi.com
www.mdpi.com/journal/water

www.ingramcontent.com/pod-product-compliance
Lightning Source LLC
Chambersburg PA
CBHW061829210926
51850CB00006B/837

* 9 7 8 3 7 2 5 8 1 4 0 8 4 *